Contamination in Tissue Culture

Contributors

DONALD ARMSTRONG
MICHAEL F. BARILE
LEWIS L. CORIELL
ETIENNE de HARVEN
JØRGEN FOGH
NELDA B. HOLMGREN
GEORGE E. KENNY
ALEXIS J. KNIAZEFF
DONALD J. MERCHANT
F. KINGSLEY SANDERS
CYRIL S. STULBERG

Contamination in Tissue Culture

Edited by

JØRGEN FOGH

Sloan-Kettering Institute for Cancer Research
Rye, New York

ACADEMIC PRESS New York and London 1973

A Subsidiary of Harcourt Brace Jovanovich, Publishers

ACADEMIC PRESS, INC.
111 Fifth Avenue, New York, New York 10003

United Kingdom Edition published by
ACADEMIC PRESS, INC. (LONDON) LTD.
24/28 Oval Road, London NW1

Library of Congress Cataloging in Publication Data

Fogh, Jørgen.
Contamination in tissue culture.

Includes bibliographies.
1. Culture contamination (Biology) 2. Tissue culture. I. Title. [DNLM: 1. Histological technics. 2. Tissue culture. QS530 F655c 1973]
QH585.F64 574.8′2′0724 72–88363
ISBN 0–12–261850–5

PRINTED IN THE UNITED STATES OF AMERICA

Contents

List of Contributors ix

Preface xi

Chapter 1. Extrinsic Cell Contamination of Tissue Culture

Cyril S. Stulberg

I. Introduction 2
II. Interspecies Identification of Cell Cultures 4
III. Intraspecies Identification of Cell Cultures 16
IV. Current Status of Cell Contamination 22
References 23

Chapter 2. Methods of Prevention of Bacterial, Fungal, and Other Contaminations

Lewis L. Coriell

I. Source of Infection 30
II. Procedures within the Transfer Room 41
III. Laminar Flow Transfer Room 43
IV. Aseptic Technique 44
V. Antibiotics 45
VI. Summary 47
References 48

Chapter 3. Contamination of Tissue Culture by Bacteria and Fungi

Donald Armstrong

I. Introduction 51
II. Materials and Methods 52

III. Results 53
IV. Discussion 56
V. Summary 63
References 64

Chapter 4. Contaminants Demonstrated by Microscopy of Living Tissue Cultures or of Fixed and Stained Tissue Culture Preparations

Jørgen Fogh

I. Preparation of Cell Cultures for Microscopy 66
II. Methods of Microscopy 68
III. Photomicrographs Illustrating Various Types of Contaminants in Cell Cultures 69
IV. Review of Data Obtained by Microscopy of Mycoplasma-Infected Cultures after Hypotonic Treatment and Orcein Staining 72
References 76

Chapter 5. Contamination of Mammalian Cells in Culture by Mycoplasmata

George E. Kenny

I. Introduction 108
II. Biology of the Mycoplasmatales 108
III. Epidemiology of the Contamination of Cell Cultures with Mycoplasmata 110
IV. Host–Parasite Relationship of Mycoplasmata and Mammalian Cells 117
V. Effect of Mycoplasmata on Cell Cultures 119
VI. Elimination of Mycoplasmata from Cell Cultures 123
VII. Conclusions 126
References 126

Chapter 6. Mycoplasmal Contamination of Cell Cultures: Mycoplasma–Virus–Cell Culture Interactions

Michael F. Barile

I. Introduction 132
II. Classification 133
III. Detecting Cell Culture Contamination 135
IV. Mycoplasmal Contamination 140
V. Interactions between Mycoplasmas and Animal Cells 150
VI. Mycoplasma–Virus–Cell Interactions 154
References 156

Chapter 7. Reversible and Irreversible Alterations of Cultured FL Human Amnion Cells after Experimental Mycoplasmal Infection

Jørgen Fogh

I. Effects of Mycoplasmal Infection on the *in Vitro* Growth of FL Cells 175
II. *In Vitro* Growth and Susceptibility to Mycoplasmal Infection of Previously Mycoplasma-Infected Cells from which Mycoplasmas Have Been Eliminated 176
III. Chromosome Changes following Mycoplasmal Infection of FL Cells 178
IV. Reduced Tumor-Producing Capacity of Mycoplasma-Modified FL Cells 183
V. Effects of Mycoplasmas on SV40 Transformation of Human Amnion Cells 188
VI. Decreased Resistance of Transformed Human Cells to the Destructive Effects of Mycoplasmal Infection 189
VII. Cell-Associated Mycoplasmas: Preferential Attachment to Transformed Cells and Cells of Malignant Origin 190
VIII. Discussion of Results 192
References 193

Chapter 8. Contamination in Tissue Culture by Parasites

Nelda B. Holmgren

I. Introduction 195
II. Identification of the Amoebae 196
III. Source of Contamination 197
IV. Methods of Detection, Prevention, and Cure 198
V. Summary 202
References 203

Chapter 9. Identification of Tissue Culture Contaminants by Electron Microscopy

Etienne de Harven

I. Introduction 206
II. Examples of Tissue Culture Contaminations Observed under the Electron Microscope 206
III. Normal Particulate Cell Organelles of Viral Size 217
IV. Abnormal Cell Structures of Possible Viral Significance 221
V. Methods of Electron Microscopic Identification of Viral Contaminants 226
References 229

Chapter 10. Endogenous Virus Contaminants in Fetal Bovine Serum and Their Role in Tissue Culture Contamination

Alexis J. Kniazeff

I. Introduction 233
II. Target Viruses 234
III. Procedures for Isolating and Identifying Viral Contaminants 237
IV. Results 240
V. Discussion and Conclusions 241
References 241

Chapter 11. The Presence of Viruses in Uninoculated Tissue Cultures: Sources and Methods of Detection

F. Kingsley Sanders

I. Introduction 243
II. Passenger, Indigenous, and Cryptic Viruses 244
III. The Relationship of Viruses to an Ability of Cells to Grow in Culture 248
IV. Significance of the Presence of Viruses Inherent to Cultured Cells 250
V. Detection of Viruses Indigenous to Cultured Cells 252
References 254

Chapter 12. Summary

Donald J. Merchant

I. Introduction 257
II. Types and Sources of Contamination 258
III. Sterility Tests of Media and Solutions 259
IV. Detection and Identification of Contaminants 261
V. Prevention of Contamination 264
VI. Treatment of Contaminated Cultures 266
VII. Use of Antibiotics 267
VIII. Effects of Contaminants on Cell Cultures 268
IX. Concluding Remarks 268
Reference 269

Author Index 271

Subject Index 285

interspecific cell contaminations. In addition to those reported in 1963, Simpson and Stulberg (1971) have since examined many other lines for interspecies contamination; they are listed in Table V.

The fluorescent antibody test also expresses species characteristics by a different mechanism than does hemagglutination or mixed agglutination. The hemagglutination test involves a reaction in which antibodies are made against cell cultures and are then reacted with red cells of the corresponding species. If hemagglutination occurs, it indicates that there is a species antigen in common between the red cells and the tissue culture cells used to evoke the antibody. With mixed agglutination, antibodies are prepared in animals red cells of a given species and the antibody thus obtained is used to sensitize cultured cells. When red cells are added, there appears to be a complex formed, first between the species antibody of the red cells and the tissue culture cells and then between homologous red cells and the attached antibody. This results in clumping of the red cells at the surface of the cultured cells. In both tests, species antibodies between red cells and tissue culture cells have been related. In the fluorescent antibody test, antibodies are made against a cell culture of known species and the antibodies react with

TABLE V
Summary of Interspecies Contamination of Cell Lines Detected by Fluorescent Antibody[a]

Number of lines tested	Species claimed	Actual species found	Procedure used
1	Monkey	Human	Fluorescent antibody
1	Human	Mouse and human	Fluorescent antibody
10	Human	Mouse	Fluorescent antibody
1	Mouse	1% Human, 99% mouse	Fluorescent antibody
1	Dog	Human	Fluorescent antibody
3	?	Human	Fluorescent antibody
1	Mouse	Human	Fluorescent antibody
1	Monkey	Human	Fluorescent antibody
1	Human	Monkey	Fluorescent antibody and karyology
1	Human	5% Human, 95% mouse	Fluorescent antibody
1	SV40 transformed human	Hamster	Fluorescent antibody and karyology
1	Rabbit	Monkey and human	Fluorescent antibody and karyology

[a]Data from Simpson and Stulberg (1963, 1971).

the cultured cells at their surface but no red cells are involved. However, the end results in all three of these interspecies tests are the same.

G. Other Immunologic Methods

1. Immunodiffusion

Although immunodiffusion is one of the most widely used and one of the most definitive methods in present day immunology, it did not play a large role in interspecific identification of cells and in detecting contamination of cultures by cells. Very recently, however, Greene and Charney (1970) and Ibrahim *et al.* (1970) used immunodiffusion procedures to differentiate between moth and mosquito lines. Each of these groups came to opposite conclusions. Greene and Charney found that the presumed mosquito cells and the moth cells gave identical precipitin bands and on the basis of other studies (chromosomal and cytotoxicity) concluded that the original mosquito cultures had been contaminated with moth cells. Thus, the testing of the presumed mosquito cell cultures by immunodiffusion alone (Ibrahim *et al.*, 1970) revealed lines of identity between two moth cell extracts, i.e., the contaminant.

2. Complement Fixation

Like immunodiffusion, complement fixation has not been fully utilized for detecting interspecies contamination of cell cultures. Early investigations by Coriell *et al.* (1958) used complement fixation. However, a comparison of complement fixation and mixed agglutination in tissue culture was carried out by Milgrom *et al.* (1964), and they found that whereas complement was bound by cultured cells, complement sensitivity was lower than that obtained with the mixed agglutination procedure.

H. Stable and Variable Antigenic Markers

1. Histocompatibility

The most widely studied histocompatibility antigenic markers in cultured cells have been those of the L cell line from mice—the H-2 antigens. These have been examined by many authors, using a variety of methods that have been summarized by Franks (1968). Gangal *et al.* (1966) and Merchant *et al.* (1967) have shown that all of the H-2K antigens in this antigenic complex are present in the L-M cell line. However, although these methods offer a promising approach, the results are limited at the present time and can be applied only to identifying certain inbred mouse lines.

2. Forssman Antigen

Franks (1968) also cites a number of experiments that he and others have carried out with Forssman and heterophil antigens (see Coombs *et al.*, 1961c; Spear, 1962; Franks *et al.*, 1964; Fogel and Sachs, 1964a,b). These antigens in cultured cells are unlikely to be used in cell contamination work because of the variations in their expression. For example, the mechanisms operating to produce Forssman-positive or Forssman-negative cells are important in themselves but are not adaptable to interspecies contamination work.

I. Isoenzymes

At the same time that isoenzymes were used to demonstrate intraspecies differences in cells (Section III), one enzyme, phosphoglucomutase (PGM), was shown to have specific isoenzymes in different species of cultured cells (Gartler, 1967) that could serve as useful and rapid checks on interspecies contamination. Of six (man, rabbit, mouse, rat, hamster, and cow), species-derived cultures, each had a distinctive PGM zymogram.

In an extensive study of polymorphic enzymes in animal cell cultures, Montes De Oca *et al.* (1969) used the isoenzymes of G6-PD and lactose dehydrogenase (LDH) and were able to establish a profile for identifying animal cells from 20 of 22 different species. This was predicted earlier in studies of various isoenzymes in different species (Yasin and Goldenberg, 1966; Ruddle *et al.*, 1968).

J. Malignancy

Malignancy in cultured cells is of prime importance and both antigens and transplantation of malignant lines have an extensive history (see Klein, 1966; Old and Boyse, 1964; Law, 1970; Tennant, 1970; Hellstrom and Hellstrom, 1970). However, extraneous cell contamination of lines has not been focused upon in the literature. In contrast, contamination is a primary consideration for such cultures that are reference lines in the Animal Cell Culture Collection (American Type Culture Collection, 1971, 1968, 1967, 1965, 1964).

The identification and contamination of human cell lines present a different problem. Foley *et al.* (1965, 1968) have used heterologous transplantation as a criterion of malignancy and have also given consideration to the problem of interspecies contamination of such cells. In other types of studies, certain oncogenic DNA and RNA viruses confer characteristic properties on normal cultured cells; e.g., it is well known that certain viruses of the papova and adenovirus groups can transform normal cell cultures of human and other species to malignant cells. When this occurs, the transformed cells produce several new identifiable antigens, one located in the nucleus (T antigen) and others associated with the cell membranes, e.g., tumor specific and transplantation antigens (see Habel, 1966; Rapp, 1966). Tests for interspecies contamination by extraneous cells are criteria for such of these cultures that have been prepared as reference lines (American Type Culture Collection, 1971, 1968, 1967, 1965, 1964).

III. Intraspecies Identification of Cell Cultures

A. Cytogenetic Analysis

It is generally recognized that the normal karyotype of human cells does not significantly vary but that of other species may. Hsu and Benirschke (1967–1971), have prepared atlases of the euploid karyotypes of most of the classes of the animal kingdom and among many purposes served is the investigation of chromosomal variations useful in interspecies identification of cell cultures. Cytogenetic systems have been of great value in distinguishing between near-diploid human lines derived from individuals with definite chromosomal abnormalities. As is now well known, persons having the chromosomal aberrations of Down's, Klinefelter's, Turner's, or other comparable syndromes readily yield fibroblastic cultures from skin that show the characteristic cytogenetic abnormality for as many transfers as the culture will maintain its integrity. However, such differences cannot be observed in normal diploid cultures or in heteroploid lines, and in these cases chromosomal characteristics do not distinguish cultures within a species.

Exceptions to these general findings, which have occurred mainly in nonhuman species, are illustrated by a mouse line and a pig kidney line. Hsu and Merchant (1961) have found that a variant of the L strain, designated L-M, contains marker chromosomes that identify this line and distinguish it from all other variants of the L strain from mice. Ruddle (1962) has found that a spontaneous transformation occurred in a pig kidney cell line designated PK-13 and, after cloning, it too contained marker chromosomes the incidence and type of which did not alter during transformation. Harris (1964) has discussed these and other similar chromosomal events.

B. Metabolic Markers of Genetic Origin

Krooth (1964, 1965, 1966) and Krooth and Sell (1970) have listed some 37 genetic abnormalities that can be detected biochemically in cultured human cells. Beginning with the demonstration by Krooth and Weinberg

(1962) that cultures of diploid human fibroblasts from a galactosemic individual were unable to metabolize galactose because they lacked the enzyme uridyl transferase, many authors (see Krooth and Sell, 1970) subsequently developed cultures with other metabolic defects. Like human chromosomal aberrations, such cell systems can be used to identify a particular heritable characteristic, and a marker characteristic that distinguishes the line can be established. In this respect, certain histochemical techniques are applicable as genetic markers (Danes and Bearn, 1966). It is difficult to state whether some histochemical methods (see Farnes, 1967; Melnick, 1971) are actually genetic in origin or whether they reflect nongenetic variations. Therefore, unless the cultures contain chromosomal or biochemical deficiency markers, it has not formerly been possible to carry out intraspecies identification.

C. Isoenzyme Profiles

The development of isoenzymes directed by the cell genome (Harris, 1966) has created the possibility of using some of these as stable genetic markers. Isoenzyme polymorphisms are often well expressed in cell cultures and, therefore, their use for marking cultures derived from different individuals within a species offers another parameter of intraspecific identification.

Gartler (1967, 1968) initially found that of 18 well-known human heteroploid cell lines of presumably independent origin, all contained the type A (electrophoretically fast moving) isoenzyme of G6-PD. These results were quite astonishing because it was also known that the A type had been found only in approximately 30% of American Negroes, whereas the slower moving B type could be found in all races. Furthermore, Gartler (1967) observed that PGM type 1 occurred in 100% of these cultures, whereas the frequency of the latter isoenzyme in the general population was approximately 60%. After excluding some illogical reasons for his findings, he concluded that because the HeLa line was of Negro origin and because it grew rapidly and had been widely distributed before the other lines were developed, the latter were really HeLa contaminants. Because these findings were of such importance, three laboratories (see Peterson *et al.*, 1968; Peterson, 1968; Stulberg *et al.*, 1970) independently verified and extended the number and kinds of human cell cultures and examined and analyzed them by three different methods, i.e., starch-gel, acrylamide-gel, and sucrose-gel electrophoresis. They not only verified Gartler's original observations (1967) but where the individual laboratories investigated the same lines, they obtained the same results. Their data are summarized in Table VI and illustrated in Figs. 4 and 5.

These findings are consistent with those of Gartler (1967); in addition, Peterson *et al.* (1971) have reported that initially and after 120 passages *in*

TABLE VI
Intraspecies Analysis of Human Cell Lines by Isoenzyme Type of Glucose 6-Phosphate Dehydrogenase[a]

	Racial origin of donor		G6-PD type		
Type of cell line	Negro	Caucasian	A	B	AB
Heteroploid epithelial-like, permanent	3	17	+		
Diploid or near-diploid,		8		+	
fibroblastlike, finite	1		+		
	1			+	
	1				+
Near-diploid, epithelial-like, permanent		1		+	
Primary, diploid		1		+	
SV40 transformed		1		+	
Lymphocytoid	1	1		+	

[a]Data from Stulberg *et al.* (1970); Peterson *et al.* (1968, 1971).

vitro, a human, heteroploid, permanent cell line retained the B type of G6-PD. This provides experimental evidence that clarifies the stability of these isoenzymes *in vitro*. Recently, other evidence has been reported that shows that even after ten irradiations in the course of 35 transfers *in vitro*, a heteroploid tumor cell line also retained its type B G6-PD and PGM isoenzymes unchanged (Auersperg and Gartler, 1970).

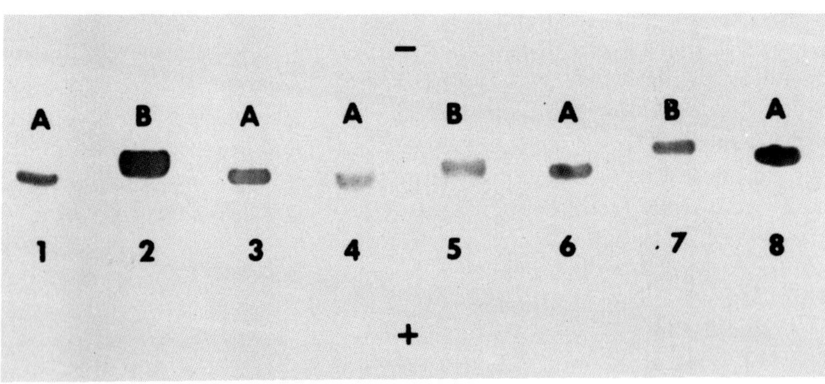

Fig. 4. Glucose 6-phosphate dehydrogenase isoenzyme pattern of cell lines by sucrose-gel film electrophoresis. (A) Fast-moving isoenzyme; (B) slower moving form. (1) MA-160; (2) IMR-32; (3) HeLa; (4) Hep-2; (5) Det. 562; (6) Det 98; (7) WI-38; (8) Det. 6. (MA-160 was supplied by Mr. M. Vincent, Microbiological Associates, Bethesda, Md., and IMR-32 was supplied by Dr. L. L. Coriell, Institute of Medical Research, Camden, N.J.)

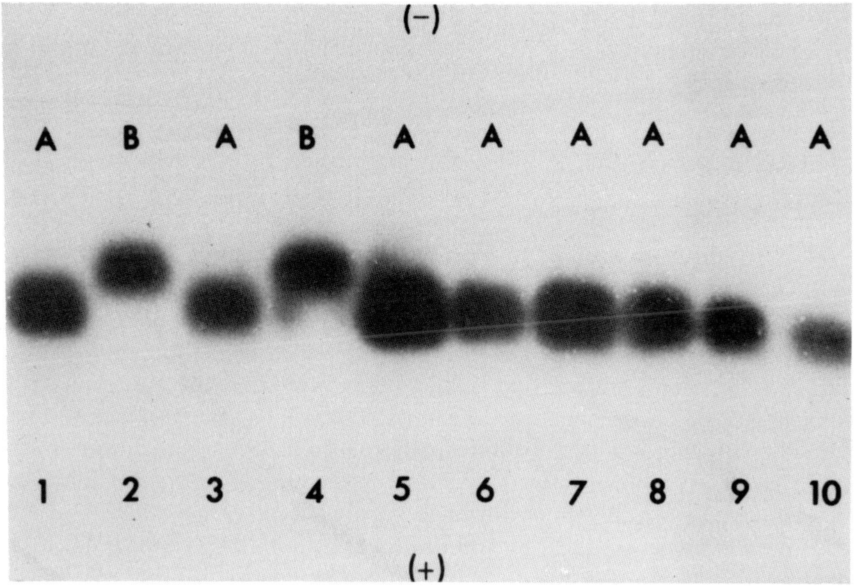

Fig. 5. Glucose 6-phosphate dehydrogenase isoenzyme patterns of Animal Cell Culture Collection lines by starch-gel electrophoresis. Origin is at the cathode (−). (A) Fast-moving isoenzyme; (B) slower moving form. (1) HeLa; (2) Det. 532; (3) Chang conjunctiva (Wong-Kilbourne derivative); (4) Dempsey; (5) AV_3; (6) FL; (7) NCTC 3075; (8) NCTC 2544; (9) WISH; (10) WISH (CDM derivative). See Stulberg *et al.* (1970).

Attention has been focused on isoenzymic variants as genetic markers and for the genetic analysis of somatic cell populations by Ruddle (1968). He has shown that certain properties of isoenzymes readily lend themselves for marker purposes because they are usually highly distinctive between and within different species and also because they occur at a high frequency within a species. Ruddle has demonstrated the *in vitro* stability of isoenzymes and has examined malate, isocitrate, and glucose 6-phosphate dehydrogenases, as well as phosphoglucomutase isoenzymes. Each possesses two variant alleles that are not closely linked. In strains of inbred and hybrid mice and tissue culture populations derived from them, he has demonstrated that isoenzyme analysis identifies the cell populations unequivocally, whereas it would be extremely difficult to determine their origins by other means, including karyotypes.

The isoenzyme findings furnish great impetus for acquiring systematic genetic profiles of any line. It should be possible to utilize other enzyme polymorphisms, i.e., dehydrogenases, kinases, esterases, and transferases, in addition to those that have been discussed. Such profiles allow changes in cells to be followed, thereby avoiding interspecific and intraspecific contaminations. An underlying factor stresses the adequate monitoring of lines by means of preservation by freezing and routine testing for these genetic markers.

D. Immunology

1. Cell Antigens and Isoantigens

For many years, Brand and his associates (Brand, 1965; Silberman-Ziv *et al.*, 1967; Brand *et al.*, 1967) studied the antigens of a number of cell lines with the hope that they could use them to differentiate cultures within a species. Antiserums were prepared against homogenized cultures and in different experiments were tested by double immunodiffusion either against a variety of cell homogenates or against red cells. These investigators obtained analytical data that exhibited differences between human lines in a dual form, i.e., soluble antigens reacting by immunodiffusion or structural elements that reacted with erythrocyte membranes.

However, in view of the interpretations of Gartler's later work (1967), the data referring to human heteroploid lines may have to be reevaluated. Fogel (1968) points out the phenotypic variability of liver-specific antigen in cultures of hamster embryonic liver cells; the subject of antigenic heterogeneity in cultured cells has been discussed in detail by Franks (1967). He points out that the main reasons for studying antigens of cell lines are to establish genetic markers for identification and to distinguish one cell line from another where contamination by cells of extraneous origin has occurred. After illustrating differences between human cell lines as determined by antigens in leukocytes and also by H antigens, he concludes that antigenic differences are no guarantee that cell lines have or do not have a common origin. Dissimilarities between his (Franks and Gurner, 1965) and Metzgar's *et al.* (1965) work may have been caused by different techniques or different antigens that were present. Thus, it is important to examine antigenic heterogeneity that may occur in isoantigens or alloantigens (Franks and Dawson, 1966; Franks, 1966, 1967; Kano and Milgrom, 1967).

2. Immunoglobulin Determinants and Immunoglobulin Production

Lymphoid human cell lines in tissue culture, both malignant and normal, have been shown to produce immunoglobulins. For example, Fahey *et al.* (1966) have demonstrated a variety of immunoglobulins synthesized by lines derived from individuals with Burkitt lymphoma, myeloma, leukemia, and other cells. Similar findings have been reported by Matsuoka *et al.* (1967) and by Finegold *et al.* (1967). These earlier results, as well as subsequent reports, have highly significant implications. However, the kind of immunoglobulins produced depends on determinants or HL-A alloantigens on the cell surfaces (see Coombs *et al.*, 1969; Mann *et al.*, 1969; Klein *et al.*, 1970). These phenomena potentially offer means for distinguishing between cultures of this nature but have not yet been widely used to verify the identification of a line for contamination purposes.

E. Fusion and Hybridization

The discovery by Barski *et al.* (1960) that the mating of somatic cells was possible was followed by the classical studies of Ephrussi and his associates (Ephrussi and Sorieul, 1962; Ephrussi *et al.*, 1964; Weiss and Ephrussi, 1966) and by Littlefield (1966), showing that cell hybridization is an important means for the study of gene mapping in mammalian cells. Following the production of artificial heterokaryons from different animal species through the mediation of inactivated viruses by Harris and his colleagues (see Harris *et al.*, 1966; Watkins and Grace, 1967), hybridization was more commonly employed because the use of fusing viruses allowed the phenomenon to progress more rapidly and predictably. Hybridization not only became a tool for gene localization but also lent itself to studies on virus replication and virus rescue (see Neff and Enders, 1968; Weiss and Green, 1967; Koprowski *et al.*, 1967). Knowles *et al.* (1968) investigated tumor viruses by this means and by 1971 (see Defendi, 1971; Koprowski, 1971; Ruddle, 1971; Green *et al.*, 1971), hybridization of somatic mammalian cells had developed to the point where it had rapidly become a valuable experimental means for investigation in these areas. The number and use of hybrid cells had begun to parallel the development of individual cell lines. With methodological advances that have been rapidly introduced, it now becomes of great importance to be able to clearly identify hybrid cells by such genetic markers as they are capable of producing. Hybrids of mouse–mouse, man–mouse, man–hamster, man–rat, etc., are examples of interspecies hybrid cells.

Although such internal genetic markers as chromosomes and isoenzymes have been studied at length, there appear to be few data on the surface antigens of interspecies hybrid cells (Kano *et al.*, 1969). The latter studies utilize the mixed agglutination reaction and whereas cytogenetic studies indicate that human chromosomes disappear more rapidly than do chromosomes of other species in hybrids, histocompatibility antigens have been detected for a longer period; species-specific antigens are always demonstrated by mixed agglutination in hybrids of mouse lines and are present in man–hamster and man–rat hybrids during the period of testing.

Those who work with hybrid cell cultures have been especially concerned with the characterization of their cell populations and the utilizaion of their techniques. The study of hybrid cells is relatively recent, and newer applications of cytogenetic and other techniques are now areas of intense study (see Miller *et al.*, 1971; Caspersson *et al.*, 1970; Migeon, 1970; Matsuya *et al.*, 1968; Weiss and Green, 1967). However, it can be seen that the possibilities for inter- and intraspecies contaminations are extensive.

IV. Current Status of Cell Contamination

Although interspecies contamination has been recognized for many years, it is still apparently occurring (Simpson and Stulberg, 1971). In view of the numerous methods available for detecting such contamination of cell lines, it is a routine matter to identify any cell line as far as species of origin is concerned and at the present time there is no reason to work with cells in which the species has been unverified or in which there has been contamination of cells from other species (see Stulberg, 1968). The work of Herrick *et al.* (1970), in which species contamination has been recognized and reported, is the only recent example of its kind to the author's knowledge.

Intraspecies contamination is more difficult to verify, although methods for doing so have become available. It is true that there are many instances in which intraspecies identification does not have a bearing on the work reported. In other cases, particularly where malignancy is concerned, some authors have taken precautions to insure the integrity of the cultures they are working with (see Tumilowicz *et al.*, 1970; McAllister *et al.*, 1971; Adams *et al.*, 1971; Peterson *et al.*, 1971; Fogh *et al.*, 1971).

Another aspect of contamination is also of concern. With the advent of human lymphoid cultures, it is interesting that of those cultures that have been tested for isoenzymes of G6-PD, most have turned out to have the type B variant of this enzyme, although there are instances in which the type A isoenzyme has been found (Peterson, 1971). It would seem that unless more isoenzyme polymorphisms are sought, the problem of intraspecies contamination with such cells might very well turn into the same kind of problem that has existed with the human heteroploid lines of the past. This can become a serious hazard, especially as the type B variant of G6-PD occurs in all races. At least one investigator (Gerber, 1971) has looked into such possibilities. In applicable situations, it is important for investigators to be cognizant of extrinsic cell contamination of cell cultures. One way that such an event can be avoided is to have significant cell lines carefully characterized as reference cells and constantly tested for all types of contamination. Therefore, it is important for both commercial and noncommercial organizations to make known and to confirm the nature of cell lines to all individuals who utilize them. An example has been set by the Animal Cell Culture Collection and although I admit to a bias in its favor in view of my long-standing association with the Collection, the Collection serves well as an example of the possibilities of standardization in the field of cell culture (see Stulberg *et al.*, 1970).

Extraneous cell contamination is of concern in many rapidly expanding fields employing tissue culture, and examples of investigative work have been cited. However, the principal objective of this chapter has been to reemphasize inherent problems, both past and present.

Acknowledgment

My research was supported by USPHS Grant CA-02947 from the National Cancer Institute.

References

Adams, R. A., Hellerstein, E. E., Pothier, L., Foley, G. E., Lazarus, H., and Stuart, A. B. (1971). *Cancer* **27**, 651.

American Type Culture Collection. (1964). "Registry of Animal Cell Lines Certified by the Cell Culture Collection Committee," 1st ed., *Supplements* 1965, 1967, 1968, 1971. U. S. Public Health Service.

Auersperg, N., and Gartler, S. M. (1970). *Exp. Cell. Res.* **61**, 465.

Barski, G., Sorieul, S., and Cornefert, F. (1960). *C. R. Acad. Sci.* **251**, 1825.

Bassett, C. A. L., Campbell, D. H., Evans, V. J., and Earle, W. R. (1957). *J. Immunol.* **78**, 79.

Brand, K. G. (1962). *Nature (London)* **194**, 754.

Brand, K. G. (1965). *Nature (London)* **206**, 1164.

Brand, K. G., and Syverton, J. T. (1959). *Proc. Amer. Ass. Cancer Res.* **3**, 8.

Brand, K. G., and Syverton, J. T. (1960). *J. Nat. Cancer Inst.* **24**, 1007.

Brand, K. G., and Syverton, J. T. (1962). *J. Nat. Cancer Inst.* **28**, 147.

Brand, K. G., Chiu, S. Y., Silberman-Ziv, G., Guoen, L. C., and Brand, I. (1967). *J. Immunol.* **99**, 1042.

Caspersson, T., Zech, L., Johansson, C., and Modest, E. J. (1970). *Chromosoma* **30**, 215.

Chu, E. H. Y., and Giles, N. H. (1968). *J. Nat. Cancer Inst.* **20**, 383.

Clarkson, B., Strife, A., and de Harven, E. (1967). *Cancer* **20**, 926.

Clausen, J. J., and Syverton, J. T. (1962). *J. Nat. Cancer Inst.* **28**, 117.

Colter, J. S., Defendi, V., Wallace, R. E., and Bird, H. H. (1958). *J. Nat. Cancer Inst.* **20**, 1141.

Committee on Terminology of the Tissue Culture Association. (1967). *J. Nat. Cancer Inst.* **38**, 607.

Coombs, R. R. A. (1962). *Nat. Cancer Inst., Monogr.* **7**, 91.

Coombs, R. R. A., Daniel, M. R., Gurner, B. W., and Kelus, A. (1961a). *Nature (London)* **189**, 503.

Coombs, R. R. A., Daniel, M. R., Gurner, B. W., and Kelus, A. (1961b). *Immunology* **4**, 55.

Coombs, R. R. A., Daniel, M. R., Gurner, B. W., and Kelus, A. (1961c). *Int. Arch. Allergy Appl. Immunol.* **19**, 210.

Coombs, R. R. A., Gurner, B. W., Beale, A. J., Christofinis, E., and Page, Z. (1961d). *Exp. Cell Res.* **24**, 604.

Coombs, R. R. A., Beinstein, A., and Wilson, A. B. (1969). *Lancet* **2**, 1157.

Coriell, L. L., Tall, M. G., and Gaskill, H. (1958). *Science* **128**, 198.

Danes, S., and Bearn, A. (1966). *J. Exp. Med.* **123**, 1.

Defendi, V. (1971). *Fed. Proc. Fed. Amer. Soc. Exp. Biol.* **30**, 912.

Defendi, V., and Colter, J. S. (1959). *J. Nat. Cancer Inst.* **23**, 411.

Defendi, V., Kamrin, A. A., and Colter, J. S. (1958). *Nature* (*London*) **182**, 1246.

Defendi, V., Billingham, R. E., Silvers, W. K., and Moorhead, P. (1960). *J. Nat. Cancer Inst.* **25**, 359.

Deinhardt, F., and Henle, G. (1957). *J. Immunol.* **79**, 60.

Eagle, H. (1955). *Science* **122**, 501.

Earle, W. R. (1943). *J. Nat. Cancer Inst.* **4**, 165.

Ephrussi, B., and Sorieul, S. (1962). *Univ. Mich. Med. Bull.* **28**, 347.

Ephrussi, B., Scaletta, L. J., Stenchever, M. A., and Yoshida, M. C. (1964). *Symp. Int. Soc. Cell Biol.* **3**.

Epstein, M. A., Achong, B. G., and Barr, Y. M. (1964). *Lancet* **1**, 702.

Epstein, M. A., Achong, B. G., Barr, Y. M., Zajac, B., Henle, G., and Henle, W. (1966). *J. Nat. Cancer Inst.* **37**, 547.

Evans, V. J., Earle, W. R., Sanford, K. K., Shannon, J. E., and Waltz, H. K. (1951). *J. Nat. Cancer Inst.* **11**, 907.

Fahey, J. L., Finegold, I., Rabson, A. S., and Manaker, R. A. (1966). *Science* **152**, 1259.

Farnes, P. (1967). *Nat. Cancer. Inst., Monogr.* **26**, 199.

Finegold, I., Fahey, J. L., and Granger, H. (1967). *J. Immunol.* **99**, 839.

Fogel, M. (1968). *Develop. Biol.* **17**, 85.

Fogel, M., and Sachs, L. (1964a). *Exp. Cell Res.* **34**, 448.

Fogel, M., and Sachs, L. (1964b). *Develop. Biol* **10**, 411.

Fogh, J., Holmgren, N., and Ludovici, P. P. (1971). *In Vitro* **7**, 26.

Foley, G. E., Handler, A. H., Lynch, P. M., Woman, S. R., Stulberg, C. S., and Eagle, H. (1965). *Cancer Res.* **25**, 1254.

Foley, G. E., Lazarus, H., Farber, S., Uzman, B. G., and Adams, R. A. (1968). *In* "The Proliferation and Spread of Neoplastic Cells," Univ. of Texas M. D. Anderson Hospital and Tumor Inst., Symp. Monogr. No. 21, p. 65. Williams & Wilkins, Baltimore, Maryland.

Ford, D. K., and Yerganian G. (1958). *J. Nat. Cancer Inst.* **21**, 393.

Franks, D. (1966). *Vox Sang.* **11**, 674.

Franks, D. (1967). *In Vitro* **2**, 54.

Franks, D. (1968). *Biol. Rev. Cambridge Phil. Soc.* **43**, 17.

Franks, D., and Dawson, A. (1966). *Exp. Cell Res.* **42**, 543.

Franks, D., and Gurner, B. W. (1965). *Exp. Cell Res.* **37**, 693.

Franks, D., Gurner, B. W., Coombs, R. R. A., and Stevenson, R. (1963). *Exp. Cell. Res.* **28**, 608.

Franks, D., Daniel, M. R., Gurner, B. W., and Coombs, R. R. A. (1964). *Exp. Cell Res.* **36**, 310.

Gangal, S. G., Merchant, D. J., and Shreffler, D. C. (1966). *J. Nat. Cancer Inst.* **36**, 1157.

Gartler, S. M. (1967). *Nat. Cancer Inst., Monogr.* **26**, 167.

Gartler, S. M. (1968). *Nature* (*London*) **217**, 750.

Gerber, P. (1971). Personal communication.

Gey, G. O., Coffman, W. D., and Kubicek, M. G. (1952). *Cancer Res.* **12**, 294.

Goldstein, M. N. (1957). *J. Immunol.* **79**, 113.

Grace, T. D. C. (1962). *Nature* (*London*) **195**, 798.

Green, H., Wang, R., Basilico, C., Pollack, R., Kusana, T., and Salas, J. (1971). *Fed. Proc., Fed. Amer. Soc. Exp. Biol.* **30**, 930.

Greene, A. E., and Charney J. *Symp. Arthropod Cell Cult. Appl. Study Viruses, 1970* (in press).

Greene, A. E., Coriell, L. L., and Charney, J. (1964). *J. Nat. Cancer Inst.* **32**, 779.

Habel, K. (1966). *In* "Malignant Transformation by Viruses" (W. H. Kirsten, ed.), p. 60 Springer-Verlag, Berlin and New York.

Habel, L., Hornibrook, J. W., Gregg, N. C., Silverberg, R. J., and Takemoto, K. K. (1958). *Virology* **5**, 7.
Harris, H. (1966). *Proc. Roy. Soc., Ser. B.* **164**, 298.
Harris, H., Watkins, J. F., Ford, C. E., and Schoefl, G. L. (1966). *J. Cell Sci.* **1**, 1.
Harris, M. (1964). "Cell Culture and Somatic Variation." Holt, New York.
Harris, M. (1968). *Nat. Cancer Inst., Monogr.* **29**, 1.
Hayflick, L. (1965). *Exp. Cell Res.* **37**, 614.
Hayflick, L., and Moorhead, P. S. (1961). *Exp. Cell Res.* **25**, 585.
Hellstrom, I., and Hellstrom, K. E. (1970). *Int. J. Cancer* **5**, 195.
Henle, G., Henle, W., and Diehl, V. (1968). *Proc. Nat. Acad. Sci. U.S.* **59**, 94.
Herrick, P. R., Baumann, G. W., Merchant, D. J., Shearer, M. C., Shipman, C., Jr., and Brackett, R. G. (1970). *In Vitro* **6**, 143.
Hsu, T. C., and Benirschke, K. (1967–1971). "An Atlas of Mammalian Chromosomes," Vols. 1–5. Springer-Verlag, Berlin and New York.
Hsu, T. C., and Merchant, D. J. (1961). *J. Nat. Cancer Inst.* **26**, 1075.
Hsu, T. C., and Moorhead, P. S. (1957). *J. Nat. Cancer Inst.* **18**, 463.
Ibrahim, A. N., Cupp, E. W., and Sweet, B. H. (1970). *21st Annu. Meet. Tissue Culture Assoc.*, Abstract 15.
Kano, K., and Milgrom, F. (1967). *Int. Arch. Allergy Appl. Immunol.* **31**, 209.
Kano, K., Baranska, W., Knowles, B. B., Koprowski, H., and Milgrom, F. (1969). *J. Immunol.* **103**, 1050.
Klein, E., Eskeland, J., Inoue, R., Strom, R., and Johanson, B. (1970). *Exp. Cell Res.* **62**, 133.
Klein, G. (1966). *Annu. Rev. Microbiol.* **20**, 223.
Knowles, B. B., Steplewski, Z., Swetly, P., Barbanti-Brodano, G., and Koprowski, H. (1968). *In* "Heterospecific Genome Interaction" (V. Defendi, ed.), Wistar Inst. Symp. Monogr. No. 9, p. 37. Wistar Inst. Press, Philadelphia, Pennsylvania.
Koprowski, H. (1971). *Fed. Proc., Fed. Amer. Soc. Exp. Biol.* **30**, 914.
Koprowski, H., Jensen, F. C., and Steplewski, Z. (1967). *Proc. Nat. Acad. Sci. U.S.* **58**, 127.
Krooth, R. S. (1964). *In* "Somatic Cell Genetics" (R. S. Krooth, ed.), p. 167. Univ. of Michigan Press, Ann Arbor.
Krooth, R. S. (1965). *Birth Defects, Orig. Art. Ser.* **1**, 21.
Krooth, R. S. (1966). *In Vitro* **2**, 82.
Krooth, R. S., and Sell, E. K. (1970). *J. Cell. Physiol.* **76**, 311.
Krooth, R. S., and Weinberg, A. N. (1962). *J. Exp. Med.* **113**, 1155.
Latta, H., and Kutsakis, A. (1957). *Lab. Invest.* **6**, 12.
Law, L. W. (1970). *Transplant. Proc.* **2**, 117.
Levan, A., and Biesele, J. J. (1958). *Ann. N. Y. Acad. Sci.* **71**, 1022.
Littlefield, J. W. (1966). *Exp. Cell. Res.* **41**, 190.
McAllister, R. M., Gardner, M. B., Greene, A. E., Bradt, C., Nichols, W. W., and Landing, B. H. (1971). *Cancer* **27**, 397.
McCulloch, E. A., and Parker, R. C. (1957). *In* "Canadian Cancer Conference" (R. W. Begg, ed.), Vol. 2. Academic Press, New York.
Mann, D. L., Rogentine, G. N., Jr., Fahey, J. L., and Nathenson, S. G. (1969). *Science* **163**, 1460.
Matsuoka, Y., Moore, G. E., Yagi, Y., and Pressman, D. (1967). *Proc. Soc. Exp. Biol. Med.* **125**, 1246.
Matsuya, Y., Green, H., and Basilico, C. (1968). *Nature* (*London*) **220**, 1199.
Melnick, P. J. (1971). *Progr. Histochem. Cytochem.* **2**, 1.
Merchant, D. J., Gangal, S. G., and Holmgren, N. B. (1967). *In Vitro* **2**, 97.
Metzgar, R. F., Flanagan, J. F., and Zmijewski, C. M. (1965). *J. Immunol.* **95**, 494.
Migeon, B. R. (1970). *Ann. N. Y. Acad. Sci.* **171**, 396.
Milgrom, F., Witebsky, E., Barron, A. L., and Millers, R. (1964). *Vox Sang.* **9**, 289.

Miller, O. J., Allerdice, P. W., Miller, D. A., Breg, W. R., and Migeon, B. R. (1971). *Science* **173**, 243.

Montes De Oca, F., Macy, M. L., and Shannon, J. E. (1969). *Proc. Soc. Exp. Biol. Med.* **132**, 462.

Moore, G. E., Ito, E., Ulrich, K., and Sandberg, A. A. (1966). *Cancer* **19**, 713.

Moore, G. E., Gerner, R. E., and Franklin, H. A. (1967). *J. Amer. Med. Ass,* **199**, 519.

Mountain, I. M. (1955). *J. Immunol.* **75**, 478.

Neff, J. M., and Enders, J. F. (1968). *Perspect. Virol.* **6**, 39.

Nichols, W. W. (1971). *Hereditas* **67**, 1.

Old, L. J., and Boyse, E. A. (1964). *Annu. Rev. Med.* **15**, 167.

Parker, R. C. (1958). "Poliomyelitis: Papers and Discussions Presented at the 4th International Poliomyelitis Conference," p. 257. Lippincott, Philadelphia, Pennsylvania.

Parker, R. C., Castor, L. N., and McCulloch, E. A. (1957). *Spec. Publ. N. Y. Acad. Sci.* **5**, 303.

Peterson, W. D., Jr. (1968). *Nat. Cancer Inst., Monogr.* **29**, 55.

Peterson, W. D., Jr. (1971). Unpublished data.

Peterson, W. D., Jr., Stulberg, C. S., Swanborg, N. K., and Robinson, A. R. (1968). *Proc. Soc. Exp. Biol. Med.* **128**, 772.

Peterson, W. D., Jr., Stulberg. C. S., and Simpson, W. F. (1971). *Proc. Soc. Exp. Biol. Med.* **136**, 1187.

Puck, T. T., and Kao, F. (1968). *Proc. Nat. Acad. Sci. U.S.* **60**, 561.

Puck, T. T. Cieciura, S. J., and Robinson, A. (1958). *J. Exp. Med.* **108**, 945.

Quersin-Thiery, L. (1959). *J. Immunol.* **82**, 542.

Rapp, F. (1966). *In* "Malignant Transformation by Viruses" (W. H. Kirsten, ed.), p. 77. Springer-Verlag, Berlin and New York.

Rothfels, K. H., and Parker, R. C. (1960). *J. Exp. Zool.* **507**, 519.

Rothfels, K. H., Axelrad, A. A., Siminovitch, L., McCulloch, E. A., and Parker, R. C. (1959). *Proc. Can. Cancer Res. Conf.* **3**, 189.

Ruddle, F. H. (1962). *Proc. Soc. Exp. Biol. Med.* **109**, 116.

Ruddle, F. H. (1968). *Nat. Cancer Inst., Monogr.* **29**, 9.

Ruddle, F. H. (1971). *Fed. Proc., Fed. Amer. Soc. Exp. Biol.* **30**, 921.

Ruddle, F. H., Shows, T. B., and Roderick, T. H. (1968). *Genetics* **58**, 599.

Sanford, K. K., Earle, W. R., and Likely, G. D. (1948). *J. Nat. Cancer Inst.* **9**, 229.

Scherer, W. F. (1953). *Amer. J. Pathol.* **29**, 113.

Scherer, W. F., Syverton, J. T., and Gey, G. O. (1953). *J. Exp. Med.* **97**, 695.

Silberman-Ziv, G., Chiu, S. Y., Buoen, L. C., Brand, I., and Brand, K. G., (1967). *Z. Immunitaetsforsch., Allerg. Klin. Immunol.* **134**, 23.

Simpson, W. F., and Stulberg, C. S. (1963). *Nature (London)* **189**, 616.

Simpson, W. F., and Stulberg, C. S. (1971). Unpublished data.

Spear, G. S. (1962). *Bull. Johns Hopkins Hosp.* **111**, 307.

Stulberg, C. S. (1968). *Nat. Cancer Inst., Monogr.* **29**, 15.

Stulberg, C. S., Simpson, W. F., Peterson, W. D., and Berman, L. (1961a). *Fed. Proc., Fed. Amer. Soc. Exp. Biol.* **20**, 150 (abstr.).

Stulberg, C. S., Simpson, W. F., and Berman, L. (1961b). *Proc. Soc. Exp. Biol. Med.* **108**, 434.

Stulberg, C. S., Coriell, L. L., Kniazeff, A. J., and Shannon, J. E. (1970). *In Vitro* **35**, 1.

Sweet, B. H., and McHale, J. S. (1970). *Exp. Cell Res.* **61**, 51.

Swim, H. E. (1959). *Annil. Rev. Microbiol.* **13**, 141.

Szybalski, W., and Szybalska, E. H. (1962). *Univ. Mich. Med. Bull.* **28**, 277.

Tennant, J. R. (1970). *Transplant. Proc.* **2**, 104.

Tjio, J. H., and Levan, A. (1956). *Hereditas* **42**, 1.

Tumilowicz, J. J., Nichols, W. W., Cholon, J. J., and Greene, A. E. (1970). *Cancer Res.* **30**, 2110.

Watkins, J. F., and Grace, D. M. (1967). *J. Cell Sci.* **2**, 193.

Weiss, M. C., and Ephrussi, B. (1966). *Genetics* **54**, 1111.
Weiss, M. C., and Green, H. (1967). *Proc. Nat. Acad. Sci. U.S.* **58**, 1104.
Westwood, J. C. N., Macpherson, I. A., and Titmuss, D. H. J. (1957). *Brit. J. Exp. Pathol.* **38**, 138.
Yasin, R., and Goldenberg, G. J. (1966). *Nature (London)* **211**, 1296.
Yasumura, Y., Tashjian, A. H., Jr., and Sato, G. H. (1966). *Science* **114**, 1186.

CHAPTER 2

Methods of Prevention of Bacterial, Fungal, and Other Contaminations

LEWIS L. CORIELL

I. Source of Infection 30
A. Cells 31
B. Glassware and Apparatus 33
C. Culture Media 36
D. Airborne Modes 38
II. Procedures within the Transfer Room 41
III. Laminar Flow Transfer Room 43
IV. Aseptic Technique 44
V. Antibiotics 45
VI. Summary 47
References 48

Tissue culture media provide all the essential nutrients required for the multiplication of mammalian cells. The basic ingredients of culture media contain the essential inorganic salts, amino acids, and vitamins but for the growth of most cells this must be supplemented by the addition of whole serum to provide essential nutrients not yet identified. Chemically defined media without serum are adequate to support growth of some heteroploid cell cultures and many investigators have devised special media that favor the growth and survival of certain cell types. All media share one common characteristic: they provide an excellent culture medium for the growth of microbial contaminants and under average cell culture procedures there are ample opportunities for contamination, i.e., frequent opening of flasks for feeding or subculture and incubating at temperatures that favor microbial growth. Furthermore, cell cultures are virtually devoid of the physical and chemical defenses that protect cells from contamination in the living animal. For example, some characteristics of the human skin that prevent infection are (*1*) dehydrated surface cells, (*2*) multilayering, and (*3*) continuous sloughing, (*4*) acid pH in the range of pH 4.0–6.0, averaging 4.85, (*5*) an oil coating to resist wetting and maceration, (*6*) scaling to prevent abrasion, (*7*) folding of the epidermis, (*8*) elasticity of the corium, and (*9*) the movement and padding of the subcutaneous tissue. The rich capillary bed in the subcutaneous tissue (*10*) maintains nutrition, (*11*) removes wastes, (*12*) stabilizes temperature and (*13*) pH of the deeper viable cells, and allows the delivery of (*14*) humoral and (*15*) cellular immune components that attack foreign microbes and cells not identified as self. From this brief listing, it is clear that the intact skin has multiple barriers against infection, some to exclude infection and others to deal with infections that penetrate the mechanical and chemical barriers. To avoid contamination of cell cultures, experience has shown that we must emulate nature and establish multiple barriers to infection. In the absence of immune mechanisms in cell cultures, primary emphasis must be placed on preventing infection. The use of antibiotics as surrogate immune mechanisms has proved to be disasterous in long-term cultures. The reasons for this will be presented in a later section but for now it is sufficient to note that Barile (see Chapter 6) has found that 72% of long-term cell cultures with antibiotics in the culture medium have been contaminated (46 out of 64), whereas only 7% of long-term cultures without antibiotics have been contaminated (2 out of 28).

I. Source of Infection

The routes by which cell cultures become contaminated may be grouped under four headings, as illustrated in Fig. 1: (*1*) contaminated cells, (*2*) glassware or apparatus, (*3*) culture media, and (*4*) airborne modes. The frequency of occurrence of each pathway varies among laboratories and changes within

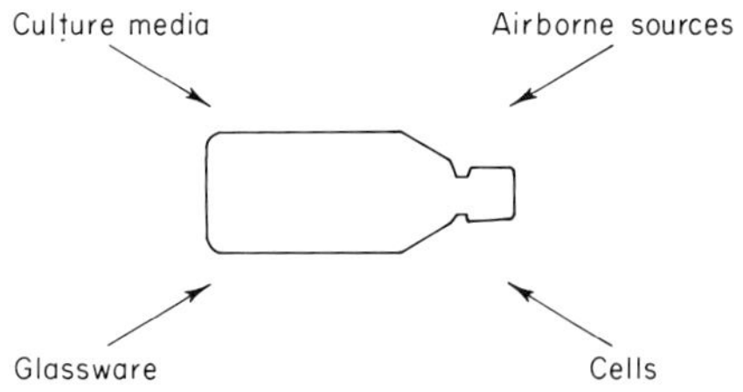

Fig. 1. Sources of microbial contamination of cell cultures.

each laboratory with changes of weather, season, personnel, procedures, housekeeping, sources of supply, etc. All four pathways of infection can be controlled or prevented by following two principles. First is the adoption of effective standard procedures that exclude or greatly reduce the opportunity for contamination and second is continuous surveillance to verify that the procedures are properly executed. This chapter will discuss the essential preventive and surveillance procedures to maintain contamination-free cell cultures. Granted that no complex system involving human participation is likely to be perfect, the suggested procedures are also designed to provide prompt evidence of contamination and to indicate its source, so that corrective measures can be taken. Consistent freedom from contamination provides evidence of the adequacy of all procedures.

The minimum barriers needed to protect against introducing contamination into the tissue culture laboratory with cell cultures are outlined in Fig. 2. These measures can be instituted in any size tissue culture laboratory with relative ease.

A. Cells

The absence of microbial contamination can never be assumed but must be proved by sterility tests on cell cultures before they can be accepted as

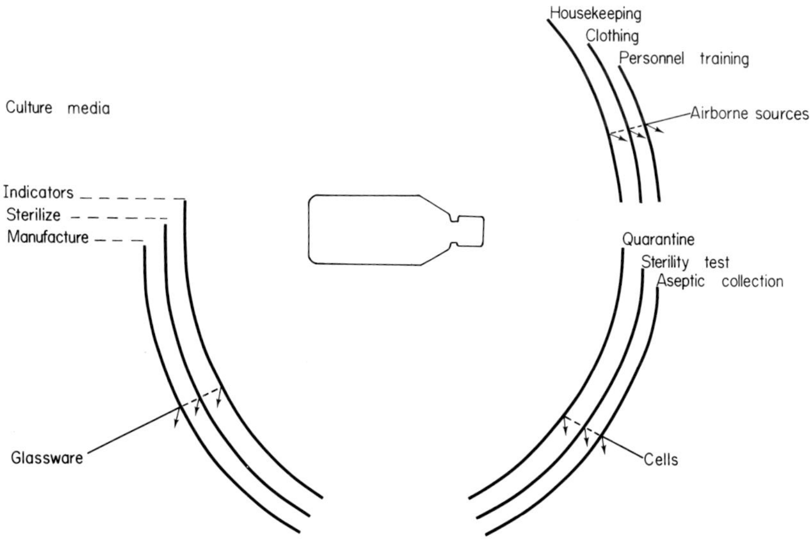

Fig. 2. Diagramatic representation of barriers to microbial contamination of cell cultures via glassware, cells, and the airborne routes.

clean. Occult, unsuspected microbial contamination is frequently transferred from one laboratory to another by exchanging contaminated cell cultures. This type of contamination is a great hazard in the cell culture laboratory because it is difficult to detect and is usually resistant to antibiotics because of previous exposure; also, its frequent presence in several different cell lines from the same laboratory suggests some ability of the organism to get from one cell culture to another. It is therefore very important to keep new cell cultures isolated until sterility tests are completed. This quarantine period should be a minimum of 3 weeks and frequently longer, during which time the cells will be refed, transferred, and incubated in a facility separate from other clean cell cultures. The hands should be washed and the work surface scrubbed with disinfectant after each cell culture is transferred in the quarantine area. If a separate area cannot be made available for handling quarantined cultures, they should be handled at the end of the day after all clean cultures have been processed. Clean the transfer area, open the quarantined cultures one at a time, and clean between each culture as above.

Sterility Tests

Sterility tests of cell cultures from another laboratory must be made the first time the flasks are opened to provide valid data on their condition upon receipt. Detection of contamination at subsequent passages may reflect contamination in the new laboratory. This precaution seems obvious but is often ignored in practice. If the cell culture contains antibiotics, they may inhibit the growth of contaminants and special procedures will be required. Centrifugation at high speed with culture and special stains of the sediment will eliminate much of the antibiotic and concentrate the contaminant. Special stains may show mycoplasmas (Fogh and Fogh, 1964) or bacteria that fail to grow on culture media. Electron microscopy has been used to demonstrate mycoplasmas and viruses (Hummeler and Armstrong, 1967; Hall *et al.*, 1967; Kniazeff, 1968). A simple and practical test is to remove the antibiotics from the culture medium for several passages and then repeat the sterility tests.

No single culture medium will support the growth of all microbial contaminants found in cell cultures. To have reasonable assurance of sterility and to save time, it is therefore necessary to use multiple culture media and procedures to encourage growth of the organisms most commonly encountered, i.e., bacteria, yeasts, molds, and mycoplasmas. Organisms that grow rapidly and produce grossly apparent changes in tissue culture media or damage to the cell culture are not a serious problem because their presence is readily apparent and the experiments or tests can be promptly terminated. Organisms that grow slowly, without producing gross changes in the culture medium or cytotoxicity to the cells, are more serious because they are not easily detected. Furthermore, these fastidious organisms frequently grow poorly or not at all in the routine culture media used for sterility tests. These contaminating organisms can invalidate many types of experiments with cell cultures because they introduce foreign antigens, enzymes, metabolic pathways, breakdown products, nutritional requirements, nucleic acids, and nucleoproteins (Stanbridge, 1971).

The efficiency of sampling depends partially on the size of the inoculum. The inoculums recommended in Table I represent compromises between maximum efficiency and ease of handling. Cultures are incubated at 37°C and at 30°C. Cells, as well as culture fluid, should be inoculated into the culture medium. Cultures should be inspected daily for evidence of growth for 14 days.

If culture tubes are cloudy or develop a precipitate, a gram stain and wet mount should be examined and the organism subcultured and identified. It is important to keep a record of the morphology and cultural characteristics of each contaminant as an aid in determining the epidemiology of culture contamination so that effective controls can be instituted.

B. Glassware and Apparatus

Three barriers to the introduction of contamination by unsterile apparatus are (*1*) selection of glassware and/or control of the manufacturing process, (*2*) effective sterilization, and (*3*) use of indicators or controls of sterilization.

Glassware

Borosilicate glass is preferred by many workers because of its resistance to heat, scratching, mechanical breakage, and solvents. It can be used over and over and this is a saving for such expensive pieces as graduated pipets and complex apparatus. Soda glass flasks and tubes are also widely used for tissue culture. They are frequently used once and discarded because they are inexpensive and because breakage is higher during cleaning and resterilization. If used for storage of solutions, soda glass tends to dissolve and form crystals of sodium silicate in the liquid. Disposable sterile flasks and pipets made of polystyrene plastic are now used extensively in cell culture. Flasks have welded seams and are brittle under impact. Occasionally, this leads to cracks that admit contamination. Occasionally, cells do no adhere to the plastic surface over portions of the flasks. This is apparently because of some defect in the manufacturing process, which applies a special nontoxic finish for tissue culture. This is not a frequent problem. Whereas borosilicate glass, soda glass, and plastic are each successfully used in tissue culture, they are not always interchangeable. Some delicate cell cultures adapted to grow in

TABLE I
Recommended Inoculums for the Discovery of Contamination in Cell Cultures

Medium	Inoculum	Incubation: Temperature, °C		Incubation: Time, days	Subculture	Stain
Brain heart infusion broth	0.5	37	30	14	+	+
Tryptose phosphate broth	0.5	37	30	14	+	+
Trypticase soy broth	0.5	37	30	14	+	+
Mycoplasma broth	0.5	37	30	14	+	+
Yeast and mold broth	0.5	37	30	14	−	−
Sabourand dextrose broth	0.5	37	30	14	−	−
Blood agar plate	0.1	37	—	7	+	+
Mycoplasma agar plate	0.1	37	—	14	+	+

one type of container will not grow or will require a period of adaptation if changed to a different type of container. The choice of container is empiric and is conditioned by cost, convenience, reliability of manufacture, type of culture vessel, method of cell culture, experience of the user, and type of cell to be cultured. Chemical contamination of cell culture vessels can prevent cell growth because the usual culture media have limited buffering capacity to neutralize, dilute, or remove toxic substances. This is most critical when cells are grown in chemically defined media with no addition of serum. For this purpose, it is essential to use borosilicate glassware and to clean it with hot acid by the following method (Earle, 1943).

Submerge flasks for 48 hours in a mixture of 80% concentrated H_2So_4, 5% concentrated nitric acid, and 15% distilled water in a container of heavy porcelain, glass, polyethylene, or stainless steel. Remove flasks with tongs, wash 2 times in tap water, and submerge in Calgolac (Calgon Corp., P.O. box 1346, Pittsburgh, Pa.), 1 tbs/gal of distilled water. Heat to 90°–95°C (not boiling) for 30 minutes, cool to 50°C, remove, and rinse 4 times in tap water and 8 times in distilled water; dry in oven. This cleaning procedure is excellent for any new glassware.

The quality of water is an important factor in cleaning glassware. Precipitates may form from impurities in water and the cleaning procedure used in one location may not be satisfactory in another. For this reason, some experimentation may be necessary. At the author's institution, the water supply has a high content of dissolved calcium and ionic salts and conditioning consists of successive passage through beds of sand, resin, carbon, and a mixed bed demineralizer. The routine glasswashing is carried out in a Better Built Turbomatic machine as follows:

Add 1 lb Simbol (Caustic soda) and [mono(trichloro)*tetra*(monopotassium dichloro)]*penta-s*-triazine trione (Wyandotte Chemical Corporation, J.B. Ford Division, Wyandotte, Michigan). There is a prewash cycle of 3 minutes, a wash cycle of 10 minutes, a series of four rinses in tap water, and one rinse in demineralized water. Total time required is 20 minutes. The cycle is repeated without Simbol for all glassware that goes through the washer. This satisfactorily removes all traces of cleaner and detergent, the equivalent of 12 rinses.

If glassware is washed by hand, the procedure is as follows: Use 2 oz per 5 gal of a mixture of two detergents: (*1*) Micro-Solv detergent (from Microbiological Associates, Inc., 4813 Bethesda Avenue, Bethesda, Maryland) and (*2*) Calgolac, using a 150 lb drum of Calgolac mixed with a 50 lb drum of Micro-Solv. The ingredients must be mixed thoroughly; then use 2 oz of this mixture per 5 gal water. Glassware is soaked and hand scrubbed. It is rinsed 14 times in tap water and four times with deionized water.

Commercially prepared plastic culture flasks and pipets are wrapped in cellophane and sterilized with ethylene oxide by the manufacturer and plastic petri dishes are sterilized by ultraviolet light by the manufacturer. Glassware can also be sterilized with ethylene oxide (ETO) but most laboratories do not have this capability and employ moist or dry heat sterilization for its convenience and speed. Ethylene oxide is corrosive to some materials and items sterilized with ETO must be aired before use to remove residual ETO.

The sterilized article must be wrapped or protected so that it will remain sterile until used. Metal cannisters for pipets and petri dishes and wrapping with cloth or paper are most frequently employed. Brown wrapping paper contains sulfides and other chemicals that can be volatilized during sterilization, leaving a chemical deposit on glassware. Patapar parchment paper (Patterson Paper Co., Bristol, Pa.), first used by Earle, is free of these residues and is preferable. Glass vessels are commonly plugged with cotton and covered with paper or cloth before sterilization. The cotton permits air or steam to enter and leave the vessel but filters out microorganisms. Cotton fibers may shed into the vessel, and if it is overheated the cotton may leave a precipitate on the glass. Wrapping the cotton plug in gauze will prevent shedding. Substitutes for cotton plugs, such as screw caps, Morton closure, and aluminum foil, are now used in many laboratories.

Indicators and Controls of Sterilization. To detect and prevent mechanical or personnel errors during sterilization, a number of controls should be included in every run. These include the use of chemical indicators in liquids or large packages to be autoclaved, autoclave tape on the outside of every package, and the date sterilized and initials of the operator on the outside of every package. Ovens and autoclaves should be equipped with a recording thermometer and the charts should be dated, filed, and signed by the person operating the equipment. These measures, if accompanied by inspection, reporting of all contaminations, and coordination with a personnel training program, will develop a group of reliable personnel, the best protection against unsterile equipment.

C. Culture Media

The four components of tissue culture medium shown in Fig. 3 are representative of all media components from the standpoint of sterilization. Basic salt solutions are not decomposed by heat and with proper allowance for evaporation they can be autoclaved. Such enzymes as trypsin, pronase, and collagenase are destroyed by heat and must be sterilized by filtration. Ethylenediaminetetraacetic acid (EDTA) in 0.25% solution is frequently used alone or in combination with trypsin to remove cells from glass. It is heat stable and may be autoclaved but when combined with trypsin must be filtered.

Basal culture media contain heat-sensitive essential amino acids and vitamins and must be sterilized by filtration. Glutamine is more unstable than other amino acids and is usually lyophilized or made up in 100-fold concentration, filtered, stored frozen, and added to the final medium just prior to use.

Serum

The use of serum in the culture medium to assure growth of many cell types is a source of recurring problems. Serum is a complex of small molecules of diverse composition adsorbed on larger proteins by ionic and hydrophobic bonding, all stabilized for ideal cellular nutrition. Its organization and composition are not well known. The growth-stimulating properties of different serum lots vary widely and a serum lot that supports excellent growth of one cell line may be poor for another. For cell culture purposes, the use of serum from pooled bleedings of many animals is cheaper, confers more uniformity, and requires less frequent switches to new serum lots at the cost of some loss of growth promotion. Baby beef, newborn, and embryo calf serum obtained in the abattoir are most widely used for culture of mammalian cells. Commercial suppliers of serum for cell culture have usually obtained their raw serum from the abattoir under conditions that invite some degree of contamination. Delivery of a sterile serum is then dependent on removal of the microorganisms by filtration, always a difficult problem with viscous liquids. Boone showed that of 125 lots of raw serum from commercial slaughterhouses all were contaminated and 10% were still contaminated after they

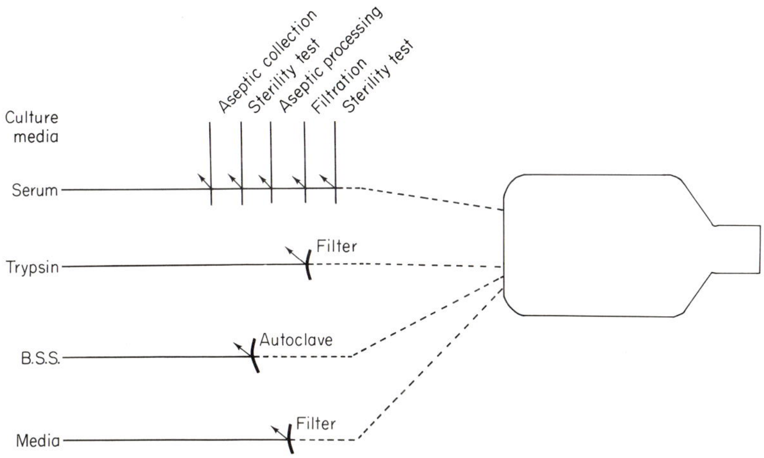

Fig. 3. Recommended procedures for ensuring sterility of cell culture media components. BSS = balanced salt solution.

had passed routine processing, filtration, and sterility tests. This illustrates the difficulty of sterilizing serum by filtration and emphasizes the importance of aseptic collection and processing. The organisms most frequently found in unsterile serum sold for cell culture are slow growing fungi, yeasts, bacteria, and mycoplasmas not detected by simple sterility tests observed for a short time. A study (Boone *et al.*, 1971) showed that the critical step in producing sterile serum is its collection under aseptic conditions. The practicality of this was demonstrated by producing 111 sterile 20 liter lots of unfiltered fetal calf serum. These serum lots were superior in growth promotion for cell cultures and, of course, more expensive to produce. The major steps in production included aseptic collection; removal of red blood cells, leukocytes, and platelets by passing through glass wool; sterility test; and frozen storage of each separate bleeding—all within the day of collection. Those bleedings that passed sterility tests were thawed, pooled into 20 liter lots, and processed aseptically into the final containers without filtration. Cultures on the final containers were all sterile. On the basis of these studies and the experience of many workers, the minimal guidelines to assure sterile serum for cell culture media are indicated in Fig. 3. To reduce the contamination probability to a minimum, the recommended steps are aseptic collection verified by a sterility test, aseptic processing, and filtration as a final step, if desired, all confirmed by a final sterility test. Sterility tests for serum are the same as those recommended for cell cultures in Table I, except for mycoplasmas. Experience has shown that *Mycoplasma arginini* and *Acoliplasma laidlarvii* present in bovine serum in very small numbers may be missed if only 0.5 ml quantities of serum are tested (Barile and Kern, 1972).

It is recommended that serum be cultured for mycoplasmas by inoculating 25 ml of serum into 100 ml of mycoplasma broth, incubated at 37°C, and subcultured onto mycoplasma agar after 3 and 7 days incubation.

D. Airborne Modes

Contamination arising from improperly sterilized culture media, apparatus, and contaminated cells is relatively easy to prevent by the use of effective sterilization procedures verified by adequate sterility tests. However, airborne and contact contamination can occur anytime the culture vessel is opened or contact is made with the culture fluid through a defective culture vessel, stopper, or poor technique. Studies in an operating room show that most airborne viable bacteria are carried on particles measuring 4–28 μm in diameter, average 12 μm (Noble *et al.*, 1963). Such particles settle at the rate of 1 ft/minute and soon reach the floor if the air is not disturbed by convection currents. From this, one would expect the air to be relatively sterile in an undisturbed room and this proves to be the case. Bacterial counts that average 10–20/ft^3 in the air of a operating room during surgery fall to 1–3/ft^3 when the patient and all personnel leave the room (Coriell *et al.*, 1968).

To provide ambient sterile air in which to manipulate cell cultures we must determine the sources of airborne bacteria and then try to eliminate the sources and/or prevent entry into the culture. The chief sources of airborne bacteria in a room may be grouped under the following: (*1*) the air supply to the room, (*2*) the clothing and shoes of the personnel in the room, (*3*) the body surfaces of the personnel, (*4*) the floor and flat surfaces, and (*5*) organisms aerosolized during manipulation and opening of microbial cultures or contaminated cell cultures. The recommended procedures to control these sources of airborne bacteria may be listed briefly:

1. Air Supply

Low-efficiency oiled Fiberglas filters usually installed in central hot-air heating systems remove coarse particles but are only 40–60% effective in removing bacteria. The recommended solution for tissue culture rooms is to use high-efficiency filters that remove 95–99.97% of particles of 0.3 μm in diameter. This is discussed further in Section III, on laminar air flow.

2. Clothing

Street clothes and shoes collect dust that may contain microbes and resistent spores and should be changed to clean gowns, scrub suits, and shoes not worn out of the laboratory. The small-animal quarters of the average institution are prone to having high microbial counts in the air and on the floor. These come from bedding, food, hair, feathers, and dandruff and are increased by cage cleaning or movement of animals (McGarrity *et al.*, 1969). If cell culture technicians must go to the animal quarters or work with animals, they should cover or change shoes, discard outer clothing, and put on clean gowns before returning to the cell culture room.

3. Personnel

One of the greatest sources of airborne organisms is from the skin and hair of the personnel. Many organisms live on the skin and the dried skin scales or scutes are shed constantly, the number being roughly proportional to the physical activity of the person. The most common bacteria from this source are species of micrococci, diptheroids, and yeasts but others may occur. In a closed, static transfer room that has been thoroughly cleaned to remove bacteria, the human source of bacteria can be documented by continuous sampling of viable organisms per cubic foot of air and correlating this with the presence and degree of activity of personnel (McGarrity and Coriell, 1971). The recommended control measures are to (*1*) reduce physical activity of the technician to a minimum, (*2*) eliminate traffic of other personnel through the cell culture laboratory, (*3*) cover the hair and skin when transferring cultures, and (*4*) wash the hands before and between cell cultures.

4. Floor

The floor of a laboratory is constantly receiving organisms as they settle from the air. Most airborne organisms in the laboratory eventually end up on the floor or other flat surfaces, from which they reenter the air when disturbed. The number of viable organisms on the floor at any time is a product of their rate of deposit and death rate from all causes. The death rate is conditioned by many factors, including the amount of protective material, i.e., dirt and organic matter, on the floor. The recommended control procedure in addition to reduction of the sources of airborne contamination is daily scrubbing of the floor. The objectives of scrubbing are twofold: to reduce the microbial count to zero at the start of each days work and to remove accumulated dirt and organic debris. To facilitate this, it is best to have a floor surface without cracks, such as sheet vinyl. The cracks around abestos or vinyl blocks cannot be properly cleaned and repeated exposure to water will dissolve the adhesive and cause the blocks to lift. Because scrubbing is designed to clean as well as to disinfect and because bacteria vary in their susceptibility to different disinfectants, the most practical disinfectants include a combination of detergents, phenols, alcohols, quaternary ammonium compounds, and halides. In use, they must be of proper concentration, and vigorous scrubbing is required to clean the floor. At least 5 minutes must be allowed for killing before the disinfectant is removed, because chemical disinfection is time and temperature dependent. The excess fluid is best removed with a wet vacuum cleaner. A less desirable alternate is to pick it up with a mop that is rinsed and squeezed in a second mop bucket. This is the two bucket technique, which is designed to avoid the spreading of bacteria with dirty scrub water in which the disinfectant has been neutralized by an excess of organic matter. It has been repeatedly demonstrated (Westwood *et al.*, 1971; Litsky and Litsky, 1968) that a mop head used day after day becomes saturated with organic matter that provides nutrition for bacteria and the mop then becomes a source of contamination rather than a cleanser. If fresh disinfectant is added to the scrub water daily, this does not solve the problem. On the contrary, it sets up a selective pressure for propagation of organisms resistent to the disinfectant. If this policy is followed, the mopping procedure will soon be adding many more organisims to the floor than it removes. A sterile mophead should be used each day. After use, the mophead should be rinsed, wrapped in heavy paper, and autoclaved. The most practical way to monitor the effectiveness of floor cleaning is by the use of Rodac plates (Fincher, 1965; Bean, 1967). The agar surface of the plates is pressed against the floor before and after cleaning, incubated, and the average number of colonies per unit area calculated.

Work benches in cell culture areas are scrubbed each morning with disinfectant and again each time a different cell culture is introduced. The purpose of this is to remove dust and viable particles, including cultured cells or microbial contaminants of cells that have settled from all the sources mentioned above. Other sources of contamination of the bench surface and ambient air of the transfer room may be important in cross-contamination of cell cultures; for example, streaking an agar plate, preparing serial dilutions, removing cotton plugs or stoppers after centrifuging, and resuspending a cell pellet after centrifuging. Such common procedures have been shown (Wedum, 1964) to be the mechanism of most infections of laboratory personnel (Reitman and Wedum, 1956; Wedum, 1961). This is easily verified by performing a series of these culture manipulations using a suspension of fluorescent dye followed by examination of the work area with a Woods light.

The choice of disinfectant for cleaning the work surface after each cell culture is processed and before the next one is introduced presents the same problems as the selection of a detergent–disinfectant for the floors. No preparation is perfect for all situations. We use the same preparation used for floors, i.e., 2 oz of RX-Concentrate per gallon of water. Its effectiveness against viruses has not been reported. According to findings (Klein and DeForest, 1965), sodium hypochlorate (200 ppm of available Cl_2) and 70–90% ethanol inactivated all 25 viruses tested after an exposure of 10 minutes at room temperature. Exposure of 10 minutes may be difficult to achieve on the desk top. Kline and Hull (1960) demonstrated that 2% peracetic acid is the most viricidal of 19 compounds tested against a broad spectrum of viruses and recommended it as the decontaminating agent of choice when the staff is working with infectious agents or trying to maintain aseptic conditions. It also inactivates a broad spectrum of bacteria and fungi (Greenspan *et al.*, 1955) and is the decontaminant of choice in the rearing of germ-free animals (Trexler and Reynolds, 1957).

II. Procedures within the Transfer Room

All the barriers against contamination discussed so far are shown in Fig. 4 as peripheral lines because they may or may not be under the direct control of the technician. For example, serum and culture media are frequently purchased from commercial sources and/or prepared for use in a central media kitchen; glassware and apparatus are usually processed in a separate

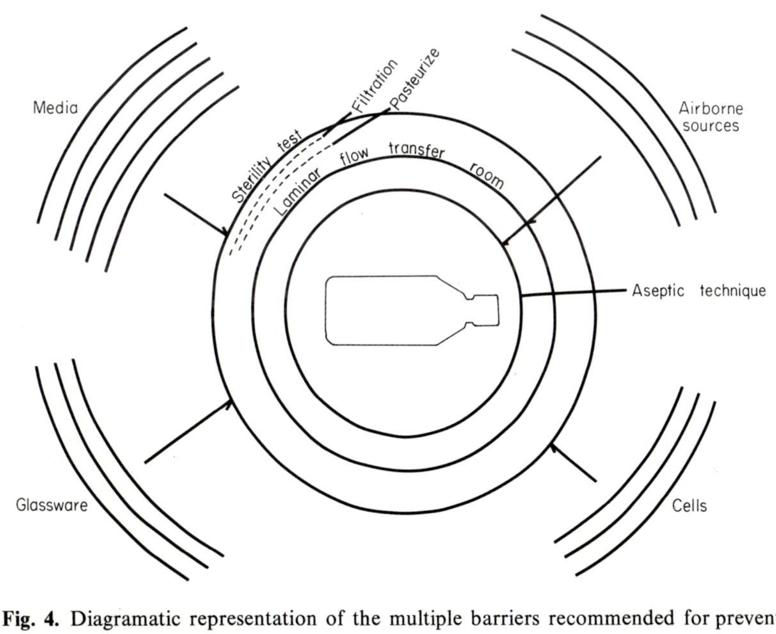

Fig. 4. Diagramatic representation of the multiple barriers recommended for preventing microbial contamination of cell cultures. The outer broken circles represent the general measures shown in Figs. 2 and 3 and described in the text. The inner complete circles illustrate the measures carried out by the technician in the transfer room and are described more fully in the text. The deflected arrows indicate the constant threat of contamination and the points of deflection identify the principal barriers to this source of contamination.

facility; cell lines or tissue specimens may come from outside the laboratory; and housekeeping–clothing–training is partially an administrative function. Figure 4 therefore shows three functions that are completely under the control of the technician, who must bear the ultimate responsibility for contamination, represented by three complete inner circles. These are the most important barriers to contamination. The technician should personally confirm the sterility of all culture media and cell cultures, examine the indicators on sterile supplies, and inspect for breaks in the protective wrapping. This means that the serum, trypsin, balanced salt solution, culture media, and other components of the medium must be cultured 2 weeks before they are used so there will be adequate time to read out the tests. If this is not done and the medium components are cultured only at the time of use, it is too late to prevent loss of the culture if one component is contaminated. However, culture at the time of use is also recommended. The volume of medium required should be prepared by mixing sterile medium, serum, glutamine, bicarbonate, and other ingredients. In use, the bottle of medium is opened several times in the course of feeding cell cultures and the residual should be discarded. The recommended culture procedure is to remove a sample at the time the bottle is prepared and a second sample from the bottom of the bottle, or at the end of the days work. A sample volume of 50 ml from a 1 liter bottle is recommended and both samples should be incubated with the cell cultures. The value of this procedure is that it enables the technician to trace the source of any contamination that occurs. If one of the medium components is contaminated, all cell cultures that received this component will have the same organism. If contamination occurs when all the medium components are sterile initially but are contaminated at the end, then technician error is indicated. If all medium samples and cells were sterile initially and the glassware passed inspection, then the techniques of the operator must be reviewed. If the contamination is spotty and by different types of organisms, operator error is suggested. When all cultures and samples are consistently sterile, the technician's confidence and skill are reinforced.

Because contaminated serum has been a recurring problem, some laboratories have found it expedient to adopt additional procedures instead of or in addition to the extensive sterility testing recommended above. These are indicated as filtration and pasteurization by dotted lines on Fig. 4.

Refiltration of serum has been practiced successfully by passing whole serum through three Millipore filters of successively smaller pore diameter, or, by Millipore filtration of the final medium containing diluted serum. The final medium containing diluted serum will usually pass through a 0.22 Millipore filter without difficulty provided it is free from gross turbidity.

Pasteurization of serum at 60°C for 30 minutes is done in a water bath that immerses the serum bottle to the neck; the time and temperature are controlled by a thermometer in a duplicate bottle filled to the same level with water. This temperature will kill mycoplasmas and some vegetative bacteria, but at a price. Both refiltration and pasteurization may reduce the cell-growth-stimulating properties of serum. If desired, this factor can be removed from the realm of the unknown by titrating the serum for growth stimulation of cells before and after pasteurization.

III. Laminar Flow Transfer Room

The importance of the airborne route of contamination requires no proof as the principle was well established many years ago by Pasteur (1862), Tyndall (1882), and Koch (1881) and by other early founders of the science of bacteriology (Bullock, 1930). The greatest advance in prevention of airborne contamination since the invention of the cotton plug is the use of high-efficiency particulate air (HEPA) filtered laminar flow air in the transfer room. This technique of filtering air through high-efficiency filters that remove dust, bacteria, and viruses, plus continuous mass movement of this clean air through a room, has been developed within the last decade (Whitfield, 1962). The filters are called *absolute* or *HEPA* and the continuous pistonlike mass movement of the filtered air is commonly referred to as *laminar flow*, although true laminar flow is not achieved. The net effect is to produce germ-free ambient air in the tissue culture transfer room. The air entering the ceiling is germ-free and aerosols generated within the room ride downstream with the air flow and move quickly out of the field. The only contaminated air in the room is directly downstream of an aerosol source and this area can be avoided for critical work. This system is being accepted by tissue culturists and microbiologists. Various modifications and applications have been described (Whitcomb and Clapper, 1966; Bodey *et al.*, 1969; Coriell *et al.*, 1967, 1968; McGarrity *et al.*, 1969; Coriell and McGarrity, 1968, 1970; Whitefield and Lindell, 1969; Whitby and Lundgren, 1965) that show that ambient dust particles in an occupied room drop from 10^6 to $< 10^2/\text{ft}^3$ air and viable bacteria from the range of 10–200 down to $< 0.1/\text{ft}^3$ air. Open agar plates or flasks of broth remain sterile if exposed in a transfer room equipped with filtered air, whereas they are 100% contaminated if the filtered air is shut off.

As a result of these many controlled studies, it is justified to recommend that no one should attempt to work with cell cultures without the use of laminar flow room or hood.

IV. Aseptic Technique

The final and most important barrier to contamination is the skill and reliability of the technician in carrying out all the recommended and approved procedures. This is represented by the inner circle labeled *aseptic technique* in Fig. 4. It includes continued observance of accepted aseptic procedures, including the use of clean clothing and hair coverings; proper cleaning of the floor and work bench; no talking or the use of a mask if talking is necessary; proper pipette handling; proper handling of open culture vessels, flames, and stoppers; the transfer of one cell culture at a time, cleaning up the bench between cultures; autoclaving or isolating known contaminated cultures; continual surveillance; and periodic review of performance. Every contamination that occurs is the result of an error and if all the recommended procedures have been followed it should be possible to identify where the break occurred and to institute corrective measures (McGarrity and Coriell, 1971).

Are all these precautions necessary and do they work? The answer is yes to both questions. The proof that they are necessary is furnished by two observations. Most laboratories do not observe all of these precautions and over 50% of cell cultures in current use are contaminated (Barile, 1968). However, laboratories that follow these procedures have virtually eliminated accidental contamination of cell cultures (Sherer *et al.*, 1964; Coriell, 1962). The most impressive evidence is furnished by the experience of the laboratories that have cooperated in attempting liquid nitrogen storage of large numbers of ampules of certified cell cultures for the Cell Culture Collection sponsored by the National Cancer Institute (Sherer *et al.*, 1964). In the beginning of the program contaminations were frequent, probably in part because of the large volume and the many manipulations required to prepare 200–300 ampules of antibiotic-free cells, each containing $2–5 \times 10^6$ cells in a volume of 0.5–0.7 ml. Techniques were gradually developed to solve these problems and eventually contaminations came to be a rare event as a result of adherence to the procedures described above (McGarrity and Coriell, 1971).

Hand and vacuum pump aspiration of spent culture fluid from monolayer cultures and centrifuged cell pellets creates airborne droplets and if the culture contains occult microbial contaminants this is an excellent way to create an aerosol that may then fall into the next cell culture transferred in the same cubicle. To avoid this, we recommend a closed aspirating system that vents the aerosol outside the laboratory, Fig. 5. Sterile pipets can be attached to the rubber tubing of the flask assembly. The spent media is collected in the flask and the air is passed through a cotton filter and a dessication flask, and then through a copper tubing to the remotely located vacuum pump, which is vented through the wall to the outside. The pump is

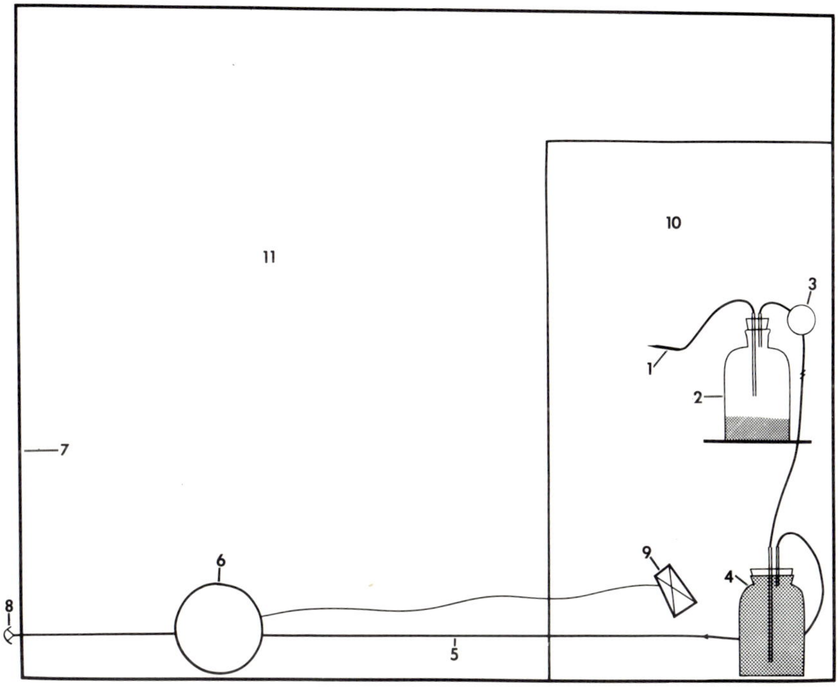

Fig. 5. Diagram of an assembly for aspirating spent tissue culture media without creating microbial aerosols in the transfer room if one of the cultures is contaminated. The numbers indicate: 1, pipet; 2, collecting flask; 3, cotton filter; 4, calcium chloride; 5, copper tubing; 6, vacuum pump; 7, building wall; 8, outside exhaust; 9, foot control; 10, transfer room; 11, laboratory.

activated by a foot control, so the technician has both hands free. The spent fluid and flask assembly can be autoclaved before washing.

V. Antibiotics

Between 1907, when Ross Harrison first grew cells *in vitro*, and 1945, when antibiotics were introduced, the basic techniques and procedures for cell and organ culture were developed by a relatively small group of investigators. However, the methods were not widely adopted because most workers were unable to avoid loss of cell cultures through contamination with bacteria and fungi. Only those who were willing to devise and observe rigid aseptic techniques were successful. The incorporation of penicillin and other antibiotics in the culture medium in 1945 (Gey *et al.*, 1945; Pomerat, 1945; Rose *et al.*, 1945; Early and Morgan, 1946) and the use of replicate cultures for quantitative studies (Evans *et al.*, 1951) opened the door for mass cultivation and quantitative measurements in many disciplines of biology (Chang, 1954; Dulbecco and Vogt, 1954; Sanford *et al.*, 1948; Bang and Gey, 1951; Coriell *et al.*, 1967; Whitefield and Lindell, 1969). As a result, thousands of investigators were able to work with cell cultures, and, as occurs with the introduction of any new experimental method, there was a great flowering of new information in all branches of biology, including such practical applications as the poliomyelitis vaccine. Moreover, the whole field of molecular biology was made possible.

A mixture of antibiotics can be selected that is nontoxic for mammalian cells but kills bacteria and fungi so that cell cultures can be initiated even from frankly contaminated specimens, such as biopsies of skin, cervix, or bowel. Because of these successes, many workers have relied heavily on the magic of antibiotics and aseptic techniques have lapsed.

There are several pitfalls to the generalization that the judicious use of antibiotics will control all contaminations. (*1*) Antibiotics do not kill all organisms; some are only suppressed or their metabolism is slowed to the point where they do not cause gross changes in the culture medium but still cause biochemical and antigenic changes (Stanbridge, 1971). (*2*) Contamination is encouraged by the relaxing of aseptic techniques if the technician feels that the use of antibiotics makes aseptic techniques unnecessary. (*3*) Many cell cultures are carried for hundreds of passages and the statistical probability of contamination increases each time the culture flask is opened. (*4*) Although antibiotics kill most contaminating organisms they also exert a selective pressure for the growth or development of resistant organisms in much the same way that widespread use of DDT exerts a selective pressure for the development of resistant mosquitoes and flies. If contaminants are constantly added to a cell culture as a result of poor technique, it is inevitable that eventually an organism will be added that likes to grow in the presence of antibiotics. (*5*) Penicillin is a recognized cause for the conversion of certain bacteria to L forms, some of which grow rapidly but cannot be detected by turbidity of broth or by visible growth on solid media. (*6*) Because of the suppressing effect of antibiotics on growth of contaminants, the best and most obvious indices of contamination are lost, i.e., gross changes in pH, turbidity, and microscopic evidence of death of the cell culture. (*7*) Some of the most troublesome and most frequent contaminants of antibiotic-containing cell cultures are not toxic to the cells, cause no gross changes in the medium, and do not grow out when subcultured in standard sterility tests. This latter point is exactly what we might predict as the end result of the foregoing considerations and a good example is the widespread prevalence of *Mycoplasma* sp. in cell cultures (see Chapter 6 by Barile). (*8*)

This chapter deals only with microbial contamination of cell cultures but it is fair to conclude that the cross-contamination of cell cultures with other cell cultures, so prevalent between 1950 and 1965, was the result of antibiotic-inspired relaxation of aseptic techniques.

In summary, antibiotics should not be routinely added to cell culture media. If they are omitted, reliance for maintaining sterile cultures must be placed on consistent observation of aseptic techniques. Contamination, when it occurs, can then be quickly recognized and the emergence of L forms will not occur. As an added inducement to omit antibiotics, if a valuable cell culture does become contaminated, it may be possible to eliminate the contaminant by appropriate antibiotic treatment.

There are legitimate uses for antibiotics, i.e., in preparation of cell cultures from contaminated specimens, such as skin biopsies. Here, the principle is to use high concentrations of bactericidal antibiotics for a short time and then delete them. First, wash the specimen several times in media containing the selected antibiotic mixture; then chop the specimen in fresh antibiotic media (a trypsin step may be added here if desired); then culture in fresh media with antibiotics. At the time of first subculture, usually 1 or 2 weeks, antibiotics should be removed because if the bacteria are not killed by then it is most unlikely that they ever will be.

Some workers believe that antibiotics are essential for large-scale cell culture and it is a valid position if the large-scale cultures are primary monkey kidney or other primary cells derived from animal or human sources. Even here, the antibiotics could be removed after the second or third refeeding. If, however, the large-scale production is from established cell lines free of contamination, the use of antibiotics is not necessary in a well-designed laboratory with properly trained personnel.

VI. Summary

The multiple barriers to contamination represented by arcs and circles in Fig. 4 are effective in preventing contamination of cell cultures, and the use of antibiotics for this purpose is not effective. In fact, it leads to massive and widespread infection of a degree that can properly be termed epidemic. The epidemic can be halted by removing the antibiotics and returning to reliance on aseptic and no-touch procedures. This system has been successfully adopted in many laboratories. Furthermore, it is the only reliable system to minimize contamination and, finally, it prevents a great deal of expense by preventing contaminations that waste culture media and time, lose cell lines, waste personnel effort, and generate erroneous data that cannot be repeated with clean cells.

References

Bang, F. B., and Gey, G. O. (1951). *Trans. N.Y. Acad. Sci.* [2] **13**, 321–327.
Barile, M. F. (1968). *Nat. Cancer Inst., Monogr.* **29**, 201–204.
Barile, M. F., and Kern, J. (1971). *Proc. Soc. Exp. Biol. Med.* **138**, 432–437.
Bean, H. S. (1967). *J. Appl. Bacteriol.* **30**, 6–16.
Bodey, G. P., Freireich, E. J., and Frei, E. (1969). *Cancer* **24**, 972–980.
Boone, C. W., Mantel, N., Caruso, T. D., Jr., Kazam, E. and Stevenson, R. E. (1971). *In Vitro* **7**, No. 3, 174–189.
Bullock, W. (1930). "History of Bacteriology," Medical Research Council, London.
Chang, R. S. M. (1954). *Proc. Soc. Exp. Biol. Med.* **87**, 404–443.
Coriell, L. L. (1962). *Nat. Cancer Inst., Monogr.* **7**, 33–53.
Coriell, L. L., and McGarrity, G. J. (1968). *Appl. Microbiol.* **16**, 1895–1900.
Coriell, L. L., and McGarrity, G. J. (1970). *Appl. Microbiol.* **20**, 474–479.
Coriell, L. L., McGarrity, G. J., and Horneff, J. (1967). *Amer. J. Pub. Health Nat. Health* **57**, 1824–1836.
Coriell, L. L., Blakemore, W. S., and McGarrity, G. J. (1968). *J. Amer. Med. Ass.* **203**, 1038–1046.
Dulbecco, R., and Vogt, M. (1954). *J. Exp. Med.* **99**, 167–182.
Earle. W., (1943). *J. Nat. Cancer Inst.* **4**, 131–133.
Early, R. L., and Morgan, H. R. (1946). *J. Immunol.* **53**, 151–156.
Evans, V. J., Earle, W. R., Sanford, K. K., Shannon, J. E., and Waltz, H. K. (1951). *J. Nat. Cancer Inst.* **11**, 907–927.
Fincher, E. L. (1965). *Annu. Tech. Meet. Exhib., Amer. Ass. Contam. Contr., 4th, 1965.*
Fogh, J., and Fogh, A. (1964). *Proc. Soc. Exp. Biol. Med.* **117**, 899–901.
Gey, G. O., Gey, M. K., Inui, F., and Vedder, H. (1945). *Bull. Johns Hopkins Hosp.* **77**, 116–131.
Greenspan, F. P., Johnsen, M. A., and Trexler, P. C. (1955). *Proc. Annu. Meet. Chem. Speci. Mfr. Ass., 42nd* pp. 59–64.
Hall, W. T., Andersen, W., Sanford, K., and Evans, V. (1967). *Science* **156**, 85–88.
Hummeler, K., and Armstrong, D. (1967). *Ann. N.Y. Acad. Sci.* **143**, 622–625.
Klein, M., and DeForest, A. (1965). *Fed. Proc., Fed., Amer. Soc. Exp. Biol.* **24**, 319.
Kline, L. B., and Hull, R. (1960). *Amer. J. Clin. Pathol.* **33**, 30–33.
Kniazeff, A. J. (1968). *Nat. Cancer Inst., Monogr.* **29**, 123–132.
Koch, R. (1881). zm *Zuecht. Pathog. Mikroorgani. Kaiserl. Gesundhartsomte* **1**, 1–48.
Litsky, Y. B., and Litsky, W. (1968). *Amer. J. Pub. Health Nat. Health* **58**, 534–543.
McGarrity, G. J., and Coriell, L. L. (1971). *In Vitro* **6**, 257–265.
McGarrity, G. J., Coriell, L. L., Schaedler, R. W., Mandle, R. J., and Greene, A. E. (1969). *Appl. Microbiol.* **18**, 142–146.
Noble, W. C., Lidwell, O. M., and Kingston, D. (1963). *J. Hyg.* **61**, 385–391.
Pasteur, L. (1862). *Ann. Chim. Phys.* **64**, 5–110.
Pomerat, C. M. (1945). *Proc. Trans. Tex. Acad. Sci.* **28**, 130–132.
Reitman, M., and Wedum, A. G. (1956). *Pub. Health Rep.* **71**, 659–665.
Rose, H. M., Malloy, E., and O'Neill, E. (1945). *Proc. Soc. Exp. Biol. Med.* **60**, 23–25.
Sanford, K. K., Earle, W. R., and Likely, G. D. (1948). *J. Nat. Cancer Inst.* **9**, 229–246.
Sherer, W. F., Coriell, L. L., Hsu, T. C., King, D. W., Madin, S. H., Meryman, H. T., Morgan, H. R., Rose, H. M., Sanford, K. K., Shannon, J. E., Stevenson, R. E., and Stulberg, C. S. (1964). *Science* **146**, 241–243.
Stanbridge, E. (1971). *Bacteriol. Rev.* **35**, 206–227.

Trexler, P. C., and Reynolds, L. I. (1957). *Appl. Microbiol.* **5**, 406–412.
Tyndall, J. (1882). "Essays on the Floating Matter of the Air in Relation to Putrefaction and Infection," Appleton, New York.
Wedum, A. G. (1961). *Bacteriol. Rev.* **25**, 216.
Wedum, A. G. (1964). *Amer. J. Pub. Health Nat. Health* **54**, 1669–1673.
Westwood, J., Mitchell, M., and LeGacé, S. (1971). *Appl. Microbiol.* **21**, 693–697.
Whitby, K. T., and Lundgren, D. A. (1965). *Trans. Amer. Soc. Agri. Eng.* **8**, 342–352.
Whitcomb, J. G., and Clapper, W. E. (1966). *Amer. J. Surg.* **112**, 681–685.
Whitefield, W. J. (1962). *Proc. Inst. Environ. Sci.* p. 529–534.
Whitefield, W. J., and Lindell, K. F. (1969). *Contam. Contr.* **8**, No. 11, 10–21.

CHAPTER **3**

Contamination of Tissue Culture by Bacteria and Fungi

DONALD ARMSTRONG

I. Introduction 51
II. Materials and Methods 52
III. Results 53
A. Bacteria 53
B. L Forms of Bacteria 55
C. Fungi 55
D. Mycoplasmas 56
IV. Discussion 56
A. Detection of the Source of Contamination 58
B. Contaminated Tissues 58
C. Contaminated Inoculums 61
D. Treatment of Contaminated Cell Cultures 62
E. Effects of Antibiotics on Cells and Inoculum 62
V. Summary 63
References 64

I. Introduction

Contamination of cell cultures by bacteria and fungi almost invariably results in (*1*) ill humor in the laboratory lasting from hours to days; and (*2*) one or many discarded cell cultures. Accompanying these two major reactions are the following: (*a*) soul searching; (*b*) culturing of all media components, usually in thioglycolate broth alone; (*c*) scrubbing of the laboratory; (*d*) tightening of techniques; (*e*) no further contamination (temporarily) because of (*c*) and/or (*d*).

A number of papers have been written on the incidence of contaminated cell cultures received in reference laboratories, usually stressing the role of mycoplasmas. Bacteria and fungi have been identified and their sources pinpointed in few instances (Coriell, 1962; Barile and Schimke, 1963; Fogh *et al.*, 1971; McGarrity and Coriell, 1971). The importance of avoiding antibiotics in media and systems of preventing contamination are described in detail by Dr. Coriell in Chapter 2. In some laboratories, it can be anticipated that the source of the cells to be cultured or the biological specimens to be inoculated into cell cultures will be contaminated, and thus antibiotics will be necessary.

This chapter will be devoted to (*1*) describing the type of organism isolated in contaminated cell cultures; (*2*) deducing the origin of the contaminant from its natural habitat; and (*3*) evaluating the antibiotics and concentrations necessary to prevent or treat infections with commonly contaminating bacteria and fungi.

II. Materials and Methods

The submission of specimens to the clinical Microbiology Laboratory of Memorial Hospital and the Sloan-Kettering Institute depended upon the willingness of the investigator to admit contamination, the knowledge that he would not be charged for the culture, and his interest in the identification of the offending organism. These factors varied with investigators and, from time to time, with the same individual so conclusions as to the incidence of infection are not drawn for the various types of laboratories. The choice of specimen was felt to be important. Investigators were urged to submit the obviously contaminated or suspected cell culture in its original container so that it could be opened sterilely in the Microbiology Laboratory and so that both cells and supernatant fluid could be obtained. A gram stain was then performed on the specimen, followed by cultures using the following methods.

1. A blood agar plate
2. A deoxycholate plate
3. Thioglycolate broth
4. A cooked meat infusion broth incubated anaerobically in Gaspak jars
5. A Sabaurad's agar plate was used if fungi were seen on gram stain or otherwise suspected

L Form (bacteria without cell walls, structurally limited by an osmotically unstable cell membrane) cultures were not routinely done; if this type of organism was suspected the specimens were cultured on a hypertonic medium previously described (Young and Armstrong, 1969) based on a brain-heart infusion containing 15% sucrose and 20% horse serum. Plates were cultured in 5% CO_2 and anaerobically (Gaspak jars) and a diphasic tube containing 4 ml of agar and 4 ml of broth was also inoculated.

Cultures for mycoplasmas were done if these organisms were suspected or if there was no bacterial or fungal growth. The methods have been previously described (Armstrong *et al.*, 1970) and as with the L form cultures included an aerobic (5% CO_2) and anaerobic plate (Gaspak jar) and a diphasic broth. Briefly, the mycoplasma broth was made with a fresh beef heart infusion (50 g/liter) and Bacto peptone (10 g/liter), supplemented with 20% unheated horse serum and 10% fresh yeast extract (Fleischmann's type 2040). Similarly, the mycoplasma agar was made with a fresh beef heart infusion, Bacto peptone, and Bacto PPLO agar. Horse serum (20%) and fresh yeast extract (10%) were added prior to use. Freshly made media components were found distinctly superior to commercial completed media (Armstrong and Yu, 1967) and were used throughout.

Antibiotic sensitivity testing was performed using disks, either by a modified triple disk (Kagnoff *et al.*, 1972) or the Kirby-Bauer method (Bauer *et al.*, 1966).

Confirmation or identification of some of the unusual organisms was done by the Center for Disease Control, Atlanta, Georgia.

III. Results

Of 48 grossly contaminated cell cultures submitted to the Microbiology Laboratory, two thirds contained bacteria and the remainder contained fungi. One outbreak of cell culture contamination was caused by an L form. Most specimens contained only one organism, but one contained three and another two different bacteria, and two more each contained two bacterial species and a fungus.

The majority of these organisms were isolated during late 1970 and 1971. Mycoplasmas were isolated and identified over a 5 year period as indicated in Table III. All bacteria and fungi were seen when cell cultures were smeared and studied after gram staining unless otherwise indicated in the tables. The *Aspergillus* species were consistently not seen after gram stain but when wet mounts were done, septate mycelia were apparent.

A. Bacteria

The various bacteria isolated are recorded in Table I along with the type of culture from which they were isolated when this information was avail-

TABLE I
Bacteria Contaminating Tissue Cultures

Organism	Number isolated	Type of cell culture: Continuous	Primary	Selected antibiotic sensitivities[a]: P	Ox	Te	Cl	G	K
Pseudomonas species	8	5	2[b]	R	R	S	R	S	S
Pseudomonas aeruginosa	1	1	—						
Micrococci	4	3	—	R	S	S	S	R	S
Klebsiella–enterobacter	3	—	2	R	R	R	S	S	S
Staphylococcus albus	2	2	—	S	S	S	S	S	S
Bacillus subtilis[c]	2	2	—						
Escherichia coli	2	—	2	R	R	R	R	S	R
Flavobacterium species	1	1	—						
Serratia species	1	—	1	R	R	R	R	S	S
Bacteroides species	1	—	1						
Hebicola–Lathyri	1	1	—						
Streptococcus viridans	1	—	—						
Diphtheroids	1	—	1						
Gram-negative rods[d]	3	3	—						

[a]Where isolates were multiple, sensitivities reflect reactions in the majority of tests: S = sensitive, R = resistant, P = penicillin G, Ox = oxacillin, Te = tetracycline, Cl = chloramphenicol, G = gentamicin, K = kanamycin.
[b]One culture inoculated with a clinical specimen.
[c]In one case, gram stain of the cell culture did not show organisms.
[d]Not further identified.

able. There was no consistent pattern as to the type of cell culture contaminated with certain bacteria. A particular laboratory showed contamination with the same bacteria for a few days running, but after this was cleared up, subsequent contamination was usually with different bacteria. The type of bacteria isolated suggested two predominant sources:

1. The inanimate environment, represented by saprophytic and particularly water-borne organisms, such as *Pseudomonas* species and *Flavobacterium* species.

2. The animate environment, e.g., man, represented by normal skin (micrococci, *Staphylococcus albus*), oral (diphtheroids), and fecal (*Enterobacter–Escherichia* species and *Pseudomonas aeruginosa*) flora.

Antibiotics to which the bacteria were sensitive are listed in the third column of Table I. The *Klebsiella–Enterobacter* and *Pseudomonas aeruginosa* isolates were broadly resistant to most antibiotics except for kanamycin, gentamicin, and colistimethate in the former case, and gentamicin and colistimethate in the latter. The *Pseudomonas* species other than *P. aeruginosa* varied in their sensitivity. Both Micrococcus species tested were resistant to penicillin.

One isolate was a *Bacteroides* species, an obligate anaerobic organism. Unfortunately, this was not speciated, nor were sensitivity studies done.

B. L Forms of Bacteria

In one instance an L form (of a *Pseudomonas* species) was isolated and the source was found to be phosphate buffered saline bought from a commercial media supply house. This company was informed of the finding but regrettably did not call in their distributed stock and the organism appeared in other laboratories in the institution using the saline. Sensitivity tests were not performed on the parent bacterium; presumably the L form would have been sensitive to an antibiotic that inhibited protein synthesis, such as tetracycline, chloramphemicol, erythromycin, and lincomycin.

C. Fungi

The fungi isolated are listed in Table II. As with the bacteria, the nature of the isolates indicates either an inanimate or animate environmental source. Antibiotic sensitivity testing was not done, but all of the organisms listed are known to be sensitive to amphotericin B and less so to nystatin.

TABLE II
Fungi Contaminating Tissue Culture

Organism	Number isolated	Type of cell culture[a]	
		Continuous	Primary
Aspergillus fumigatus[b]	1	1	—
Aspergillus niger	1	1	—
Aspergillus species[c]	1	1	—
Mucor species[b]	1	1	—
Oospora species	2	—	2
Penicillum species[c]	1	—	1
Candida albicans	3	—	3
Candida stellatoidea	1	—	1
Candida krusei	1	1	—
Candida species	1	1	—
Yeasts	4	—	4

[a]None were inoculated with clinical specimens.
[b]Seen on wet mount, not on gram stain.
[c]Negative on gram stain.

D. Mycoplasmas

The numbers of various types of mycoplasmas isolated according to the year are recorded in Table III. It is evident that *Mycoplasma hyorhinis*, a common contaminant in 1966, was subsequently replaced by *M. orale* type 1 and the latter by *Acholeplasma laidlawii* and *M. arginini* as the most common. mon.

IV. Discussion

The source of bacterial and fungal contamination in established cell cultures is usually the environment because of break-downs in sterile technique. Less often, tissue culture media components may be contaminated before use. This can be detected, however, by culturing the components and waiting for a 72–96 hour period before using them. Provided the correct isolation procedures are used, all but the most slow growing organisms will become apparent. Bacterial and fungal isolation media for detection should be those with which the microbiologist is familiar but should include at least those outlined in Section II. It would be more complete to include a hypertonic

TABLE III
Mycoplasmas Contaminating Tissue Cultures

Mycoplasma isolated	Number of isolates per year					
	1966	1967	1968	1969	1970[a]	1971[a]
M. fermentans	—	—	4	—	—	—
M. hominis	—	1	1	—	—	—
M. hyorhinis	1	22	17	3	—	—
A. laidlawii	—	—	—	20	2	15
M. orale Type 1	—	6	43	14	3	—
M. pulmonis	—	—	—	—	—	—
M. salivarium	—	—	6	—	1	4
Others[b]	1	3[c]	—	—	14[d]	—
Total	2	32	71	37	20	19

[a]During 1970 there were 29, and during 1971, 79 isolations from tissue cultures but attempts were made to identify only the isolates recorded in the table.
[b]Unidentified unless otherwise noted.
[c]One isolate was *M. arginini*, identified by Dr. Joseph Tully, National Institutes of Health, Bethesda, Md.
[d]All *M. arginini*, identified by Dr. Joseph Tully, National Institutes of Health, Bethesda, Md.

(sucrose or saline or both) medium for L forms and a Sabaurad's medium culture for fungi at room temperature as well as 37°C, although contamination picked up by these methods alone is probably rare. Certainly, thioglycolate broth alone is insufficient, for some organisms will simply not grow in this medium.

Environmental contamination is more difficult to control, requiring sustained, scrupulous techniques, which appear to be aided by special isolation rooms containing laminar flow systems. These systems are expensive and frequently not available to many investigators. The decision must be made whether to use antibiotics in the medium or not, according to (*a*) the types of tissue specimens cultured, (*b*) whether or not contaminated inoculums are used, and (*c*) the skills of the team working and the equipment in the tissue culture laboratory. Most authorities prefer to work without antibiotics when culturing noncontaminated tissues. If antibiotics are to be used routinely in established cell lines, the question is which one(s). The general use of a combination of penicillin and streptomycin is as antiquated in a tissue culture laboratory, as it is in clinical infectious diseases. A specific indication in either discipline would be the treatment of Group D streptococci or enterococci. Otherwise, streptomycin is woefully inadequate in trying to control common gram-negative bacillary infections because of widespread bacterial resistance to this antibiotic. From the data presented here and from our knowledge of the antibiotic sensitivities of the majority of aerobic gram-negative rods comprising normal human flora or environmental water-borne saprophytes, the antibiotic promising the broadest spectrum would be gentamicin. This agent is also effective against many strains of *S. aureus* and mycoplasmas. It is probably worthwhile to include penicillin in the medium also to cover organisms such as *Streptococcus viridans*, micrococci, *S. albus*, and diphtheroids. Concentrations of 100 units/ml of penicillin will also kill certain gram-negative bacilli, such as *E. coli* and *Proteus mirabilis*. As noted in our sensitivity testing, some of these organisms, plus *S. aureus*, coagulase positive, may be resistant to penicillin because of the production by the bacteria of penicillinase. A penicillinase-resistant semisynthetic penicillin, such as oxacillin, would be more appropriate in these circumstances and in sufficient concentrations would eradicate the other gram-positive cocci and bacilli as well. Concentrations of the antibiotics will depend on the cell type used and the ranges that might be employed are listed in Table IV (Perlman *et al.*, 1961; Perlman and Brindle, 1963, 1965). Probably the best combination to use would be either penicillin, 100 units/ml, or oxacillin, 20 μg/ml, and gentamicin, 50 μg/ml. This, however, would not necessarily cover gram-negative anaerobic bacteria, fungi, or all mycoplasmas. Tetracycline, 10 μg/ml, and amphotericin B, 3.0 μg/ml, would add very broad coverage. The effects of this antibiotic regimen on a particular cultured cell metabolism must be established by trial. It must be stressed that the superiority of these regimens over the commonly used penicillin and streptomycin has not been proven by controlled studies. These straightforward yet time-consuming investigations remain to be done.

A. Detection of the Source of Contamination

A search may reveal the source of contamination in the media components, such as serum used, or even in protective environmental devices, such as the footbath described by Coriell (1962). Liquid soap dispensers, if improperly and irregularly cleaned, can support the growth of bacteria. In two cases, we have presumed that contamination resulted from low grade infections in personnel doing the cell culture, one caused by *Candida albicans*, the other by *P. aeruginosa*. Contamination with these organisms stopped as soon as the infections in the personnel were cured.

A search for the source should be conducted but, in my experience, in the majority of cases the search is fruitless; the tightening of techniques in reaction to the contamination results in prevention, at least for a while.

B. Contaminated Tissues

When the cells to be cultured come from a naturally contaminated area, then the normal flora in that area should be considered. Internal organs, such as the kidney, spleen, and liver, should be sterile but areas of the oropharynx, skin, and gastrointestinal or gynecologic tracts all contain normal bacterial and frequently fungal flora. When culturing cells from these areas, the tissue should be washed with media containing high concentrations of appropriate antibiotics (Table IV) and during the first few days of culture, the same antibiotics should be included in the medium at concentrations that will not interfere with cell growth.

The microorganisms comprising predominant normal flora of humans are listed in Table V (Rosebury, 1965). This flora varies from species to species; the normal microbial flora of most other species is not as well known and will have to be treated differently according to the predominant organisms. The same applies to plants. The safest way to determine this would be to culture the tissue for bacteria and fungi before initiating the cell culture and to perform antibiotic sensitivities on the isolates. As time and facilities usually do not permit this, an educated guess or previous knowledge (Table V) of normal flora can be used as guides.

If tissue is to be taken from an individual who has been receiving antibiotics, this should be taken into account because the normal flora will be altered according to the antibiotic used. Organisms resistant to the admin-

TABLE IV
Antibiotic Concentrations for Cell Cultures[a]

Antibiotic	Recommended concentration	Range of concentration, μg/ml[b]
Penicillin G	100 units/ml	10–1000
Oxacillin	20 μg/ml	10–200
Gentamicin	50 μg/ml	10–200
Kanamycin	100 μg/ml	10–1000
Polymyxin B	50 μg/ml	10–1000
Tetracycline	10 μg/ml	5–50
Chloramphenicol	5 μg/ml	1–20
Erythromycin	50.0 μg/ml	10–100
Amphotericin B	3.0 μg/ml	1–50
Mycostatin	50.0 units/ml	10–500
Tylosin	10.0 μg/ml	1–100

[a]Concentrations necessary will vary with the size of the bacterial or fungal inoculum and toxic concentrations will vary for the cell type in use. The recommended concentrations are those for prophylaxis, treatment doses should be worked out individually using varying concentrations when possible.

[b]Ranges of concentration tolerated by various cell cultures.

istered antibiotic tend to increase in numbers, replacing the sensitive flora that is eliminated. The best chance to get rid of the organisms is during the washing procedure, when the concentration of the antibiotic is high and mechanical washing will decrease the numbers of organisms. A minimum of three washings should be used. Thereafter, it is probably wise to include antibiotics in the culture medium for at least 3 days and longer if the tissue was heavily contaminated. As high an antibiotic concentration as possible, compatible with continued cell growth, should be used. Low concentrations encourage the selection of resistant organisms, especially if the antibiotic deteriorates with time. At 37°C, many antibiotics will deteriorate over 24 hours so that fresh antibiotics should be added at least at this interval along with a change in media. Whether antimicrobial activity remains in the media after a selected period of time can be determined by saturating a blank sensitivity disk in the media and placing this on a plate recently inoculated with the offending organism or mixing the media with an inoculated tube of broth. Tube or plate dilution measurements could be done, similar to those for antimicrobial or antibiotic assays on human serum, using these methods.

Most areas of the gastrointestinal and gynecological tracts harbor anaerobic organisms that may be penicillin resistant (*Bacteroides fragilis* in the lower gastrointestinal and gynecological tract) and may require antibiotics that primarily inhibit protein synthesis. Although anaerobes are not report-

TABLE V
Predominant Bacteria and Fungi to Be Anticipated in Human Biological Specimens

Location	Organisms
Nose and throat	*Streptococcus* sp. *Staphylococcus* sp. Micrococci *D. pneumonia* *Neisseria* sp. *Corynebacteria* sp. Coliforms[a] *Klebsiella pneumoniae* *Haemophilus* sp. *Bacteroides* sp. *Fusobacterium* sp. *Spirilum* sp. *Candida* sp.
Saliva or sputum	All organisms found in nasopharynx plus: *Actinomyces* sp. *Nocardia asteroides* *Mycobacterium* sp. *Aspergillus* sp. *Cryptococcus* sp.
Feces and gastrointestinal tract	All organisms found in nasopharynx and sputum plus: Enterococci *Lactobacilli*[b] *Bacteroides* sp.[b] Coliforms[b] *P. aeruginosa* *Salmonella* sp. *Clostridia* sp. *Vibrio* sp. *Mima–Herrellea* group
Genitourinary tract, gynecological tract, urinary tract	Most organisms found in gastrointestinal tract External, as above Internal, should be sterile but may contain same organisms as external
Skin	Streptococci Staphylococci Micrococci *Corynebacterium* sp. Coliforms[a] *Candida* sp. *Dermatophytes*

ed as common contaminating organisms (we isolated only one), cultures are usually not performed for these fastidious bacteria (thioglycolate will not support the growth of many) and it is known that predominantly anaerobic mycoplasmas, such as *Mycoplasma orale*, are common contaminants. Therefore, because mycoplasmas are normal flora in the areas under discussion along with *Bacteroides* species, the inclusion of one of the protein-inhibiting antibiotics in the wash fluids to eradicate both types of organisms (plus L forms) would seem indicated. Whether these antibiotics will inhibit cell growth if included in the growth medium at various concentrations must be determined for each cell type.

C. Contaminated Inoculums

The microbial flora to be expected in biological specimens must be considered, just as the flora of tissues to be cultured. We frequently isolated *M. orale* or *M. salivarium* from saliva specimens cultured in human embryonic lung cells for cytomegalovirus until tetracycline was added to the medium in addition to penicillin. Bacteria- and fungus-contaminated cultures were usually discarded but when isolates were identified they ranged from *P. aeruginosa* to *Candida albicans*. Normal flora of various biological specimens are recorded in Table V.

The concentration of the various organisms varies considerably if the individual contributing the specimen is on antibiotics. One incident is worth relating both as an example and a warning. Peripheral leukocytes of a patient with chronic lymphocytic leukemia were placed in cell culture, refed biweekly, and observed for growth (Armstrong, 1966). After 10–14 days, cells were noted to be multiplying but they were quite small and some were budding. These were identified as *Cryptococcus neoformans*, a human pathogen that grew well in the cell culture medium and in the presence of the patient's dying leukocytes. The patient was demonstrated to contain the same organism in the cerebrospinal fluid when this was cultured on routine media in the Microbiology Laboratory.

Tissue culture media and cell cultures themselves may be excellent systems for promoting the growth of various bacteria and fungi that may be potential pathogens to the laboratory workers who are culturing apparently innocent and sterile tissue specimens. We have demonstrated that *P. aeruginosa, Aspergillus fumigatus* (Armstrong, unpublished data), and some my-

---

Footnotes to Table V:

[a]Coliforms include *E. coli*, *Klebsiella enterobacter*, and *Proteus* sp. Other organisms, such as *Pseudomonas aeruginosa*, *Alcaligenes fecalis*, *Enterobacter cloacae*, *Aeromonas* sp., and *Serratia marcescens*, are frequently found with the coliforms.

[b]Predominant organisms; *Bacteroides fragilis* in the gastrointestinal tract is usually penicillin resistant.

coplasma species (Armstrong and Hummeler, unpublished data), grow better in cell cultures than in some of the usual microbiology laboratory nutrient broth or agar medias.

Just as with tissues from contaminated sources, the broadest possible coverage for biological specimens for inoculation into tissue culture would be as follows: penicillin, 100 μg/ml; gentamicin, 50 μg/ml; tetracycline, 10 μg/ml; and amphotericin B, 3.0 μg/ml. If penicillin-resistant gram-positive organisms, such as *S. aureus*, become a problem (these will frequently be covered by gentamicin), oxacillin can be substituted for penicillin. Tetracycline covers mycoplasmas and most penicillin-resistant *Bacteroides* species that can be expected in stool specimens, as well as some gram-negative aerobes. Amphotericin B protects against *Candida* species as well as other fungi that might be in any of the usual biological specimens submitted (Table V).

D. Treatment of Contaminated Cell Cultures

Coriell has reported successful therapy of a diphtheroid contaminant with erythromycin (1962). Usually, contamination is considerable by the time it is detected grossly and is therefore difficult or impossible to treat. I have cured hamster tumor cell lines of infection with both *P. aeruginosa* and *C. albicans* after antibiotics failed by inoculating small numbers of the cells into hamsters and trypsinizing the resultant tumor to establish another cell line, presumably (but not certainly) comprised of the same cells as the original line. This maneuver affords the investigator the tremendously important advantage of an immune system in his battle with microorganisms, something that cell cultures are lacking and that puts them at a sad disadvantage.

Whether immune serum, especially in the case of such serum-sensitive organisms as *E. coli*, might be helpful in irradicating contamination with bacteria and fungi as it has with mycoplasmas remains to be proven.

The use of antibiotics alone may suppress the microorganism but when the antibiotic is withheld the bacteria or fungus reappears. Treatment of a contaminated culture is worth attempting, especially if the organism is isolated and identified and if sensitivity tests are performed. Preventing contamination is far easier and more successful.

E. Effects of Antibiotics on Cells and Inoculum

The effect of various antibiotics on the cells in culture must be considered. Variables include the cell type and media used and the concentration of the antibiotic (Perlman and Brindle, 1963, 1965). Toxic levels have been established for some antibiotics in relation to specific cell cultures but it is unlikely that these data can be applied to cell cultures in general. Concentrations recommended for a number of antibiotics are listed in Table IV along with concentrations found toxic for various cell cultures (Perlman and Brindle, 1963). If antibiotics are to be used in a cell culture medium, a number of concentrations should be tried while observing cell growth and multiplication. Because the recommended concentrations in Table V are usually well above the minimal bactericidal concentration of the antibiotic for the target organisms, lower concentrations can be used if there is an evident effect on the cell culture. Unfortunately, the antibiotics with the closest minimal inhibitory to cytotoxic concentration are those, such as the tetracyclines and chloramphenicol, that inhibit protein synthesis. These may be necessary to prevent contamination with many gram-negative anaerobes and some gram-negative and -positive aerobes, as well as mycoplasmas and L forms. The concentrations to be used for this type of antibiotic will have to be tested over a narrower range than those that inhibit bacterial cell wall synthesis and that are bactericidal, such as the penicillins, the aminoglycosides (gentamicin, kanamycin, and streptomycin), and colymycins. A newly developed aminoglycoside antibiotic, tobramycin, has proved highly effective *in vitro* against *P. aeruginosa* and other usually resistant gram-negative organisms (Meyer *et al.*, 1971) and may have some application in tissue culture.

Two antibiotics, novobiocin and Rifampin, are known to inhibit viral replication (Hirschman, 1971) so that they should not be used in cell cultures used for virus isolation.

V. Summary

Contamination of tissue culture with microorganisms is a problem whenever cells are cultured. The types of bacteria and fungi isolated reflect the surrounding inanimate environment, normal human flora, or the flora of the tissue cultured. Less often, media ingredients contain the contaminants. Prevention by scrupulous antiseptic techniques is the hallmark of continued sterile cell culture but when facilities are unavailable or contaminated cells or inoculums are employed, antibiotics should be used. Minimum safety can be attained by using penicillin alone, moderate safety by penicillin and gentamicin, and maximum protection by penicillin, gentamicin, tetracycline, and amphotericin B. Recommended concentrations are given but this should be established by trial for each different cell culture in use whenever possible.

Acknowledgments

Dr. J. Fogh submitted many contaminated cultures that had been sent to him for examination. Miss Anne Blevins and the staff of the Microbiology Laboratory of Memorial Hospital and Mrs. Bessie H. Yu of the special studies section of that laboratory diligently persued the final identification of the microbial isolates.

References

Armstrong, D. (1966). *Proc. Soc. Exp. Biol. Med.* **122**, 475.

Armstrong, D., and Hummeler, K. (1900). Unpublished data.

Armstrong, D., and Yu, B. (1967). *Clin. Res.* **15**, 302.

Armstrong, D., Tully, J. G., Yu, B., Morton, V., Friedman, M. H. and Steger, L. (1970). *Infect. Immunity* **1**, 1.

Barile, M. F., and Schimke, R. (1963). *Proc. Soc. Exp. Biol. Med.* **114**, 676.

Bauer, A. W., Kirby, W. M. M., Sherris, J. C., and Turck, M. (1966). *Amer. J. Clin. Pathol.* **45**, 493.

Coriell, L. (1962). *Nat. Cancer Inst., Monogr.* **7**, 33.

Fogh, J., Holmgren, N., and Ludovici, P. P. (1971). *In Vitro* **7**, 26.

Hirschman, S. Z. (1971). *Amer, J, Med.* **51**, 699.

Kagnoff, M. F., Armstrong, D., and Blevins, A. (1972). *Cancer* **29**, 245.

McGarrity, G. J., and Coriell, L. L. (1971). *In Vitro* **6**, 257.

Meyer, R. D., Young, L. S., and Armstrong, D. (1971). *Appl. Microbiol.* **22**, 1147.

Perlman, D., and Brindle, S. A. (1963). *Antimicrob. Ag. Chemother.* pp. 458–461.

Perlman, D., and Brindle, S. A. (1965). *Bacteriol. Proc.* p. 120.

Perlman, D., Giuffre, N. A., and Brindle, S. A. (1961). *Proc. Soc. Exp. Biol. Med.* **106**, 880.

Rosebury, T. (1965). *In* "Bacterial and Mycotic Infections of Man" (R. Dubos and J. Hirsch, eds.), 4th ed., pp. 326–355. Lippincott, Philadelphia, Pennsylvania.

Young, L. S., and Armstrong, D. (1969). *J. Infec. Dis.* **120**, 281.

CHAPTER 4

Contaminants Demonstrated by Microscopy of Living Tissue Cultures or of Fixed and Stained Tissue Culture Preparations

JØRGEN FOGH

I. Preparation of Cell Cultures for Microscopy 66
A. Cell Cultivation and Infection 66
B. Preparation of Living Cultures for Nomarski Interference-Contrast Microscopy 66
C. Hypotonic Treatment, Fixation, Air Drying, and Orcein Staining prior to Phase-Optics Microscopy 67
II. Methods of Microscopy 68
III. Photomicrographs Illustrating Various Types of Contaminants in Cell Cultures 69
A. Mycoplasmas 70
B. Bacteria 70
C. Yeasts 71
D. Molds 71
IV. Review of Data Obtained by Microscopy of Mycoplasma-Infected Cultures after Hypotonic Treatment and Orcein Staining 72
A. Quantitative Morphological Aspects of Mycoplasma–Host Cell Relationships 72
B. Results of Tests for Mycoplasmal Contamination 74
References 76

Bacterial, fungal, and mycoplasmal tissue and cell culture contaminants in most cases can be seen and recognized by their morphology directly in tissue cultures. This chapter illustrates that bacterial and fungal contaminants can be clearly distinguished in living tissue cultures when observed by the

Nomarski interference-contrast method of microscopy. Phase-optics microscopy, when applied to tissue cultures exposed to hypotonic treatment, fixation with Carnoy's solution, air drying, and staining with natural Orcein (Fogh and Fogh, 1964), provides an excellent technique for the demonstration of mycoplasmal, bacterial, and fungal contamination. Because various structures and organelles occurring in some cell types may resemble the morphology of tissue culture contaminants, all morphological tests for evidence of contamination have, in our laboratory, been performed by inoculating uninfected cultures of indicator cells, the FL human amnion cells (Fogh and Lund, 1957), with the material suspected of contamination. The very uniform morphology of FL cells and a complete knowledge of their detailed appearance under the selected conditions of tests have removed any doubts about the presence or absence of any of the above contaminants.

I. Preparation of Cell Cultures for Microscopy

A. Cell Cultivation and Infection

Cells of the FL line of transformed human amnion cells were cultured in LY medium (Fogh and Lund, 1957) with 20% human serum, 100 units of penicillin per milliliter and 100 μg of streptomycin per milliliter. They were transferred by trypsinization to small or large Leighton tubes (Leighton, 1951) with inserted rectangular coverslips, or to Demuth cups (Demuth Glass Works, Division, Brockway Glass Co., Brockway, Pa.) with round coverslips. Five thousand cells were seeded per small Leighton tube, and 20,000 cells per large Leighton tube. At this stage, antibiotics were eliminated. Twenty-four or forty-eight hours later they were inoculated with the tissue culture material to be tested, usually in a volume of 0.1 ml. It was ascertained that the cultures from which the inoculum originated did not contain antibiotics. To assure a sufficient quantity of the contaminant at the time of observation and to avoid excessive cell destruction by the contaminant, the inoculum and the time of incubation were varied among several test cultures.

B. Preparation of Living Cultures for Nomarski Interference-Contrast Microscopy

Cultures that were to be examined without fixation and staining were mounted in a drop of medium by inverting the coverslip on a glass slide and sealing the rim with a molten mixture of 75% dental wax and 25% Vaseline (Fogh and Sykes, 1972). This rimming was necessary to avoid cytoplasmic bubbling, which occurred rapidly after mounting of many cell types. Immediately after mounting, the preparations were examined under the microscope by the Nomarski technique.

C. Hypotonic Treatment, Fixation, Air Drying, and Orcein Staining prior to Phase-Optics Microscopy

Cultures that were to be examined under phase-optics microscopy were prepared as follows: The culture fluid was poured off and was replaced by 3 ml of 0.6% sodium citrate solution. One milliliter of distilled water was added, dropwise, with a 1 ml pipet, making the concentration of sodium citrate 0.45%. After 10 minutes, 4 ml of Carnoy's fixative (1 part glacial acetic acid, 3 parts absolute ethyl alcohol) was added with a 2 ml pipet, dropwise, for gradual fixation. The fluid in the container was poured off and was replaced with 2 ml of Carnoy's fixative. After 10 minutes fixation, the slide was taken out and left until absolutely dry. The cells were now stained for 5 minutes with Orcein stain (2% natural Orcein and 60% glacial acetic acid in distilled water). Following three washes in absolute alcohol, the slide was mounted in Euparal (Flatters and Garnett, Ltd., Manchester, England).

The reasons for using FL cells as indicator cells; the choice of culture container, cell density, and culture medium; and the importance of the proper pH and temperature have been extensively discussed elsewhere (Fogh and Fogh, 1969). For the demonstration of mycoplasmas, low numbers of FL cells are preferred because the amount of FL cell associated mycoplasma increases with decreasing number of FL cells per slide (Fogh and Fogh, 1967). Cultures were usually prepared for examination 24 or 48 hours after inoculation with the test sample. Not infrequently, cell cultures received for testing in this laboratory have contained antibiotics in concentrations sufficient to mask the presence of mycoplasmas and other contaminants; for example, aureomycin, kanamycin, or streptomycin. In proper concentrations, these antibiotics sometimes can effectively eliminate mycoplasma from tissue cultures. Their presence may, therefore, result in a false negative test. Before testing, cultures should always be changed to a medium containing no antibiotics and they should be incubated in such a medium for several days before the actual tests are performed.

Although in many cases contamination with bacteria or molds may be demonstrated without exact adherence to the steps described, the demonstration of mycoplasmas by this technique requires the proper procedure (Fogh and Fogh, 1969). It is most important that the technique be carried out exactly as described. The omission or incorrect performance of individual steps of the procedure may well result in unsatisfactory preparations. Good preparations depend, for example, on a gradual expansion of the cytoplasm during the hypotonic treatment and a not too sudden fixation. The procedure as described was designed for the particular response of FL cells. These cells contain a large amount of cytoplasm, well retained for the purpose of observation of cell-associated mycoplasmas after fixation with Carnoy's solution. Different cell types vary in their response to hypotonic treatment and fixation. If carried out as described and with cells cultured under optimal conditions, this test has been found to be exceedingly reliable and reproducible. If time is important, results may be obtained within a few hours after inoculating the test sample. In connection with each series of tests, a positive control (an FL cell culture inoculated with culture supernatant from a known mycoplasma-infected culture) and a negative control (uninfected FL cells) are included.

II. Methods of Microscopy

Zeiss photomicroscopes, equipped with either phase optics or interference-contrast (Nomarski) optics, were used for all micrographs presented in this chapter.

Both types of microscopy make use of the interference effects of light. In phase-contrast microscopy (Osterberg, 1955; Clark, 1961), the light striking the specimen is separated into two components, the diffracted and undiffracted portions. Density variations in the sample cause the diffraction of the light beam. After passing through the specimen, the light goes through the phase rings, where the undiffracted portion is reduced in intensity and given a quarterwave phase shift with reference to the diffracted light. Upon leaving the phase rings, the light beams are recombined, allowing constructive and destructive interference to take place. Thus, refractile regions, such as the nucleus in living cells, produce areas of destructive interference, giving rise to their dark appearance, whereas less refractile regions, such as the cytoplasm, appear less dark than the nucleus because of constructive interference. Phase-contrast microscopy makes use of density variations within the specimen to enable the observer to see details that would be invisible under bright field microscopy.

In Nomarski interference-contrast microscopy (Nomarski, 1955; Clark, 1961; Allan *et al.*, 1969), plane polarized light passes through a birefringent plate. The emerging coherent wavefronts are mutually perpendicular and physically displaced by about 1 μm. These two wavefronts pass through the sample and are recombined at the interface of another prism. After passing through the analyzer, an interference image is produced before reaching the observer's eye. The ordinary and extraordinary wavefront are shifted in phase by geometrical thickness or relief within the specimen. Hence, visible detail results from variations in the refractive index within the sample. This gives rise to the three-dimensional quality of Nomarski photomicrographs.

III. Photomicrographs Illustrating Various Types of Contaminants in Cell Cultures

The series of photomicrographs at the end of this chapter illustrate various examples of tissue and cell culture contaminants collected during the last few years from cultures obtained from several laboratories. All the pictures show cultures of FL cells that have been inoculated with a contaminant. Both methods, phase-optics microscopy of hypotonically treated and Orcein-stained preparations or interference-contrast microscopy (Nomarski) of living, unfixed, and unstained cells, have without exception permitted the determination of the type of contaminant, i.e., bacterium, mold, yeast, or mycoplasma. A few of the micrographs illustrate Nomarski microscopy of hypotonically treated and Orcein-stained preparations. Obviously, further classification, if needed, requires studies of other special characteristics of the contaminant. Whereas mycoplasmas are best observed in the hypotonically treated and Orcein-stained preparations, Nomarski microscopy of living cultures may sometimes be the superior method for the other types of contamination. Cell and glass attachment vary considerably among bacterial, yeast, and mold contaminants. In lightly contaminated cultures, the procedures related to fixation and staining may reduce the attached amounts of the contaminant to such an extent that demonstration is difficult in the stained preparations. In heavily infected cultures, however, both techniques have very well demonstrated the morphology of the contaminants and their association with the host cells.

The distinction between mycoplasmas, bacteria, yeasts, and molds depends, in part, upon the size and morphology of the individual microorganisms but also upon their location in relation to the mammalian cells. Mycoplasmas are spherical or occasionally slightly oval. According to measurements in our material, they usually average 0.23–0.30 μm in size, but may vary from 0.18 μm to 1.6 μm (Fig. 5–10).* The size of bacteria is in the range of 0.5–2.0 μm; they can be rod shaped (Fig. 11, 13, and 19) or spherical (Fig. 15, 17, and 18). Yeasts, of course, are much larger (2–6 μm) and mostly ovoid (Fig. 21–33). Internal structure is often discernible (Figs. 27 and 32–35). The molds are generally easily distinguished by their branching hyphae (Figs. 40–45). Intercellular location of the contaminants, which is highly pronounced for essentially all strains of mycoplasmas (Figs. 3, 4, and 8), is sometimes also typical for bacterial (Figs. 11 and 17) and yeast contaminants (Figs. 24 and 30).

*See pages 77 through 106 for all figures.

A. Mycoplasmas

The easily distinguished mycoplasmal contamination shows the individual mycoplasmas located primarily at the cell borders, apparently in part attached to the peripheral parts of the cytoplasm. Within groups of cells, the mycoplasmas are predominantly in the intercellular spaces. The amounts of cell-associated mycoplasmas observed in a given preparation vary with the inoculum, strain of mycoplasma, number of FL cells, time of incubation, and temperature (Fogh and Fogh, 1967). A heavy infection of a dense cell culture gives a mosaiclike picture, resulting from the high concentration of mycoplasmas in the intercellular spaces (Figs. 3 and 4). All the cells are heavily rimmed with the mycoplasmas, including cells in mitosis. Each cell in a light cell culture, when heavily infected, is surrounded by a very dark, contrasty rim that frames the cells at the free cytoplasmic borders. The cell-associated mycoplasmas in sparsely populated cultures, when infected only to a light degree, are observed as individual entities attached to the borders and, to a minor degree, to other parts of the cell surface (Fig. 5). With light infection of more dense cell cultures, the mycoplasmas may be seen only in association with some of the cells. Glass-attached mycoplasmas can be frequently observed between the cell-covered areas of the slide (Figs. 5–7). They appear singly or in groups, sometimes showing colony morphology (Fig. 7). Their amounts vary among different mycoplasma strains. Variation in the size of the individual glass-attached mycoplasmas is very pronounced for some strains (Fig. 6). Agglutinated clumps (Figs. 8 and 9) or crystalline arrays (Fig. 5) may be seen in association with the cells or in the glass-attached colonies. Some strains have shown variations in staining intensity among the individual mycoplasmas (Fig. 6), although in most cases this intensity has been very uniform.

Nomarski microscopy of living, unfixed cultures has not proven to be a sufficiently reliable method to evaluate the presence or absence of mycoplasma in tissue cultures. In cultures exposed to hypotonic treatment prior to Orcein staining, this type of microscopy can reveal the true spherical shape of the mycoplasmas (Fig. 10).

B. Bacteria

Bacterial contaminations are easily recognized by both methods. The micrographs in this chapter illustrate the gram-negative *Pseudomonas* sp. (Figs. 11 and 19), *Escherichia coli* (Figs. 12 and 13), and a nonpigmented *Serratia* sp. (Fig. 20). Gram-positive micrococci are seen in Figs. 14 and 15 and *Staphylococcus albus* is seen in Figs. 16 and 17. In addition to the cell-attached bacteria, which at heavy infections may essentially cover the mammalian cell surfaces (Figs. 12, 16 and 20), microorganisms may attach to the glass surface (Figs. 11, 13, 18, and 19) and, of course, may float in the culture fluid. Their presence in the fluid phase may interfere with good photographic demonstration in micrographs of living cultures (Figs. 12 and 20). Depending upon the type and strain of bacterium and the length of infection, degrees of cell destruction may be observed varying from damage to the peripheral part of the cytoplasm (Figs. 11 and 15) to a nearly total destruction of the cell culture, with most of the mammalian cells having fallen off the glass surface (Fig. 19). Either method of demonstration appears to give a clear and unquestionable picture of any of these contaminants.

C. Yeasts

Different yeast tissue culture contaminants are shown in Figs. 21–39. They include several yeasts of *Candida* sp. (Figs. 21–27 and 36–39), one of *Saccharomyces* sp. (Figs. 28 and 29), and two unidentified strains (Figs. 30–35). Although the oval, spherical, or elongated yeast cells are found throughout the infected FL cell cultures, they appear to have a particular affinity for the mammalian cells. The contaminants are located at the mammalian cell surfaces, either at the free borders of the host cells (Figs. 24–28, 30–33) or on the free cell surfaces (Figs. 21, 33, and 35). Within groups of cells, intercellular location is quite common (Figs. 24 and 30). Intracellular location is indicated in several of the micrographs (Figs. 25, 27, and 30). Internal structure of the yeast cells (Figs. 23, 28, 29, 31–33, and 39) and budding (Figs. 28, 30, and 39) are apparent in many of the micrographs. Hyphalike structures are seen in Figs. 36–39, which demonstrate *Candida stellatoidea* infection. Hyphalike structures may express a double infection with a mold, but they are probably the pseudohyphae of this yeastlike species.

It is apparent from the pictures that both methods, Nomarski microscopy of living cultures and phase-optics microscopy of hypotonically treated and Orcein-stained cultures, are highly suitable for the observation of yeast contaminations.

D. Molds

As demonstrated by four contaminants, *Mucor* sp. (Fig. 40), *Aspergillus niger* (Figs. 41 and 42), *Penicillium* sp. (Figs. 43 and 44), and *Trichosporon* sp. (Figs. 45–48), molds, as they appear in cultures of FL cells, are also easily observed by either method, when long, tubular, septate hyphae can be seen (Figs. 40–45). The branching, intermingled, and often fused hyphae constitute the mycelium, which crisscrosses and is superimposed upon the mammalian cells. The septae may be seen particularly well by the Nomarski microscopy of living cultures (Fig. 41). This technique can also reveal the nuclei within the individual cells (Fig. 43). There is evidence of severe mammalian cell damage in several of the micrographs (Figs. 41–45). However, the cytoplasmic bubbling seen in Fig. 45 does not result from the mold in the culture. It also frequently occurs in living, uninfected cell cultures observed by the Nomarski method of microscopy some time after the mounting.

As shown in Figs. 46–48, *Trichosporon* sp. can occur in a yeastlike form under the conditions of tissue culture. The spherical cells may appear as chains (Fig. 46) or may produce hyphalike structures (Figs. 47 and 48).

Although certain structural details of the contaminant may be better observed by one or the other of the two methods, it appears on the basis of our experience that either method can very well demonstrate the presence of mold contaminants.

IV. Review of Data Obtained by Microscopy of Mycoplasma-Infected Cultures after Hypotonic Treatment and Orcein Staining

Hypotonic treatment and Orcein staining have been used extensively in our laboratory to test for the presence of mycoplasmas in numerous tissue cultures. Because it provides the opportunity for detailed observations of mycoplasmas attached to the host cells and for an evaluation of the amounts of mycoplasma present under various experimental conditions, the method has also been applied in quantitative morphological studies on mycoplasma–mammalian cell association (Fogh and Fogh, 1967).

A. Quantitative Morphological Aspects of Mycoplasma–Host Cell Relationships

These experiments were carried out with cultures of FL cells and with the HT strain of *Mycoplasma fermentans* (Fogh *et al.*, 1965). Prior to infection, the mycoplasma was cultured in BYE broth (Barile *et al.*, 1958) with 15% human serum or in cultures of FL cells. The concentration of colony-forming units (CFU) was determined by inoculating petri dishes containing BYE agar with X and V factor (Barile *et al.*, 1958) and 15% human serum, incubating the plates in an atmosphere of 5% CO_2 in nitrogen at 37°C, and recording the colony numbers either from total plate counts or by extrapolation of counts of randomly selected representative areas. FL-cell-associated mycoplasmas were recorded as actual counts per cell when the infection was light (Fig. 49). When the infection was too heavy for the actual number of mycoplasmas per cell to be counted, progressive grades of mycoplasma–FL cell associations ranging from + + + to + were used (Fogh and Fogh, 1967).

The results of these experiments established a correlation, independent of the incubation period (within limits), between the quantity of cell-associated mycoplasmas and the mycoplasma titer in the cell culture fluid. The amounts of each increased logarithmically with time. The maximum value of the culture fluid titer increased gradually as the number of FL cells per culture increased. Other factors equal, the amount per FL cell increased with decreasing number of FL cells, demonstrating that the mycoplasmas competed for cell association and were not randomly distributed in the culture. Cell attached mycoplasmas were also observed earliest in cultures containing low numbers of FL cells. Under such conditions, mycoplasmas could be seen in association with the host cells within 2 hours after an inoculum of 3×10^5 CFU/ml. The proportion of free to cell-associated mycoplasmas increased at temperatures lower than 37°C.

A reproducible relationship that was established between the number of host cells, the mycoplasma inoculum, the titer in the culture supernatant, and the amounts of the host-cell-associated mycoplasmas indicates the possibility of using the hypotonic treatment–Orcein staining technique to determine the concentration of mycoplasma in a sample of unknown titer.

It is clearly demonstrated in these studies that mycoplasmas are located to a large extent in connection with and associated with the mammalian host cells and that mycoplasmas may selectively favor the cell site rather than the fluid culture phase for their location of propagation. It is not a new observation that the presence of cells results in an increase in the amount of mycoplasmas in the culture supernatant (Carski and Shepard, 1961; Fogh *et al.*, 1965). Our more recent data have shown that titers in the culture supernatant increase with the number of cells per culture. It has been previously proposed that mammalian cells provide mycoplasmas with certain nutrients absent from the medium or inactivate substances that can prevent growth of mycoplasmas (Carski and Shepard, 1961). It is conceivable that better conditioning of the medium and better pH adjustment may be obtained when more mammalian cells are present.

The intimate relation between mycoplasmas and mammalian cell surfaces has also been observed by electron microscopy (Edwards and Fogh, 1960). Microvilli have been seen projecting from the mammalian cells in the neighborhood of the mycoplasmas, and the plasma membrane of the mammalian cells at the tips or on the sides of the microvilli may become continuous with the plasma membranes of the mycoplasmas, suggesting attachment. Such attachment, in some cases, is observed in nonvillated regions of the mammalian cell border. The attachment is generally by direct apposition of plasma membranes. Because data with the hypotonic treatment–Orcein staining technique have shown reductions in the proportion of cell-associated mycoplasmas to free mycoplasmas occurring after temperature reductions, it is indicated that cell attachment involves an active biological process on behalf of either host cell or mycoplasmas (Fogh and Fogh, 1967).

Because the rate of increase in cell-associated mycoplasmas with time is logarithmical and the rate of increase is similar to the rate of increase of CFU in the fluid phase, the observed cell-associated mycoplasmas are indeed living organisms. Their amount at any time is not an expression of random distribution. Our data have indicated that a high proportion of the mycoplasmas that are demonstrated in the culture fluid are derived from cell-associated replication. For quantitative determinations of cell-associated mycoplasmas, the hypotonic treatment–Orcein staining technique appears to be unique. The high fragility of mycoplasmas essentially excludes the use of physical-chemical methods applicable to virus–host cell studies for such examinations. The technique also lends itself to the studies of the effects of environmental conditions, antibiotics, mixed infections, etc. One great advantage has been that preparations obtained during many years still exist and serve as a permanent file of various cell types employed and different aspects of the work in this laboratory. They have provided the possibility of a retrospective evaluation of particular experiments carried out a long time ago.

B. Results of Tests for Mycoplasmal Contamination

Based on tests performed in our laboratory with mammalian cell cultures from approximately 350 different sources, a general evaluation of the hypotonic treatment–Orcein staining was given several years ago (Fogh and Fogh, 1969). Since then, the method has been used even more extensively to test cultured cells from many different sources. It has been used routinely in this laboratory for frequent monitoring of cell cultures used in our own work, as well as for testing cultures from a number of other laboratories. There have been no exceptions to the previous evaluation. On the basis of a comparison with other methods examined or used routinely to check the validity of the hypotonic treatment–Orcein staining method, this method is our method of choice, in terms of rapidity, ease, and reliability. There is a need, though, for a large-scale experiment in which many strains of mycoplasmas, isolated as tissue culture contaminants by several laboratories, are compared by the several available test procedures for mycoplasmal contamination. The outcome of such an experiment, in which different laboratories should participate, might establish the most efficient and practical method to recommend for general use.

Routinely, indicator cultures of FL cells have been prepared for microscopic examination 24 or 48 hours after inoculation with the suspected material. Reliable results have always been obtained within this time, whereas comparable tests using agar plates have required 3 to 10 days of incubation. Among the many different contaminants we have demonstrated by the hypotonic treatment–Orcein staining technique, some variation in the morphology of the mycoplasmas and in the mycoplasma–host cell association has been noted. However, this test has never left any doubt about their presence or absence.

During the last few years, it has become apparent that the morphology of some strains of mycoplasmas is different in the presence of heat-inactivated human serum (56°C for 30 minutes) as compared to noninactivated serum. If the serum is not inactivated, these mycoplasmas may appear agglutinated, whereas in inactivated serum they are uniformly dispersed as single units throughout the culture. Tests with heat-inactivated human serum have, therefore, often been added in our recent program.

As previously reported (Fogh and Fogh, 1969), mycoplasma contaminations on several occasions were detected in hypotonically treated and Orcein-stained cell preparations when simultaneous tests using agar plates were negative. Additional experimentation with the agar media components, however, confirmed the isolate as a mycoplasma, usually with special nutritional requirements for propagation on agar. Similar observations have been made during the last few years. In addition, we have experienced several contaminations that were clearly observed by the hypotonic treatment–Orcein staining technique without being able to culture mycoplasmas in any of the several broths or agar media used in this laboratory. As the size, morphology, and location of the contaminants in the test cultures of FL cells were typical of mycoplasmas, and as they responded to treatment with antibiotics used for mycoplasma elimination, it was concluded that the demonstrated microorganism was a mycoplasma.

It is still questionable whether our inability to compose the artificial medium necessary for propagating these particular mycoplasmas is responsible for the negative results, or whether these particular strains of mycoplasmas have an absolute requirement for living cells for their propagation. The reverse situation, in which growth of mycoplasmas could be established on agar without a positive result of the hypotonic treatment–Orcein staining test, has never been experienced.

The necessity of testing in the absence of antibiotics (except for penicillin) cannot be emphasized enough. False negative results have, indeed, been obtained if antibiotics were not removed from the culture to be tested prior to inoculating culture supernatant into the indicator cultures. Apparently, the low concentration of antibiotics, resulting from the dilution of the inoculum in the test culture medium, may be sufficient to inhibit mycoplasma propagation to a level not detectable by the test.

Depending upon the culture media and serum combinations used in the test, various strains of mycoplasmas have propagated to different titers. The amounts of cell- and glass-associated mycoplasmas that can be observed vary also with these conditions. In our laboratory, LY medium (Fogh and Lund, 1957) with 20% human serum is used routinely in all tests. Apparently, this combination supports the propagation of most mycoplasma strains to concentrations proper for a reliable demonstration by the hypotonic treatment–Orcein staining method within 24 hours after inoculation, or less.

Acknowledgments

Several tissue culture contaminants demonstrated in this chapter were supplied by Dr. Donald Armstrong of Memorial Hospital. The bacterial contaminants were all identified by Dr. Armstrong. He and Dr. Morris Gordon of the New York State Department of Health both helped us in identifying the fungi. I am grateful to Dr. Gordon and Dr. Morris Teller of the Sloan-Kettering Institute for reviewing the manuscript.

References

Allan, R. D., David, G. B., and Nomarski, G. (1969). *Z. Wiss. Mikrosko.* **69**, 193.
Barile, M. F., Yaguchi, R., and Eveland, W. C. (1958). *Amer. J. Clin. Pathol.* **30**, 171.
Carski, T. R., and Shepard, C. C. (1961). *J. Bacteriol.* **81**, 626.
Clark, G. (1961). "The Encyclopedia of Microscopy." Van Nostrand-Reinhold, Princeton, New Jersey.
Edwards, G. A., and Fogh, J. (1960). *J. Bacteriol.* **79**, 267.
Fogh, J., and Fogh, H. (1964). *Proc. Soc. Exp. Biol. Med.* **117**, 899.
Fogh, J., and Fogh, H. (1967). *Proc. Soc. Exp. Biol. Med.* **125**, 423.
Fogh, J., and Fogh, H. (1969). *Ann. N.Y. Acad. Sci.* **172**, 15.
Fogh, J., and Lund, R. O. (1957). *Proc. Soc. Exp. Biol. Med.* **94**, 532.
Fogh, J., and Sykes, J. (1972). *In Vitro* **7**, 206.
Fogh, J., Hahn, E., and Fogh, H. (1965). *Exp. Cell Res.* **39**, 554.
Leighton, J. (1951). *J. Nat. Cancer Inst.* **12**, 545.
Nomarski, G. (1955). *J. Phys. Radium* **16**, 9.
Osterberg, H. (1955). *Phys. Tech. Biol. Res.* **1**, 378.

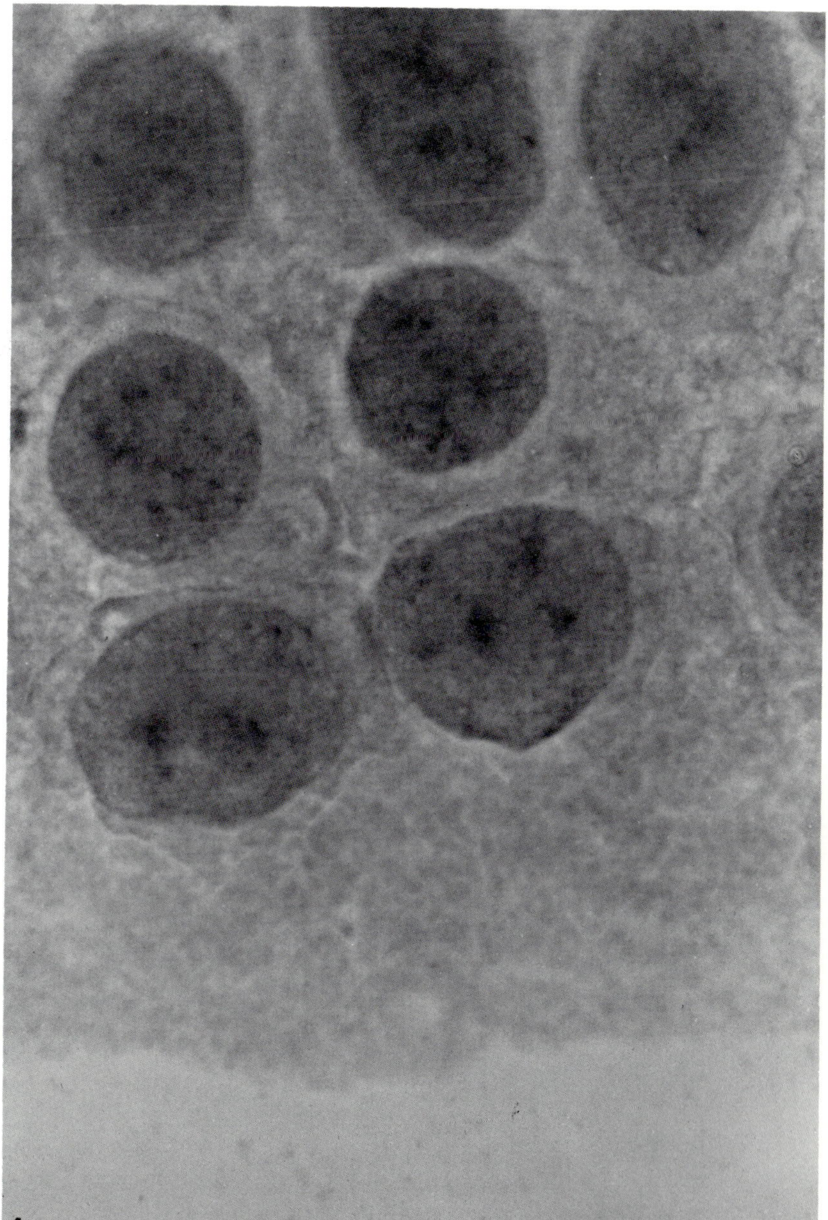

Fig. 1. Uninfected FL cells exposed to hypotonic treatment and staining with Orcein. Phase optics. × 2000.

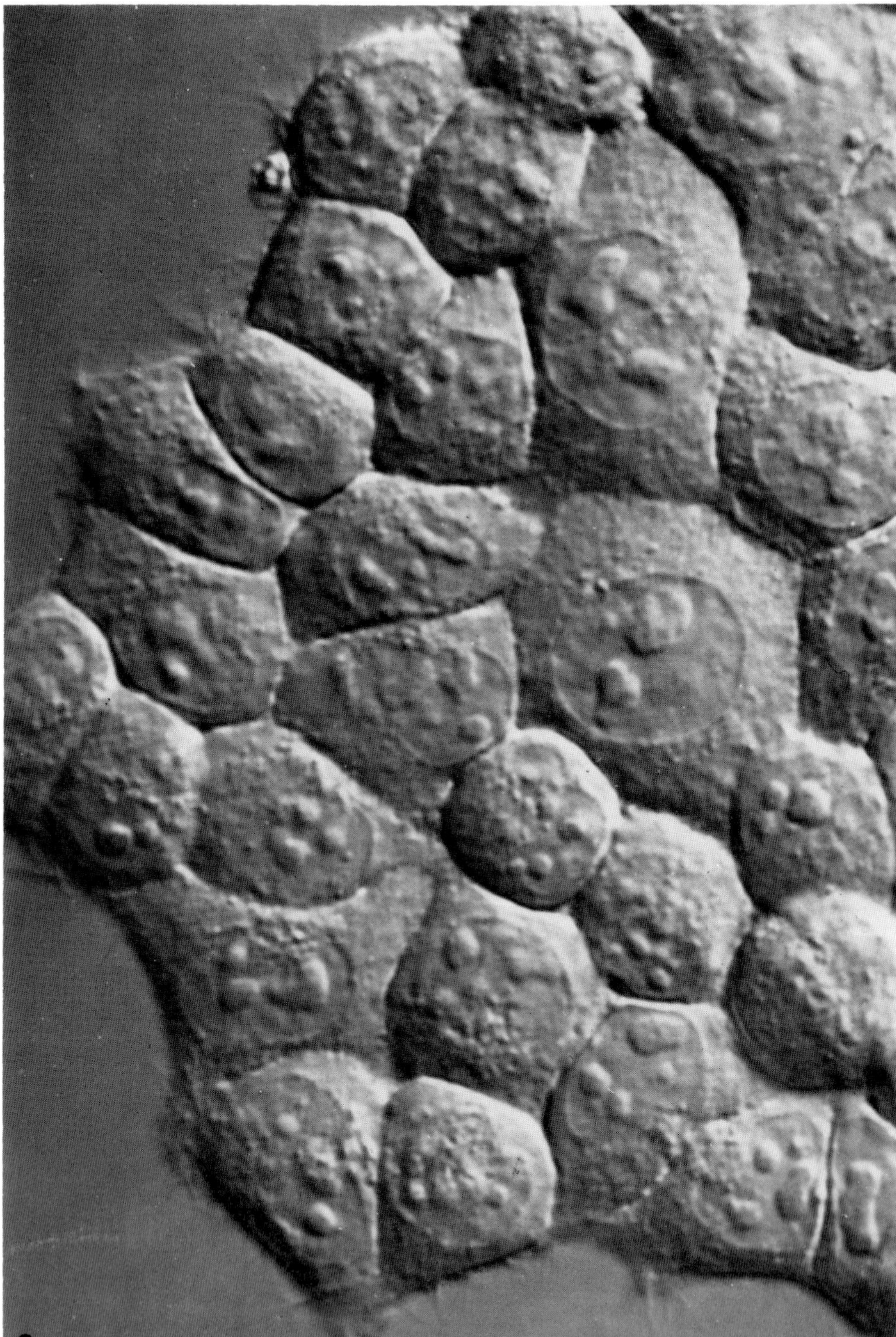

Fig. 2. Living culture of uninfected FL cells. Nomarski optics. × 1280.

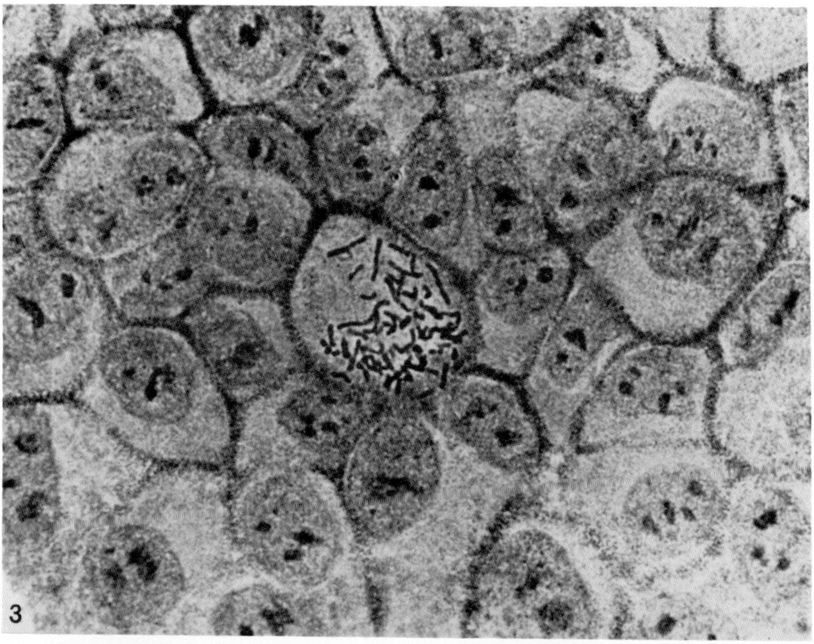

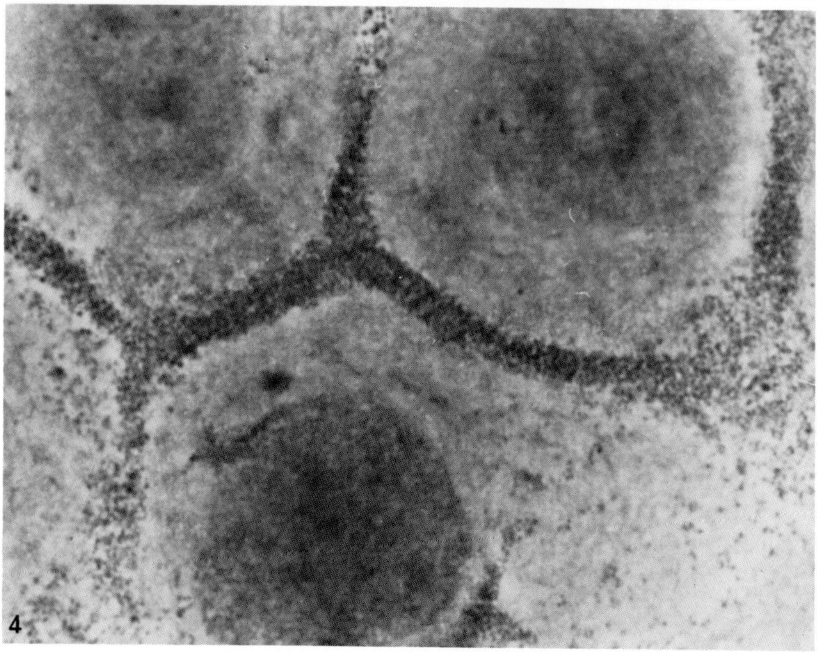

Figs. 3 and 4. Cultures of FL cells heavily infected with HT mycoplasmas. The intercellular location of the contaminants is obvious. Hypotonic treatment and Orcein staining. Phase optics. Fig. 3, × 600; Fig. 4, × 1920.

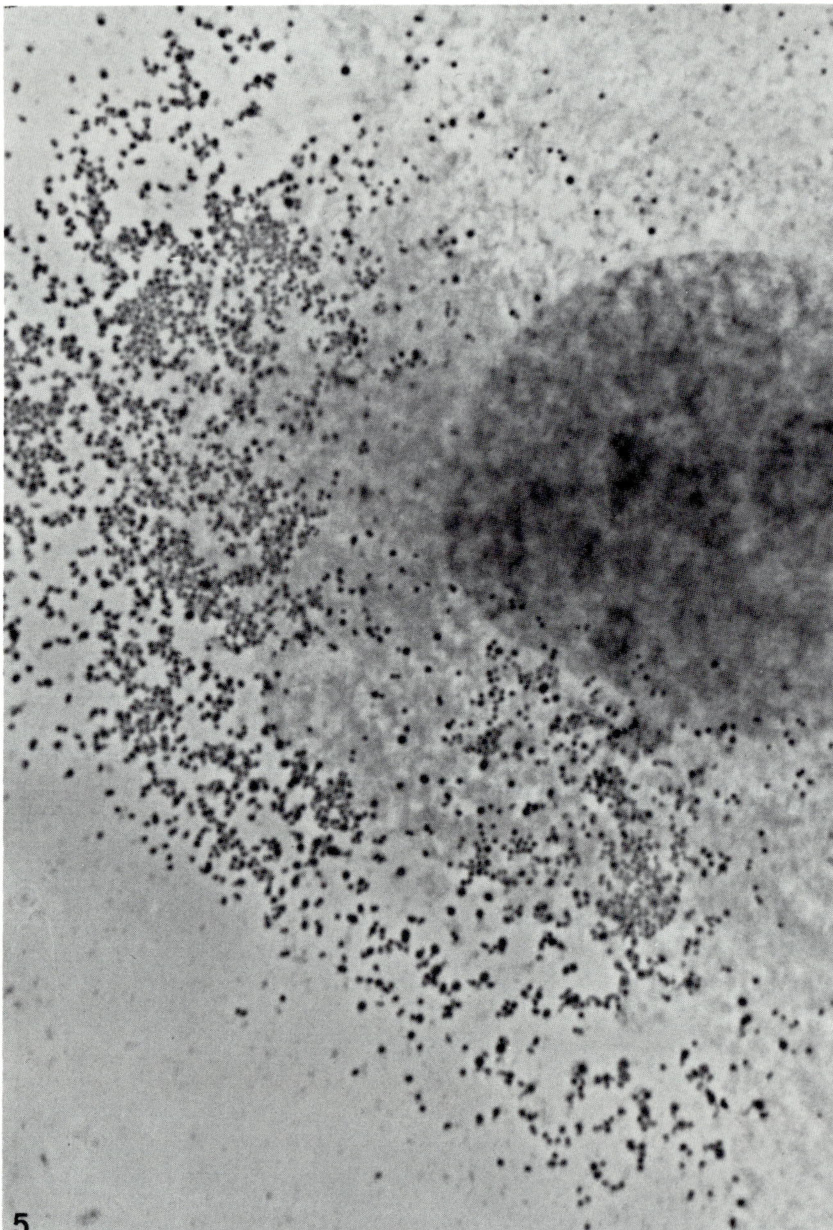

Fig. 5. High magnification of individual FL cell with cell-associated HT mycoplasmas, some of which form agglutinates. Hypotonic treatment and Orcein staining. Phase optics. × 2600.

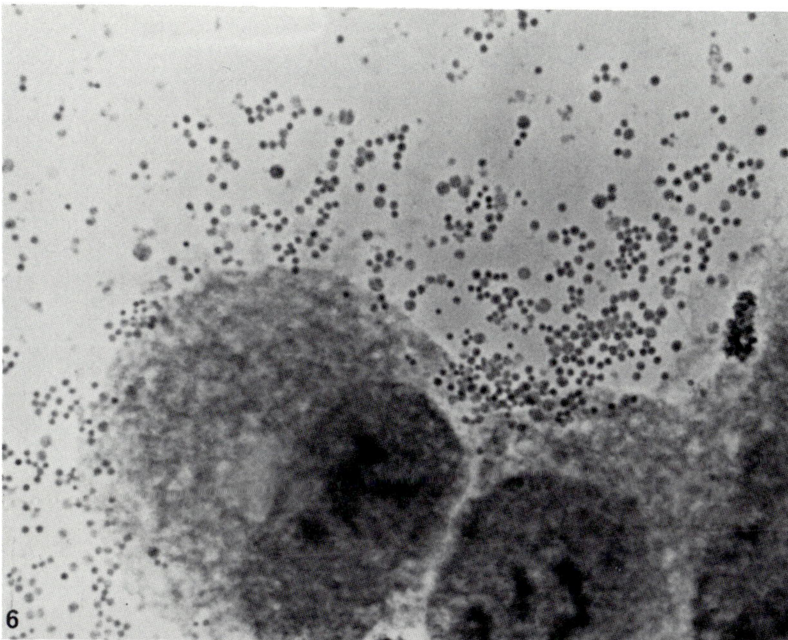

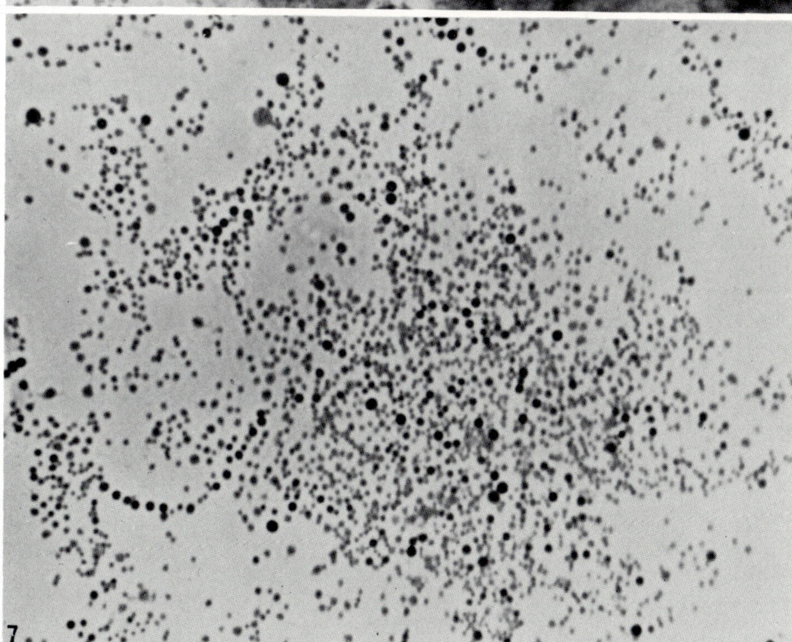

Fig. 6. Infection of FL cells with mycoplasma strain 880. Variation in size is pronounced and internal structure is noticeable in the larger mycoplasmas. Hypotonic treatment and Orcein staining. Phase optics. × 1500.

Fig. 7. Glass-attached mycoplasmas in culture of FL cells. Notable variation in size among the mycoplasmas. Hypotonic treatment and Orcein staining. Phase optics. × 1500.

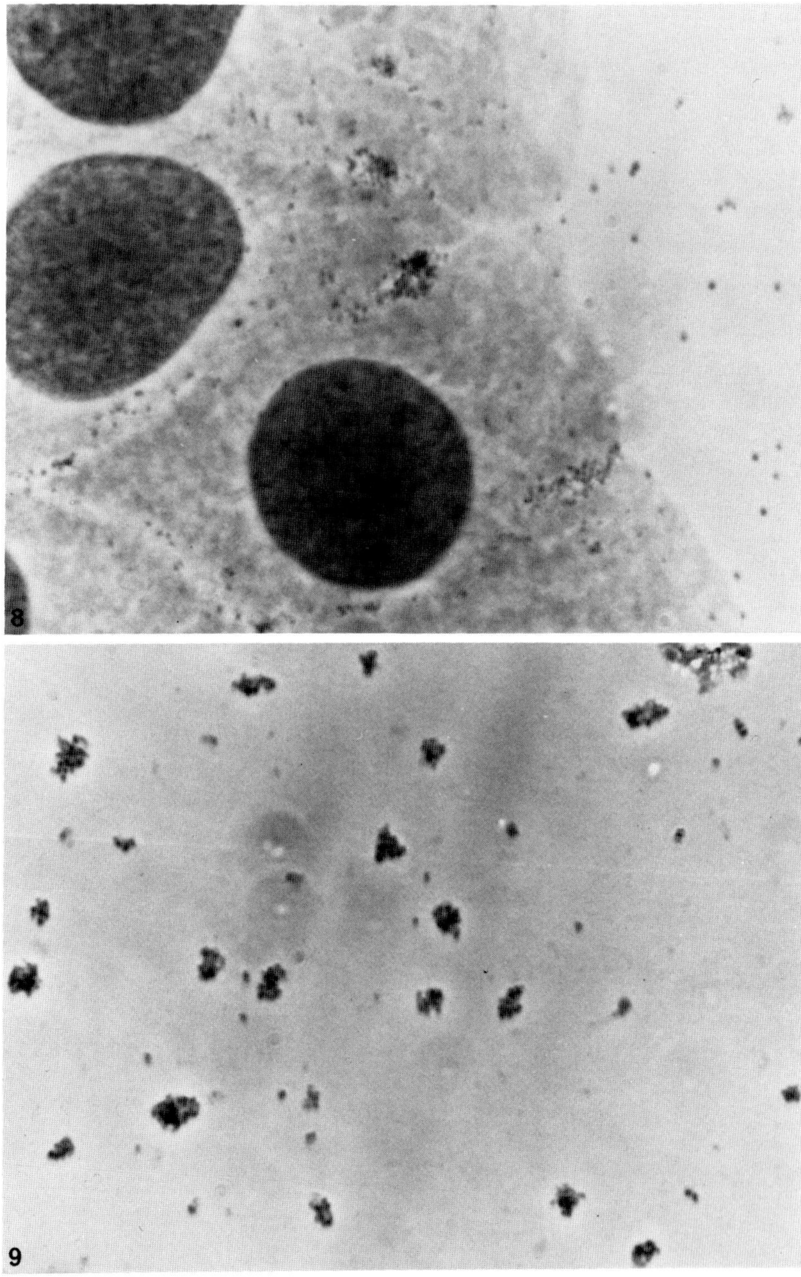

Figs. 8 and 9. Mycoplasmas isolated from ESP-1 cells, located in the intercellular spaces (Fig. 8) and attached to the glass (Fig. 9) in FL cell cultures. Hypotonic treatment and Orcein staining. Phase optics. × 1920.

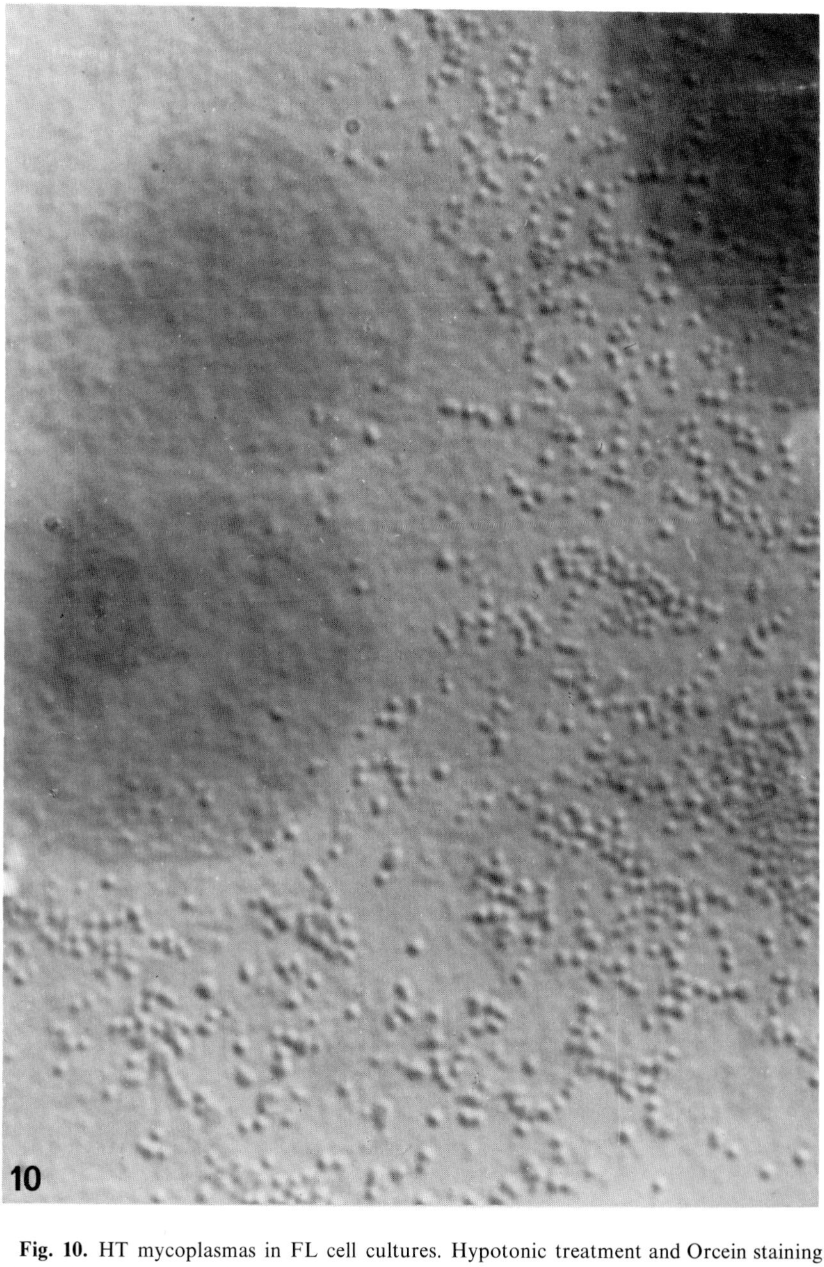

Fig. 10. HT mycoplasmas in FL cell cultures. Hypotonic treatment and Orcein staining. Nomarski optics. × 3200.

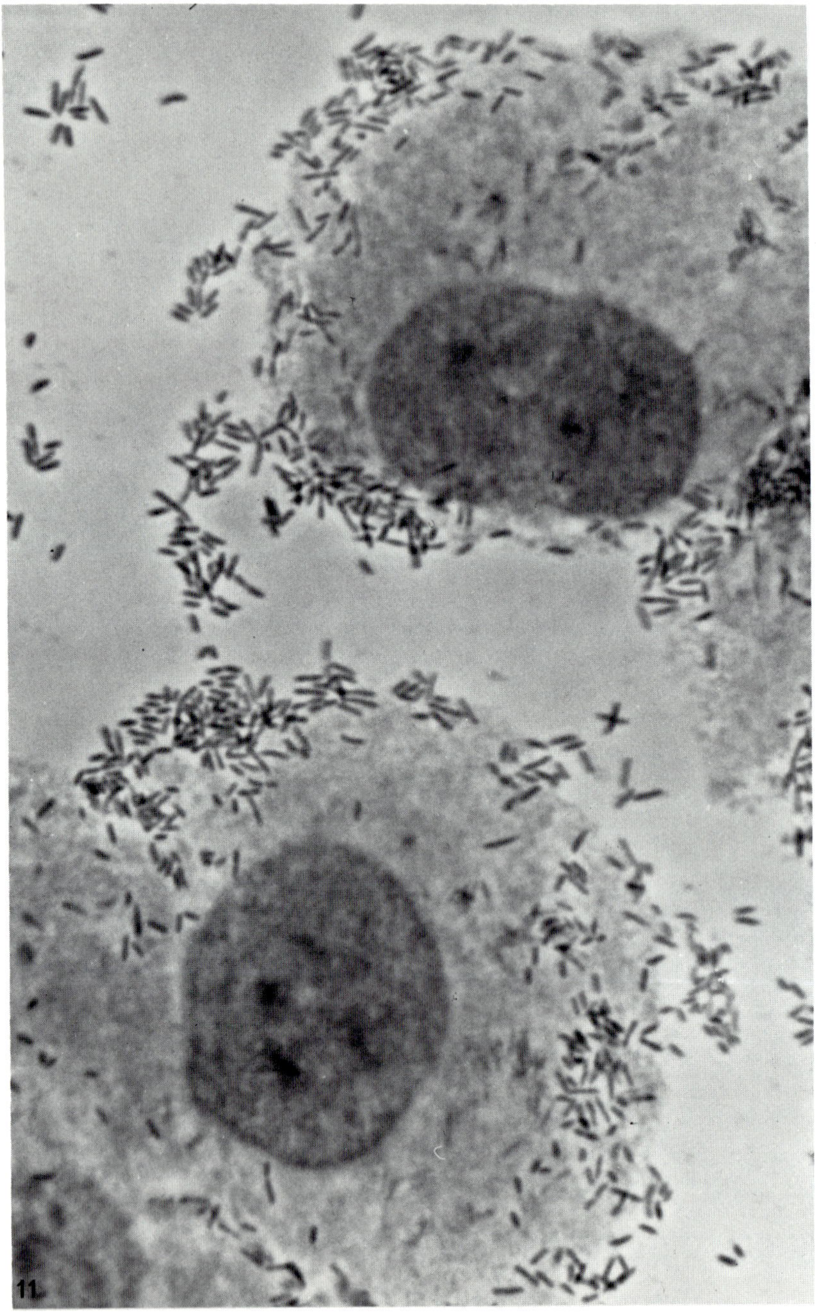

Fig. 11. *Pseudomonas* infection showing the FL-cell-associated and glass-attached rod-shaped bacteria. There is destruction in the peripheral part of the host cell cytoplasm. Hypotonic treatment and Orcein staining. Phase optics. × 2100.

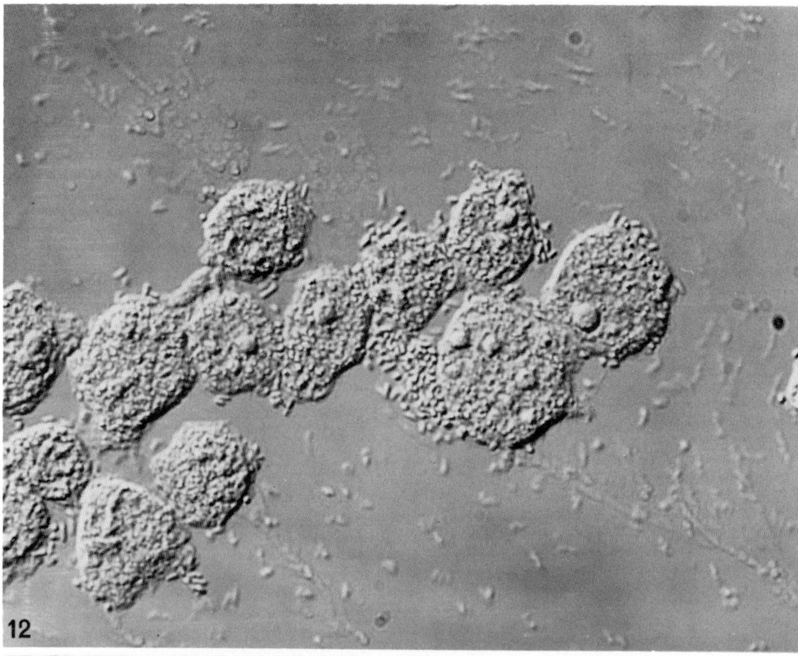

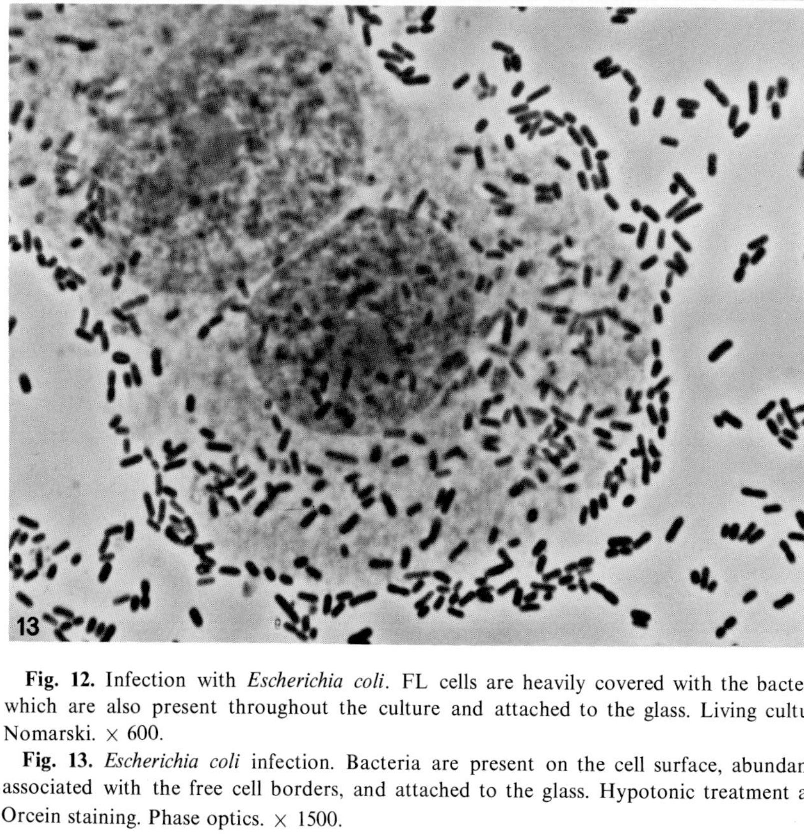

Fig. 12. Infection with *Escherichia coli*. FL cells are heavily covered with the bacteria, which are also present throughout the culture and attached to the glass. Living culture, Nomarski. × 600.

Fig. 13. *Escherichia coli* infection. Bacteria are present on the cell surface, abundantly associated with the free cell borders, and attached to the glass. Hypotonic treatment and Orcein staining. Phase optics. × 1500.

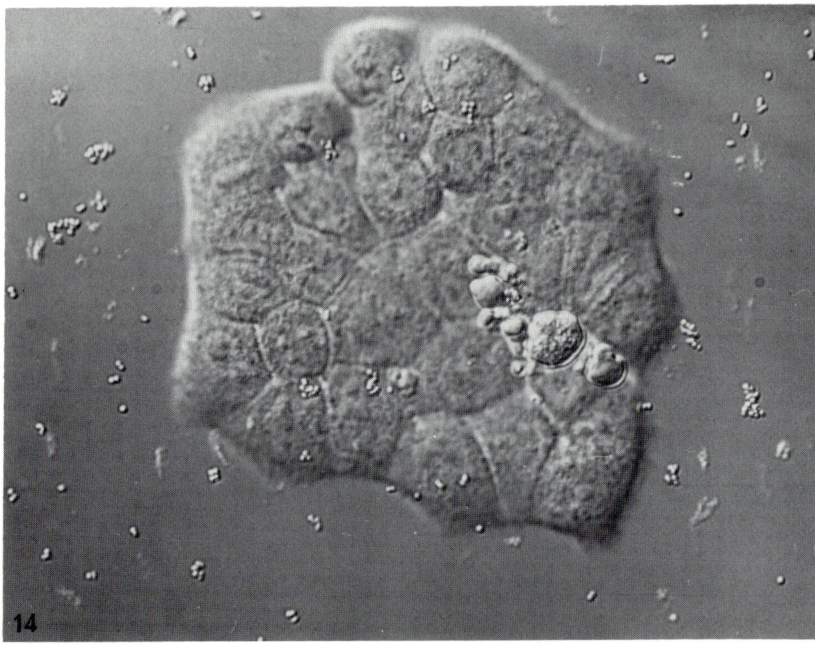

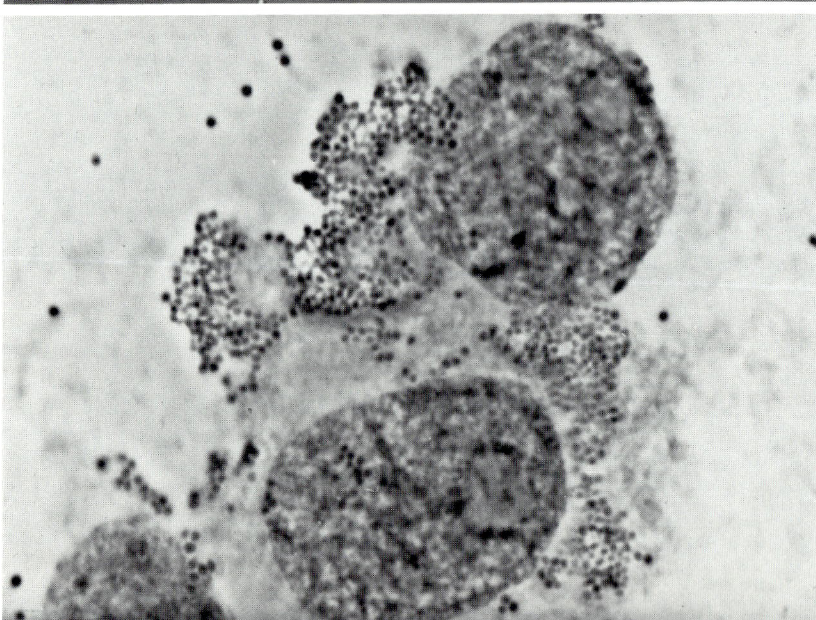

Figs. 14 and 15. Gram-positive micrococci in association with the host cells and floating in the culture medium (Fig. 14). Fig. 14: living culture, Nomarski, × 600. Fig. 15: hypotonic treatment and Orcein staining, phase optics, × 1500.

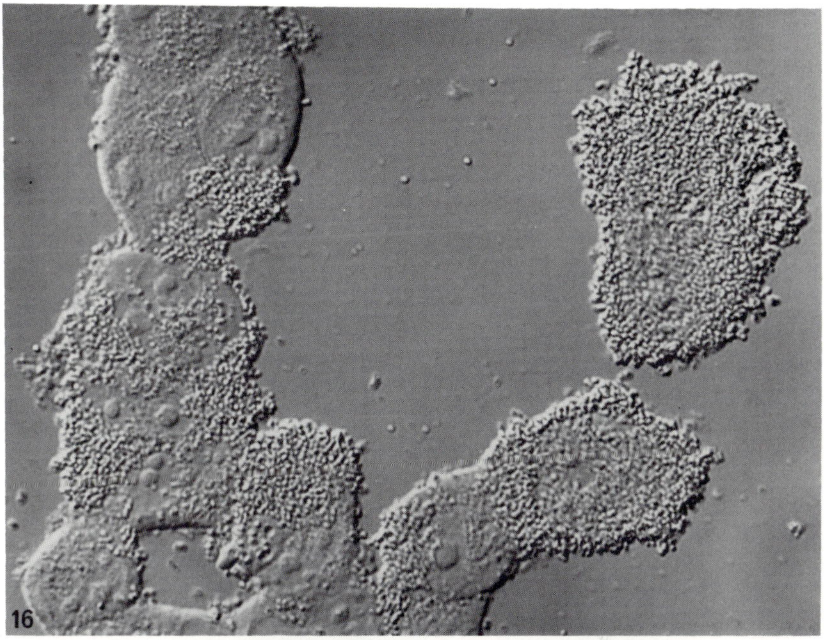

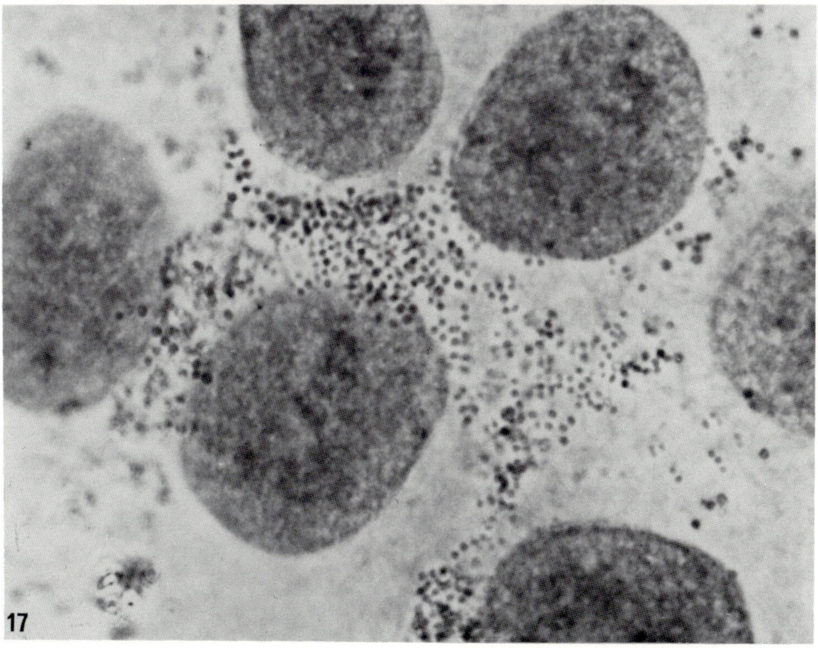

Figs. 16 and 17. *Staphylococcus albus* infection. Fig. 16: living culture, Nomarski, × 600. Fig. 17: Hypotonic treatment and Orcein staining, phase optics, × 1500.

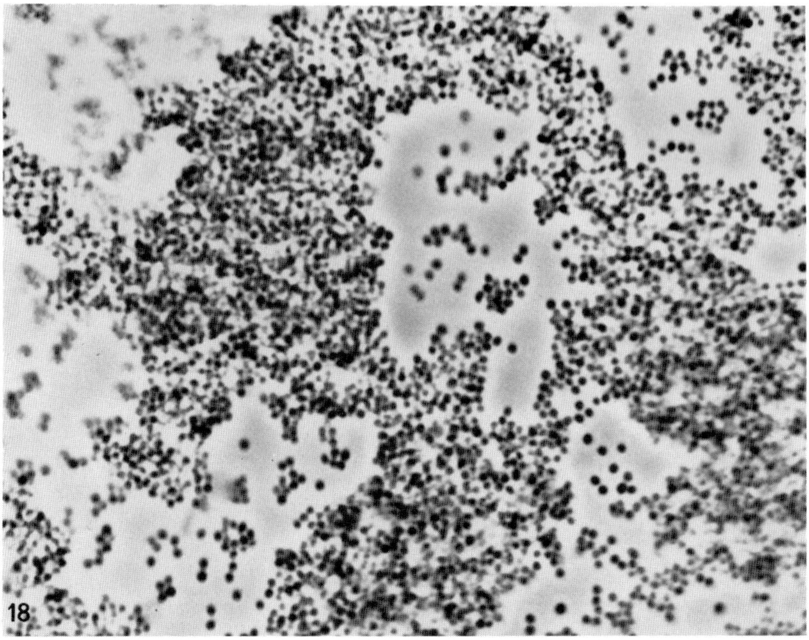

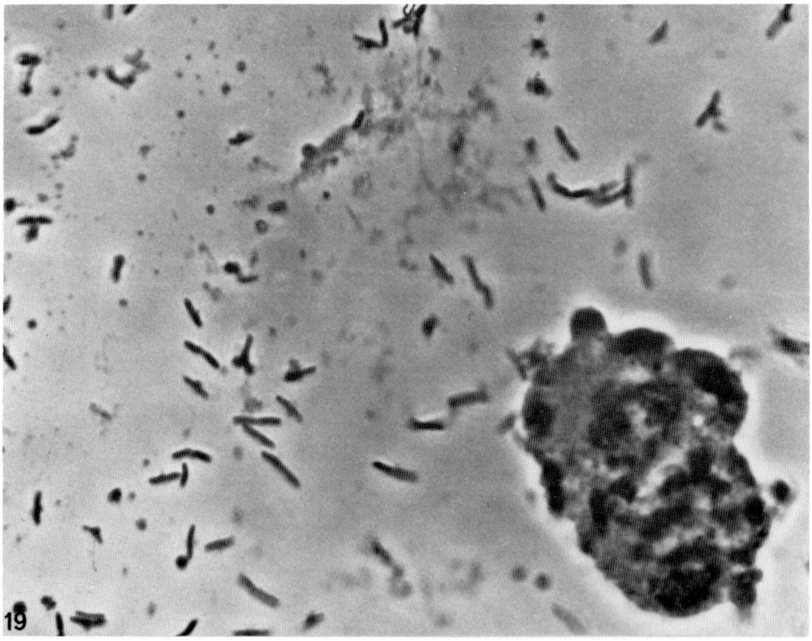

Fig. 18. Heavy infection with *Staphylococcus albus*. FL cells are totally covered with the bacteria or are destroyed. Hypotonic treatment and Orcein staining. Phase optics. × 1500.

Fig. 19. Severe culture destruction after infection with *Pseudomonas* sp. Hypotonic treatment and Orcein staining. Phase optics. × 1500.

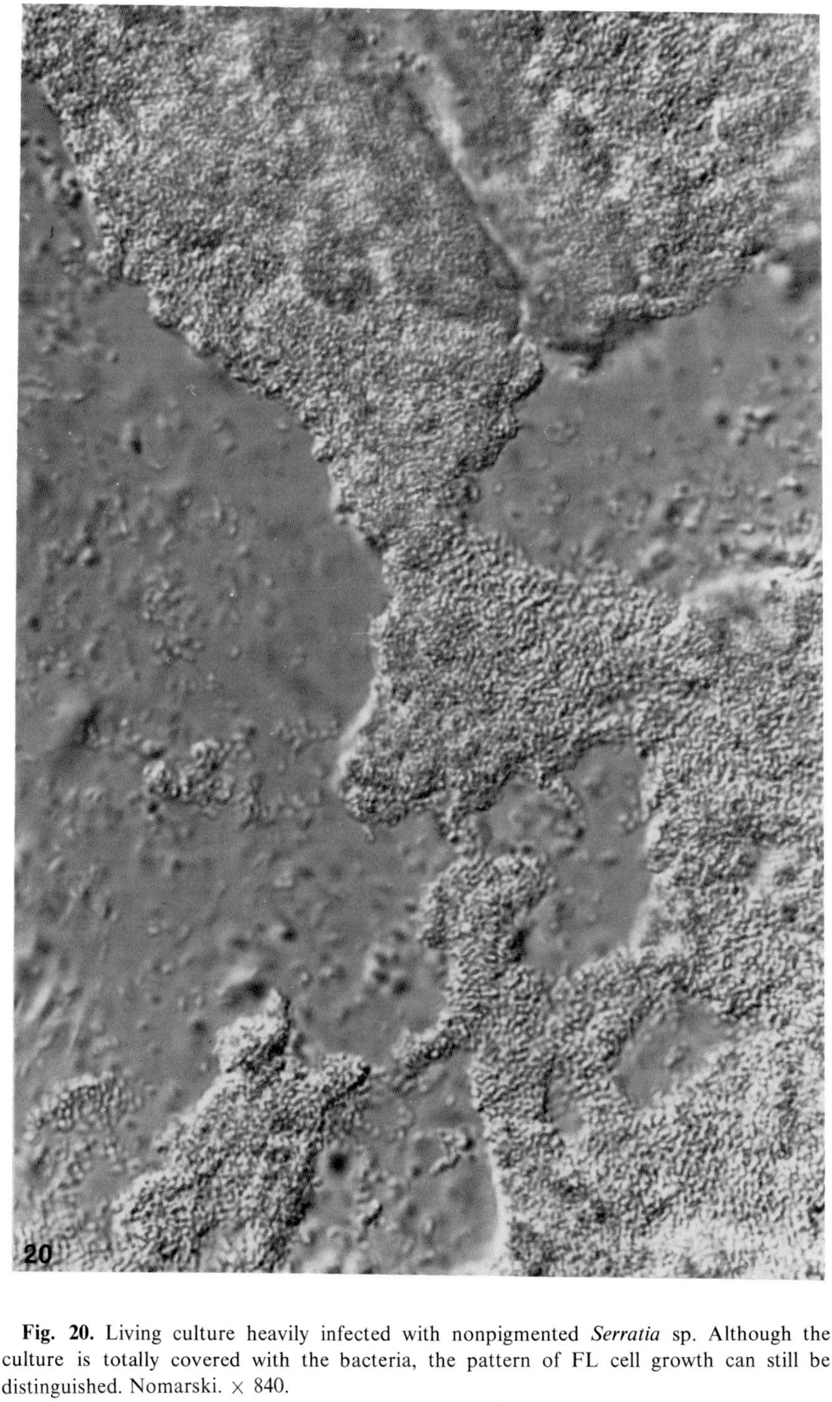

Fig. 20. Living culture heavily infected with nonpigmented *Serratia* sp. Although the culture is totally covered with the bacteria, the pattern of FL cell growth can still be distinguished. Nomarski. × 840.

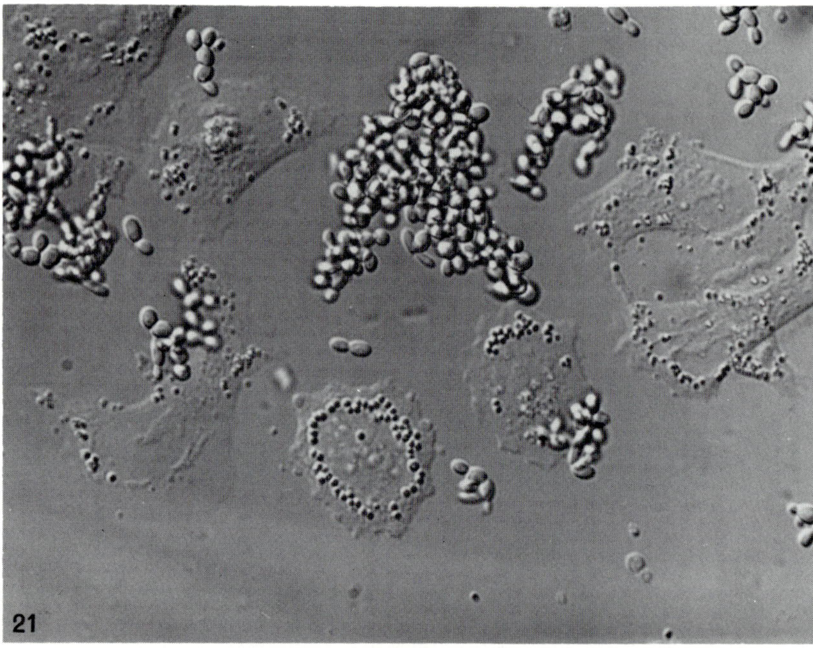

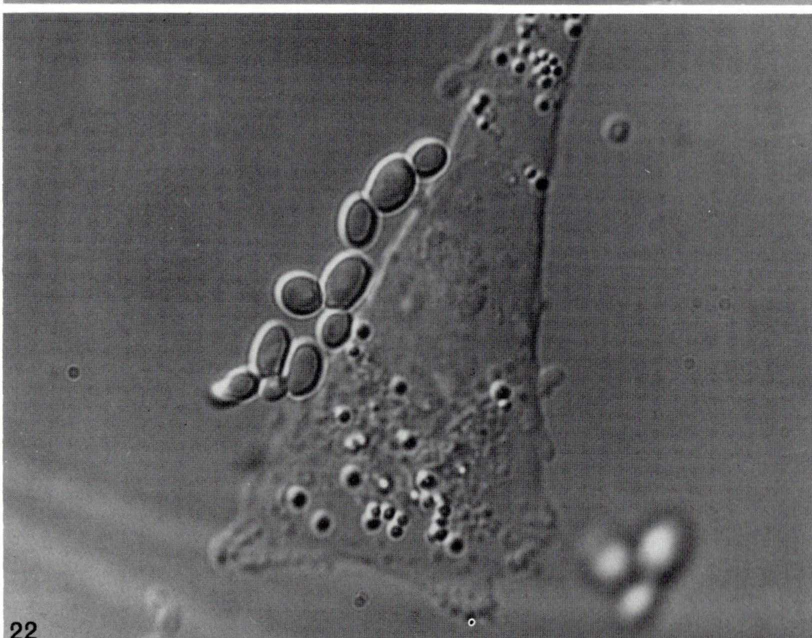

Fig. 21 and 22. Infection of FL cultures with *Candida parapsilosis*, the spherical or oval yeast cells mainly associated with the mammalian cells. Notice their large size compared with mycoplasmas and bacteria. Living cultures. Nomarski. Fig. 21, × 600; Fig. 22, × 1500.

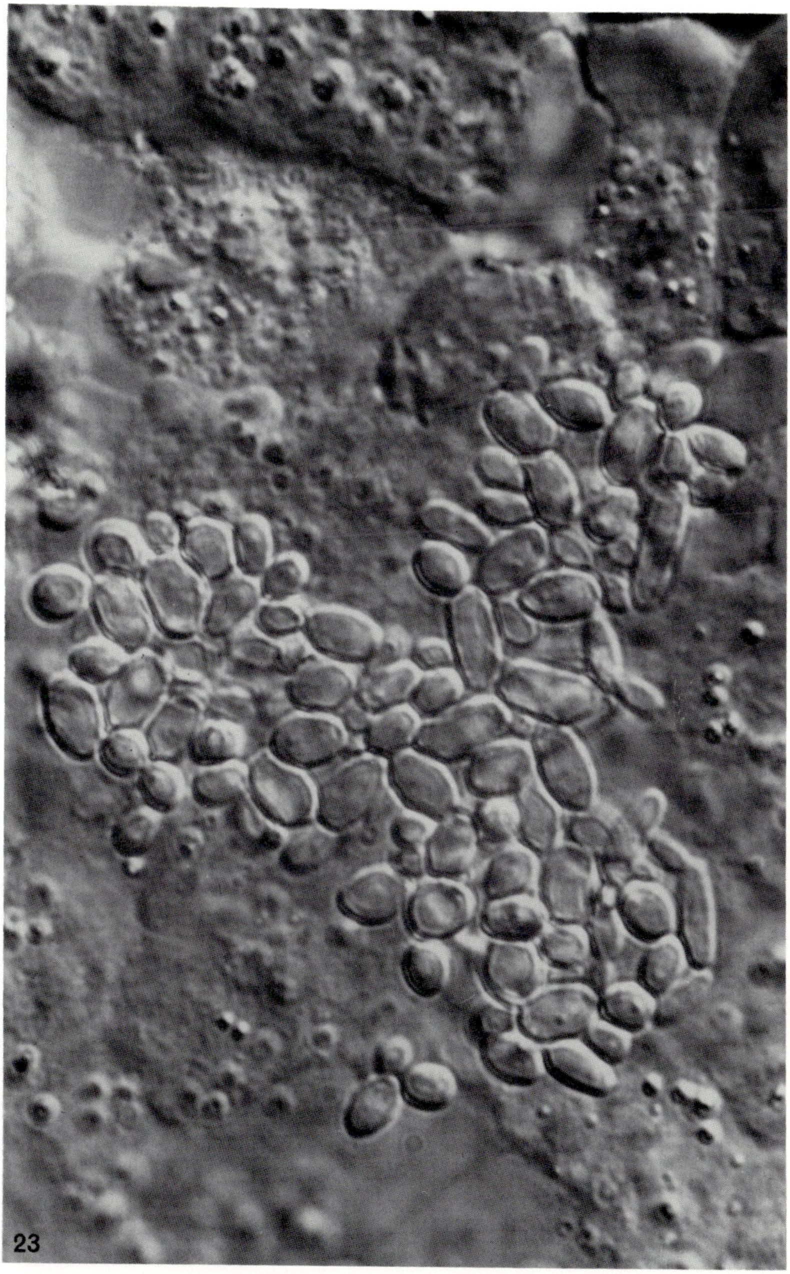

Fig. 23. Colony of *Candida parapsilosis* associated with the free surface of an FL cell monolayer. Living culture. Nomarski. × 2100.

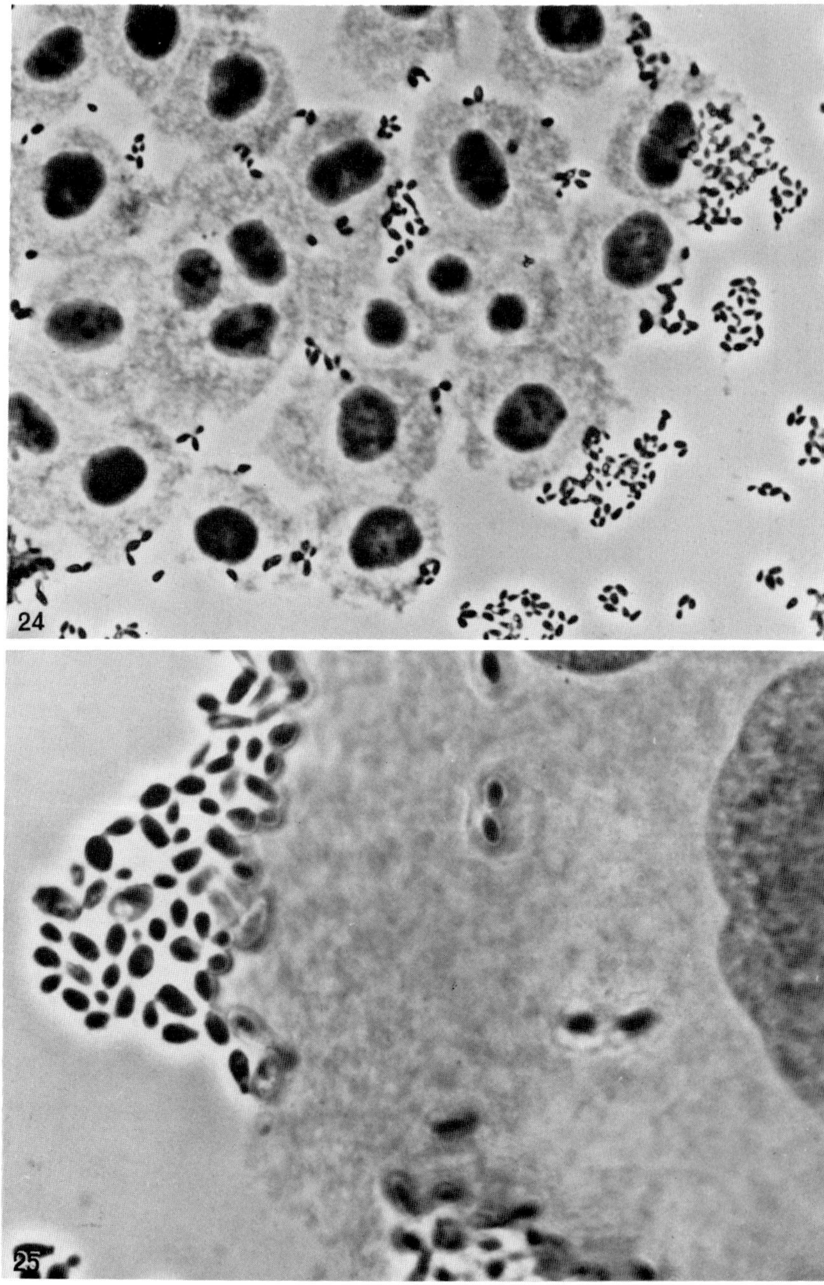

Fig. 24 and 25. *Candida parapsilosis* associated with the free borders of FL cells, located intra- and intercellularly, and attached to the glass. Hypotonic treatment and Orcein staining. Phase optics. Fig. 24, × 600; Fig. 25, × 1500.

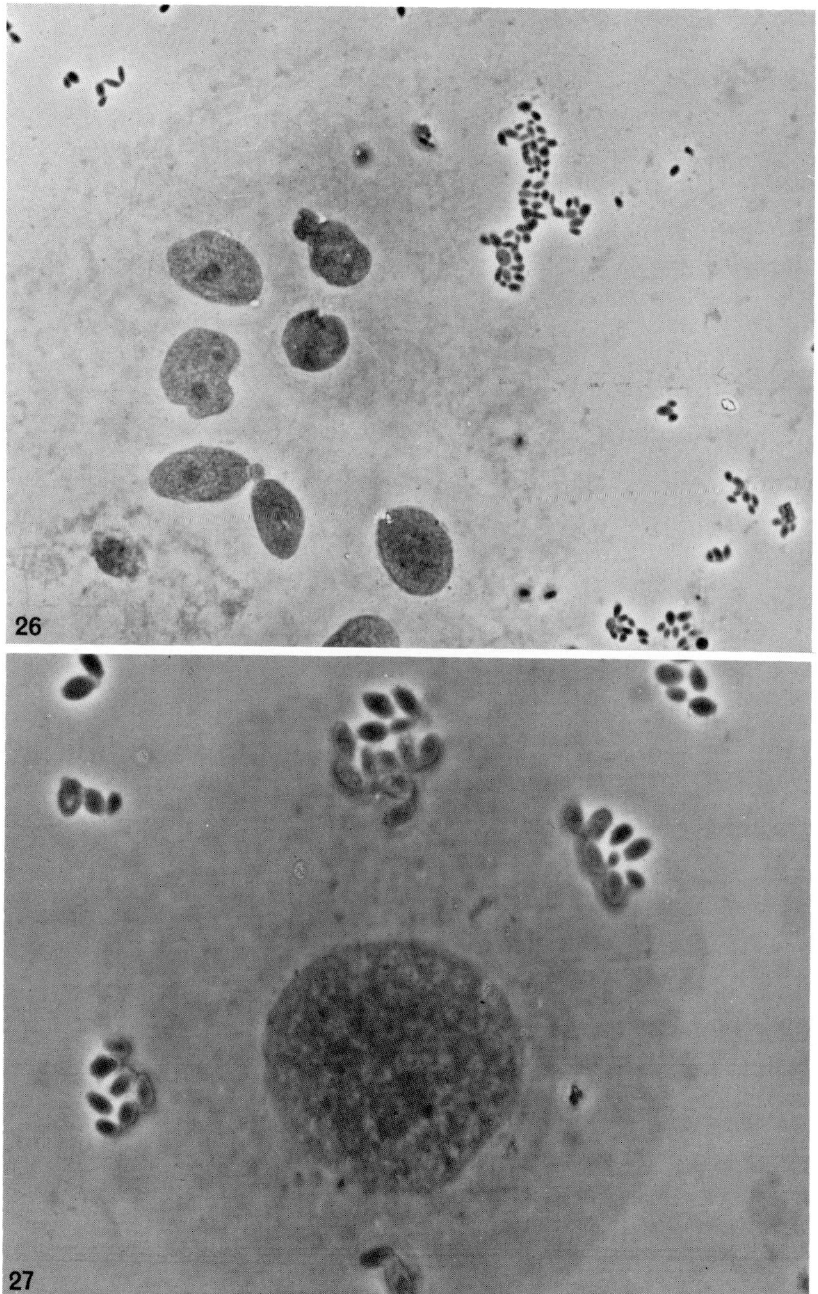

Fig. 26 and 27. Infection with a different *Candida* sp. The yeast cells, recognized by their size and shape, are closely associated with the host cell cytoplasm, causing local cytoplasmic destruction that is observed as a clear zone around the yeast cells. Other parts of the cytoplasm and the nucleus appear remarkedly unchanged. Hypotonic treatment and Orcein staining. Phase optics. Fig. 26, × 600; Fig. 27, × 1500.

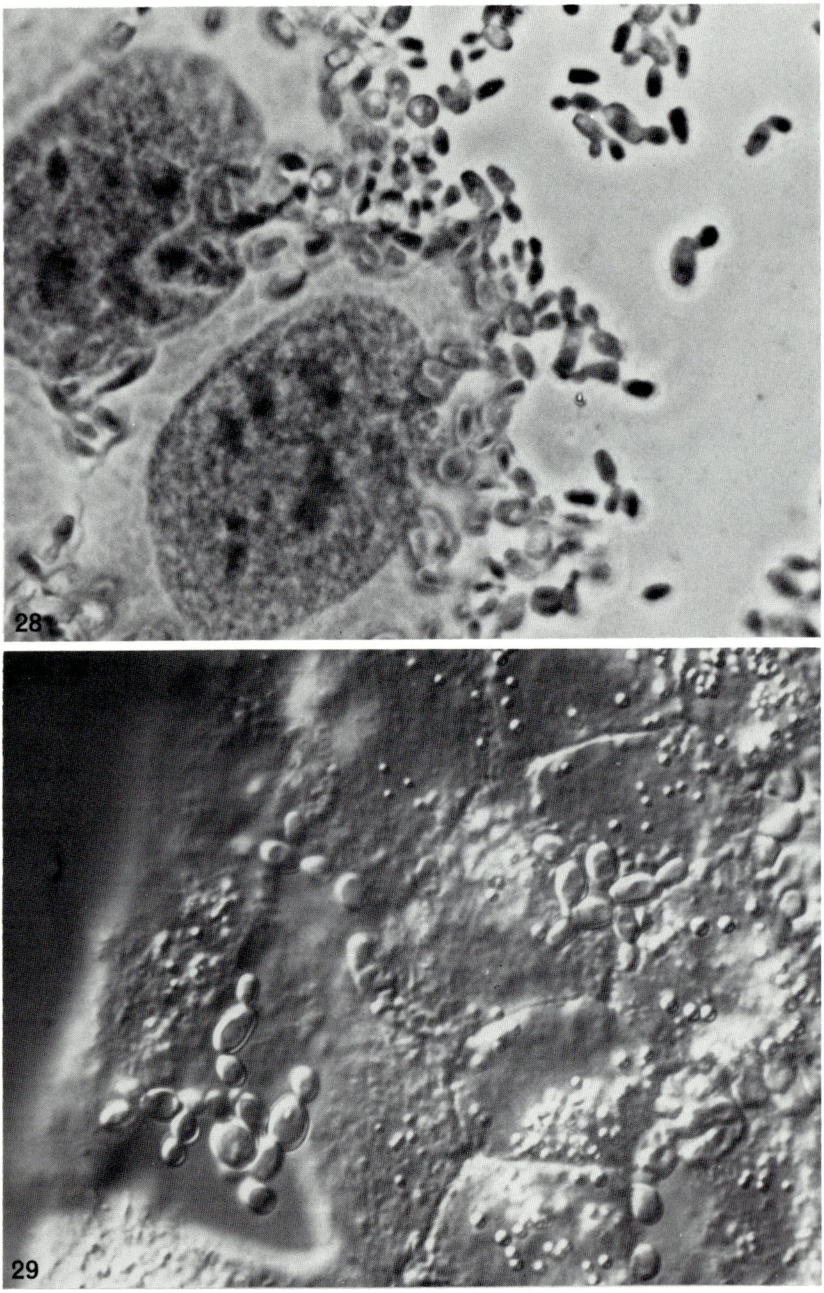

Figs. 28 and 29. Infection with *Saccharomyces* sp., which is closely associated with the host cells and causes destruction of the peripheral cytoplasm (Fig. 28). It is floating freely in the culture fluid and attached to the free host cell surfaces. Fig. 28: Hypotonic treatment and Orcein staining, Phase optics, × 1500. Fig. 29: Living culture. Nomarski. × 1500.

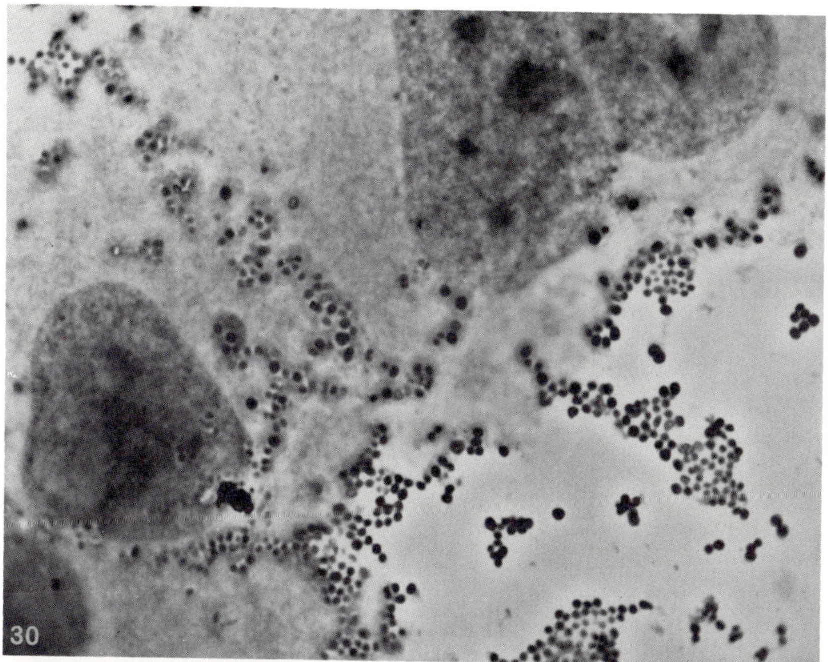

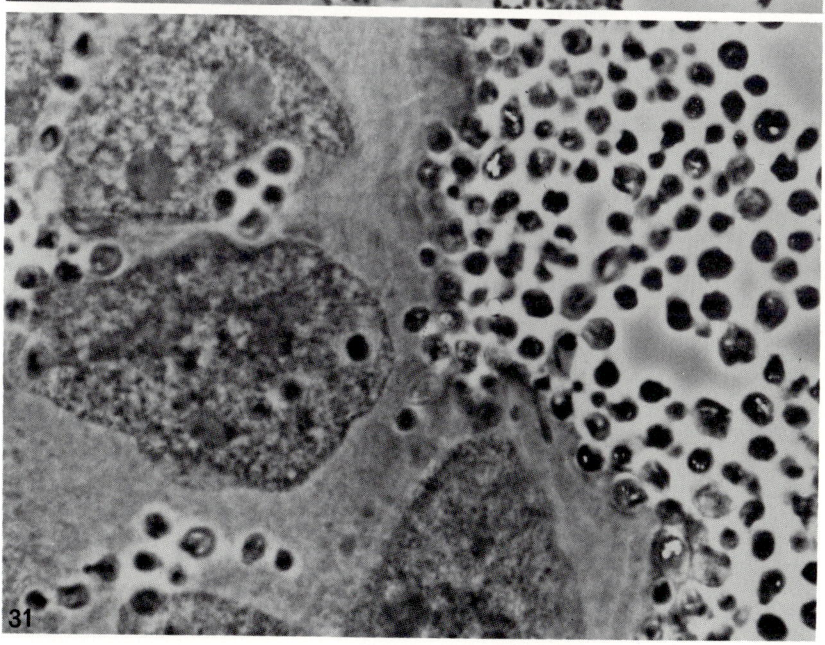

Figs. 30 and 31. Infection with unidentified yeast. The intercellular location of the yeast cells is characteristic and there is pronounced attachment to the peripheral cytoplasm. Hypotonic treatment and Orcein staining. Phase optics. Fig. 30, × 600; Fig. 31, × 1500.

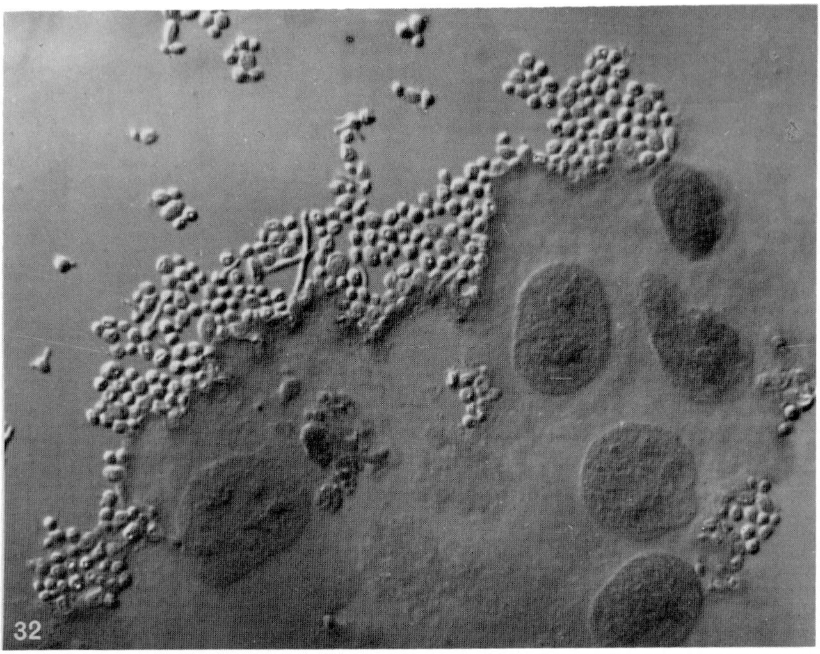

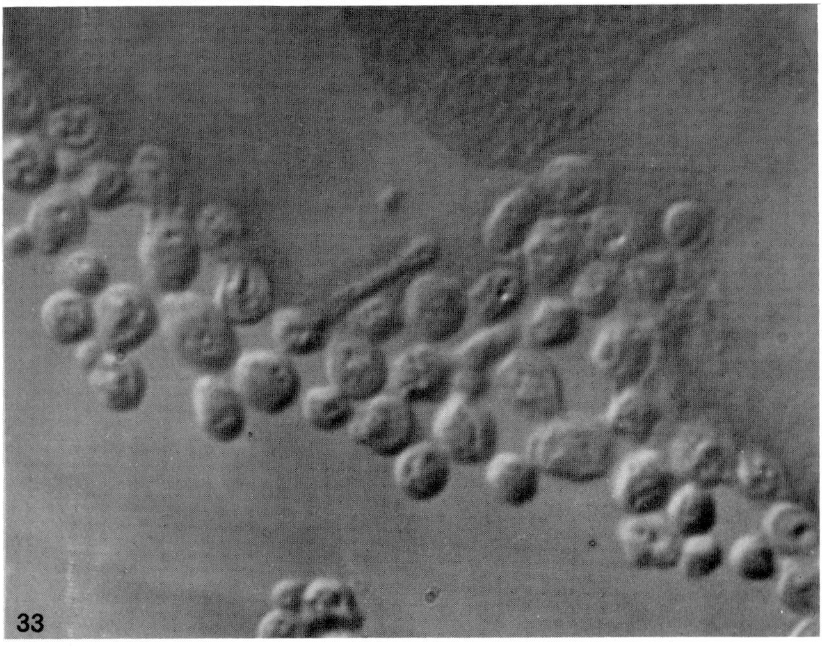

Figs. 32 and 33. Infection with the same, unidentified yeast as shown in Figs. 30 and 31. The internal structure of the yeast cells is clearly visible. Hypotonic treatment and Orcein staining. Nomarski. Figure 32, × 600; Fig. 33, × 2400.

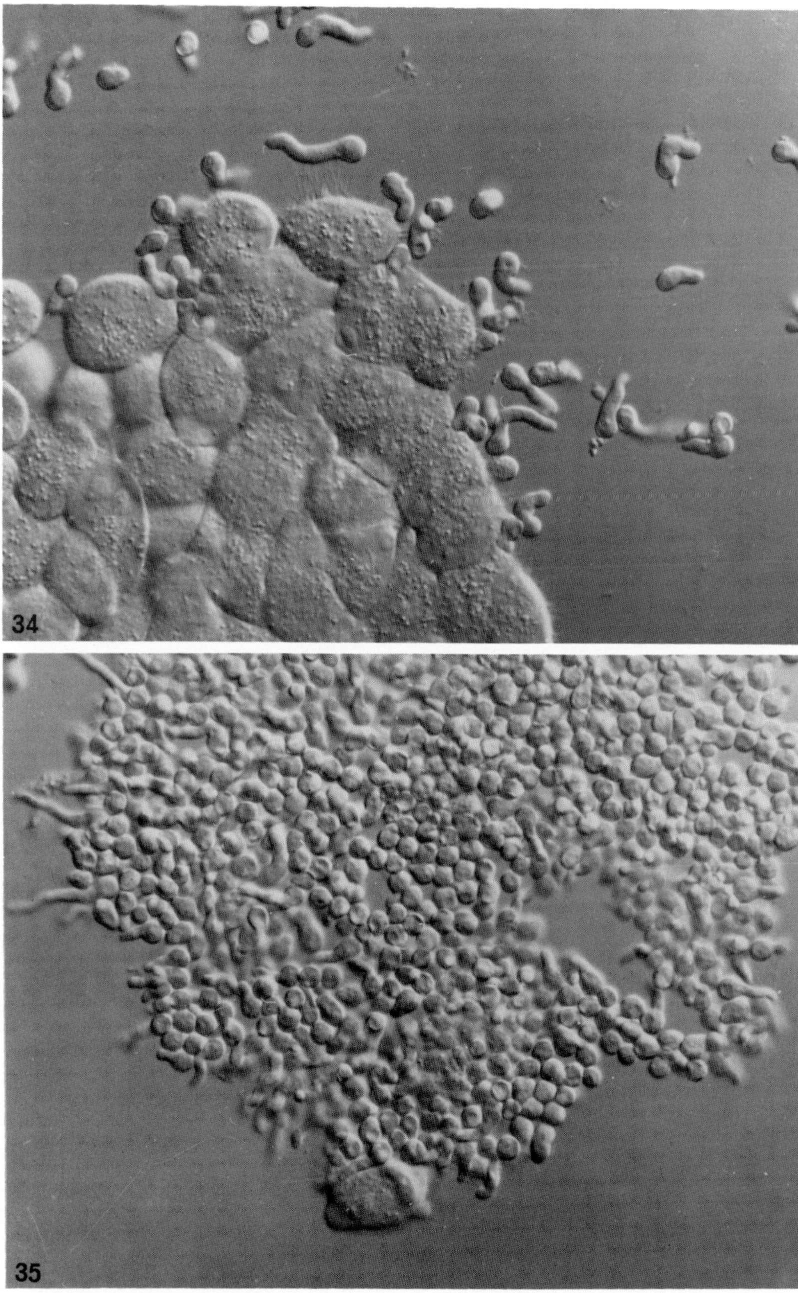

Figs. 34 and 35. Infection with a different, unindentified yeast. Fig. 35 shows a very heavy, local concentration of yeast cells, totally covering the mammalian cells. Many of the yeastlike cells have produced pseudogerm tubes. Living culture. Nomarski. × 600.

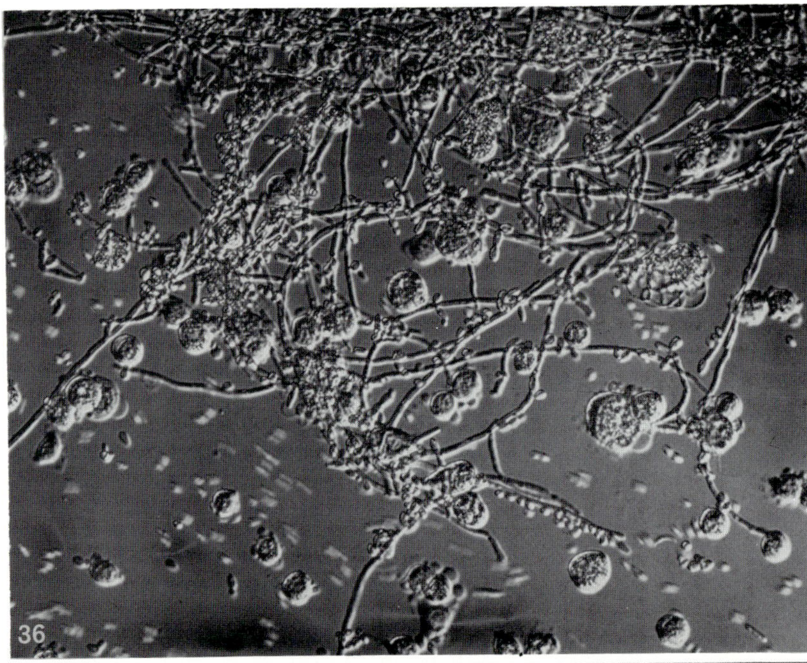

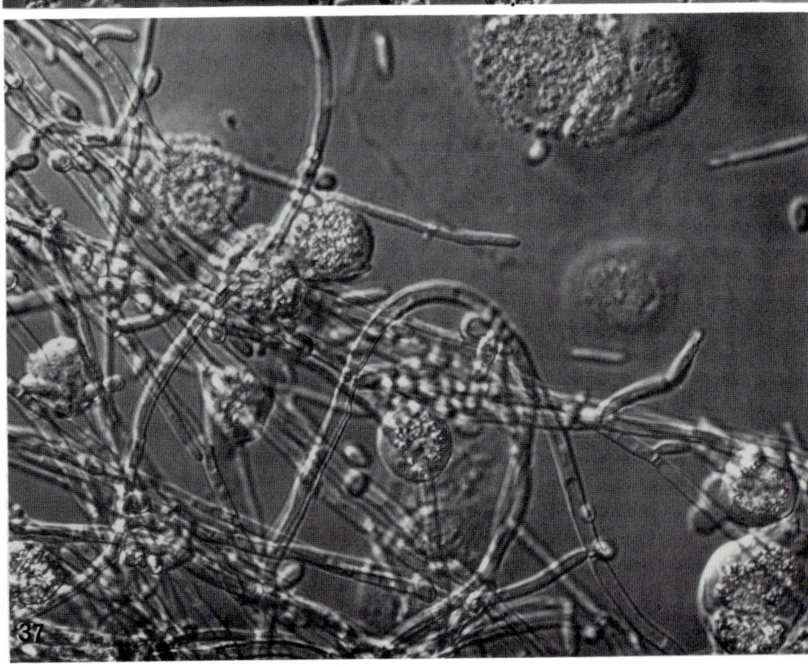

Figs. 36 and 37. Infection with *Candida stellatoidea*. Rodlike structures represent the moving ovoid microorganisms as recorded by photography. The branching hyphae might represent an unindentified mold. More likely, however, these structures are pseudohyphae of *C. stellatoidea*. Living culture. Nomarski. Fig. 36, × 240; Fig. 37, × 600.

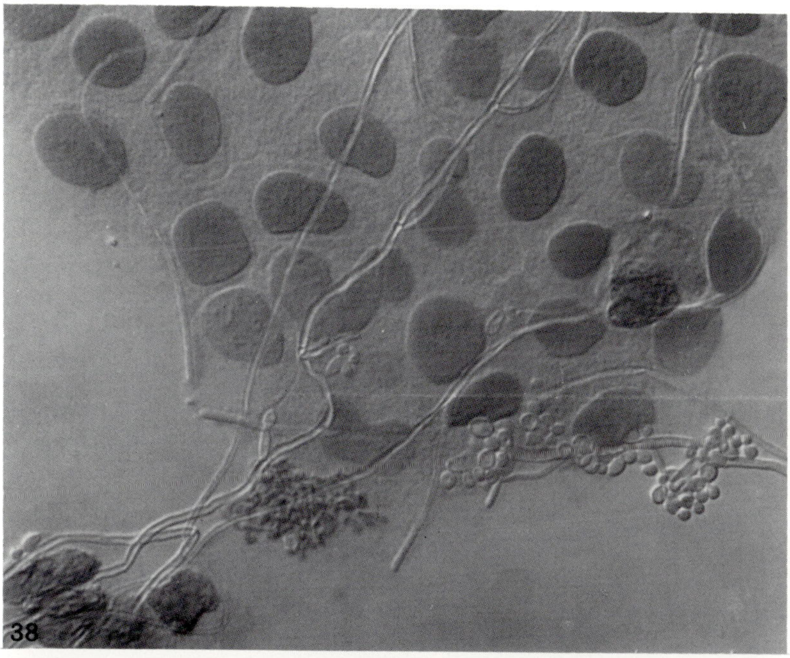

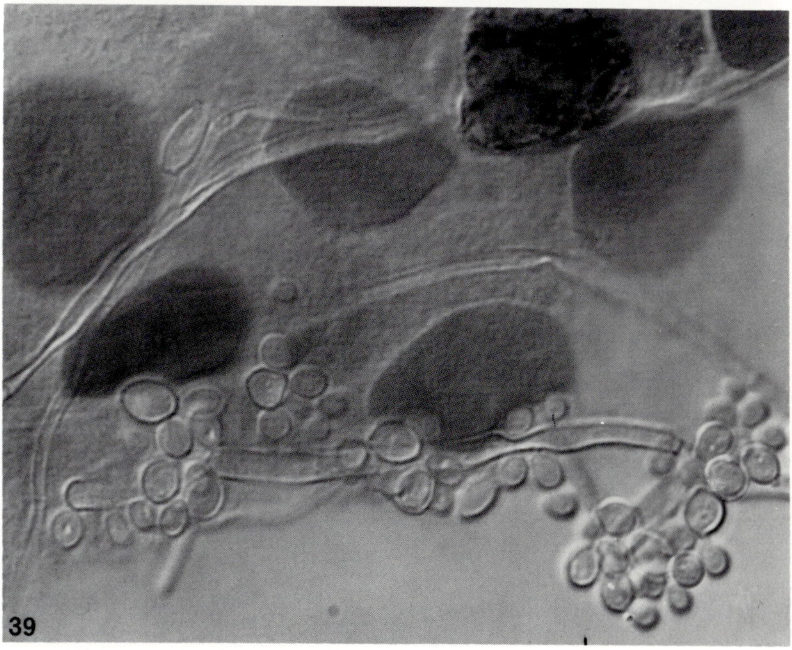

Fig. 38 and 39. Same cultures as illustrated in Fig. 36 and 37. Hypotonic treatment and Orcein staining. Nomarski. Fig. 38, × 600; Fig. 39, × 1500.

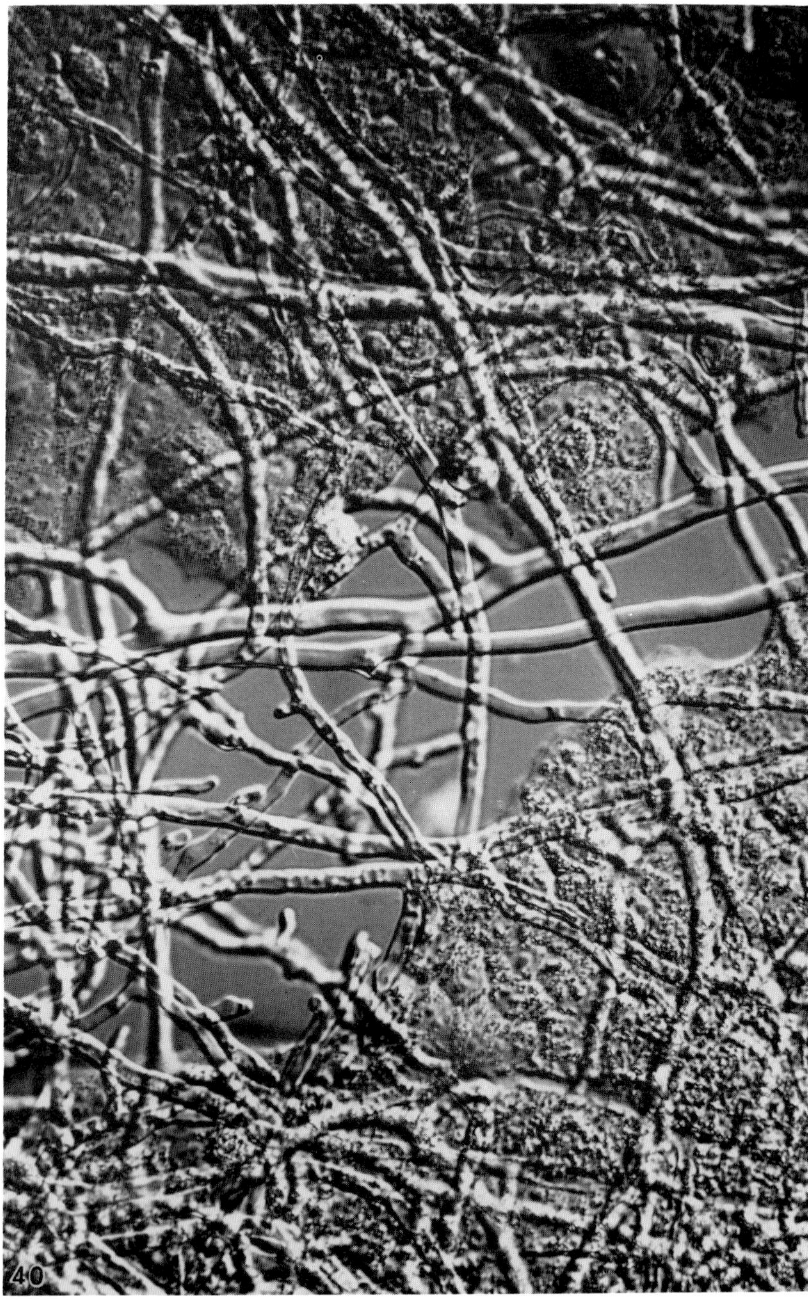

Fig. 40. Infection with *Mucor* sp. The tubular, branching hyphae as they cover the culture of the mammalian cells are clearly recognized by this technique. Living culture. Nomarski. × 341.

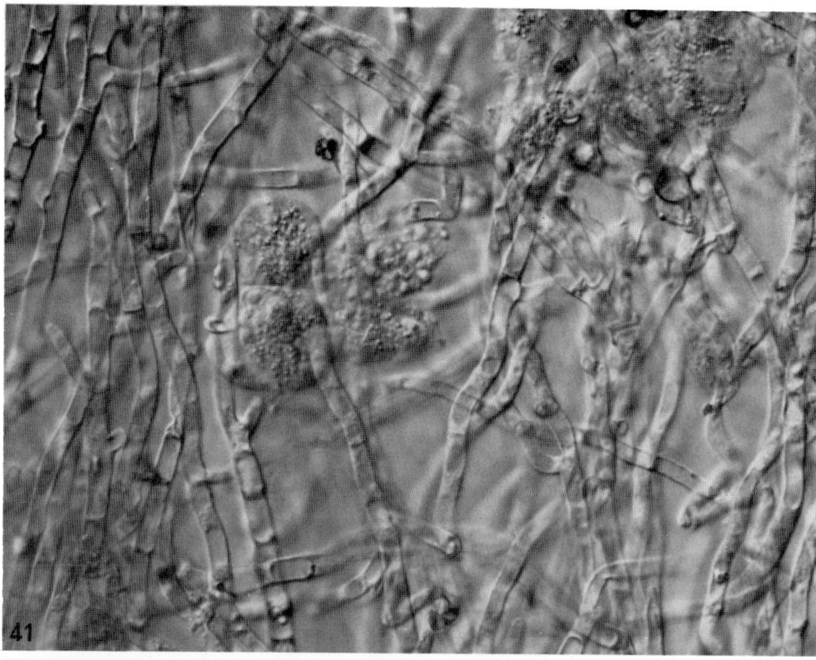

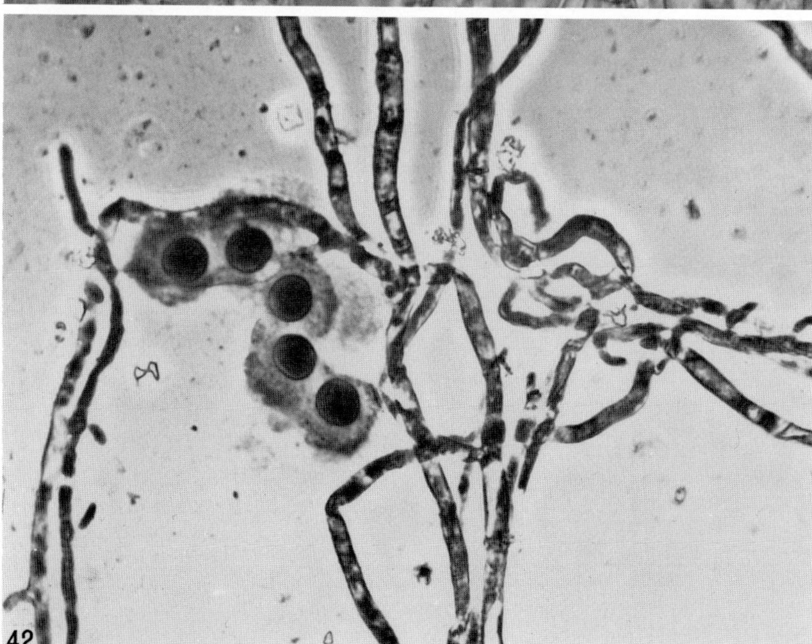

Figs. 41 and 42. Infection with *Aspergillus niger*. Fig. 41: The tubular branching hyphae, which are divided at irregular intervals by septae, are heavily covering the FL cells. There is severe cell destruction, and only a few mammalian cells are still attached to the glass. Living culture. Nomarski. × 600. Fig. 42: Hypotonic treatment, Orcein staining. Phase optics. × 600.

43

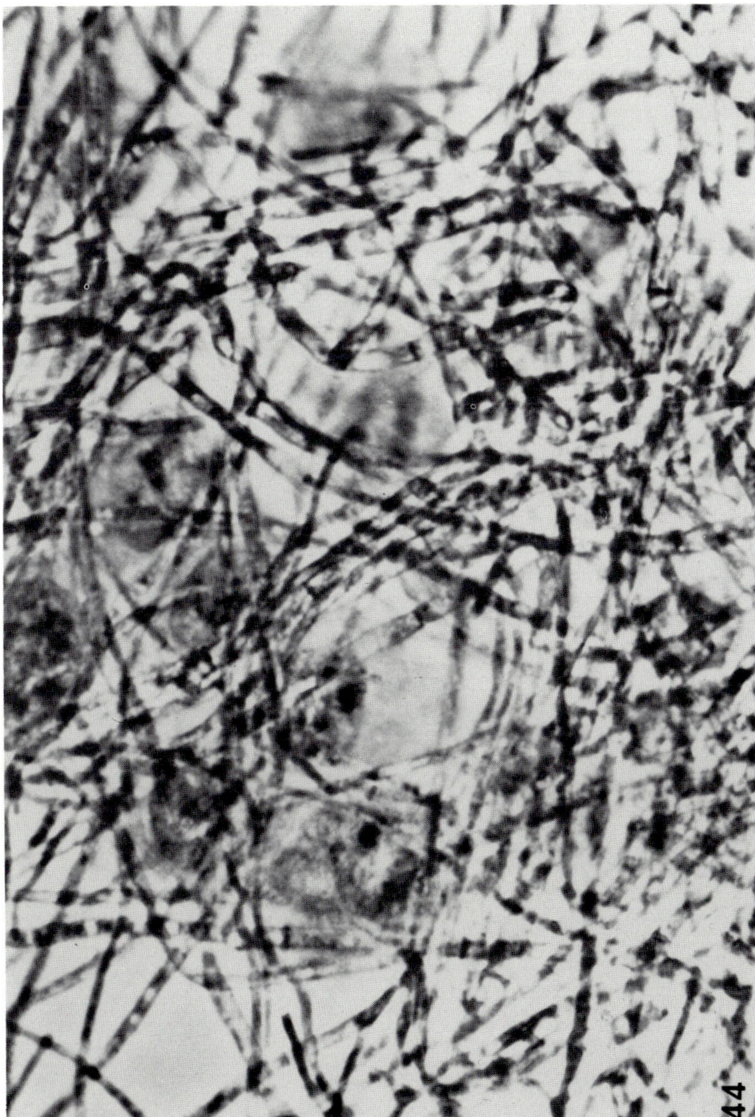

Figs. 43 and 44. Infection with *Penicillium* sp. Branching hyphae intermingle as a mycelium with a few remaining rounded FL cells. Fig. 43: living culture. Nomarski. × 600. Fig. 44: Hypotonic treatment and Orcein staining. Phase optics. × 600.

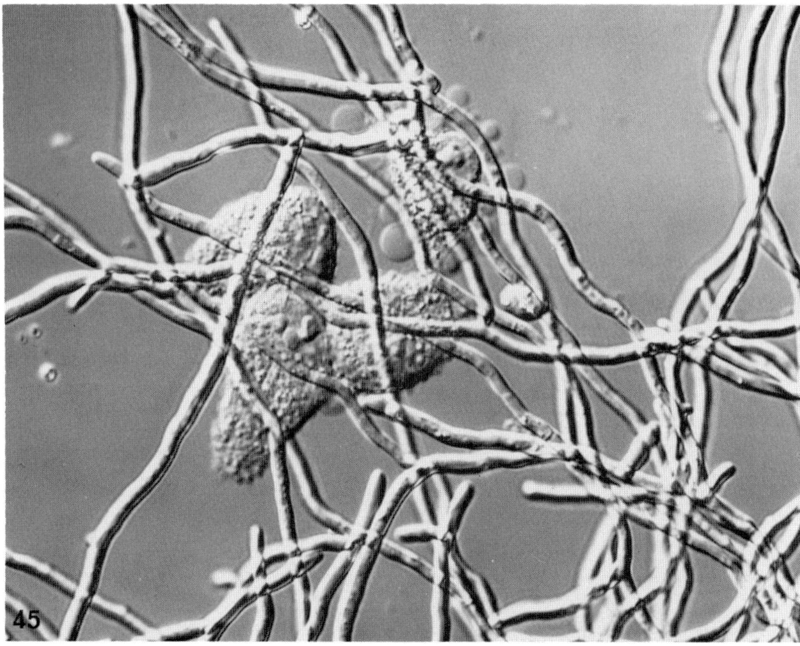

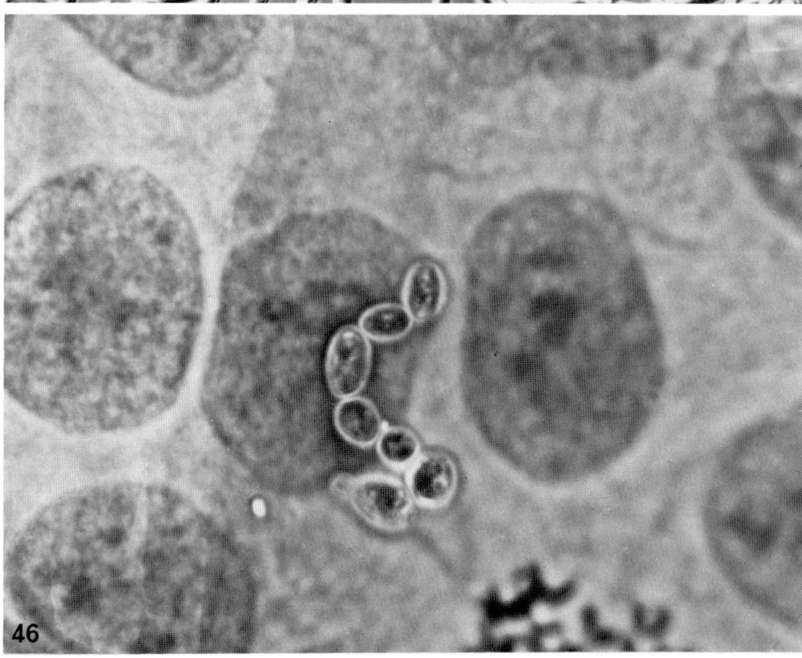

Figs. 45 and 46. Infection with *Trichosporon* sp. Fig. 45: The mycelium is crisscrossing the few remaining FL cells. Bubbles are protruding from the cytoplasm of one FL cell. Such bubbles are frequently seen in the living cultures, also uninfected, a period of time after preparation. Living culture. Nomarski. × 600. Fig. 46: Yeastlike form of *Trichosporon* sp. Hypotonic treatment and Orcein staining. Phase optics. × 1500.

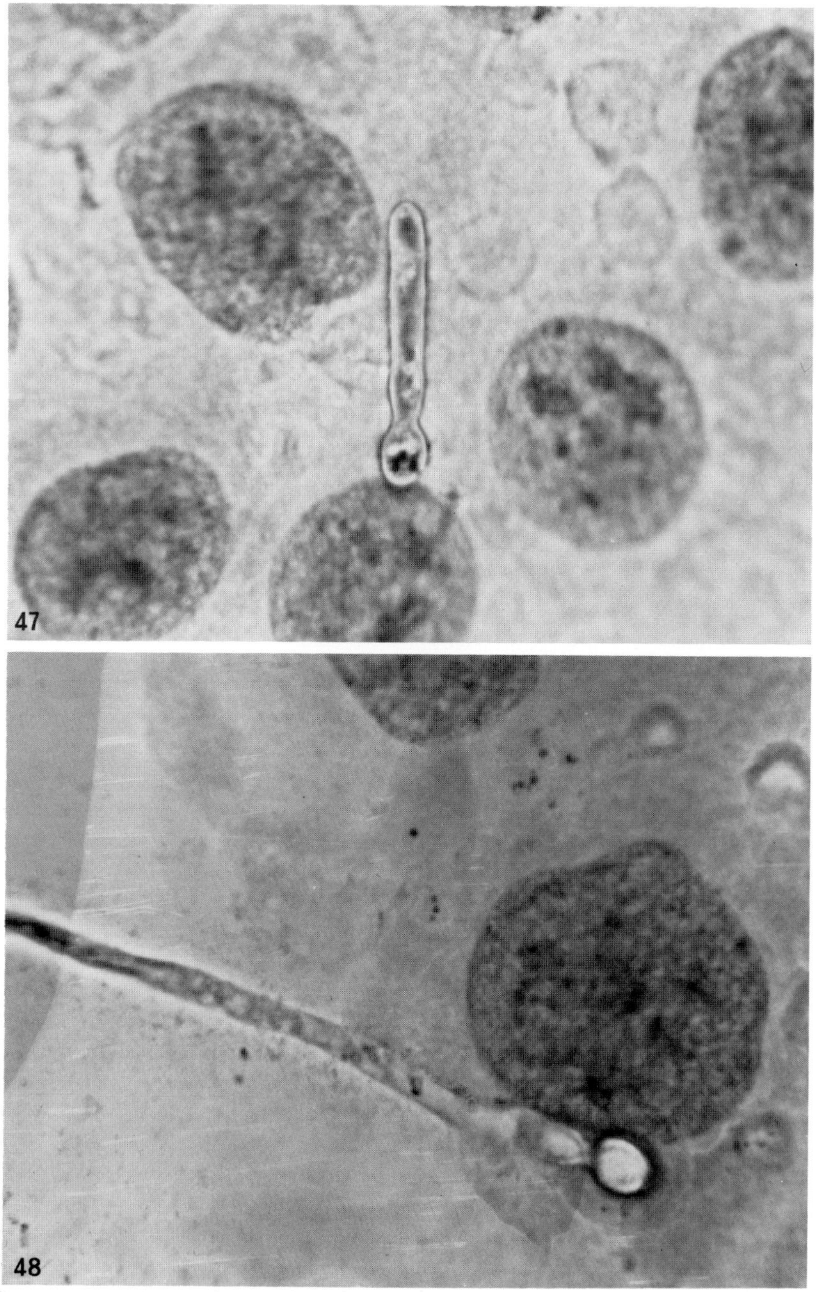

Figs. 47 and 48. Infection with *Trichosporon* sp. Special forms of this mold as observed in hypotonically treated and Orcein-stained cultures by phase optics. × 1500.

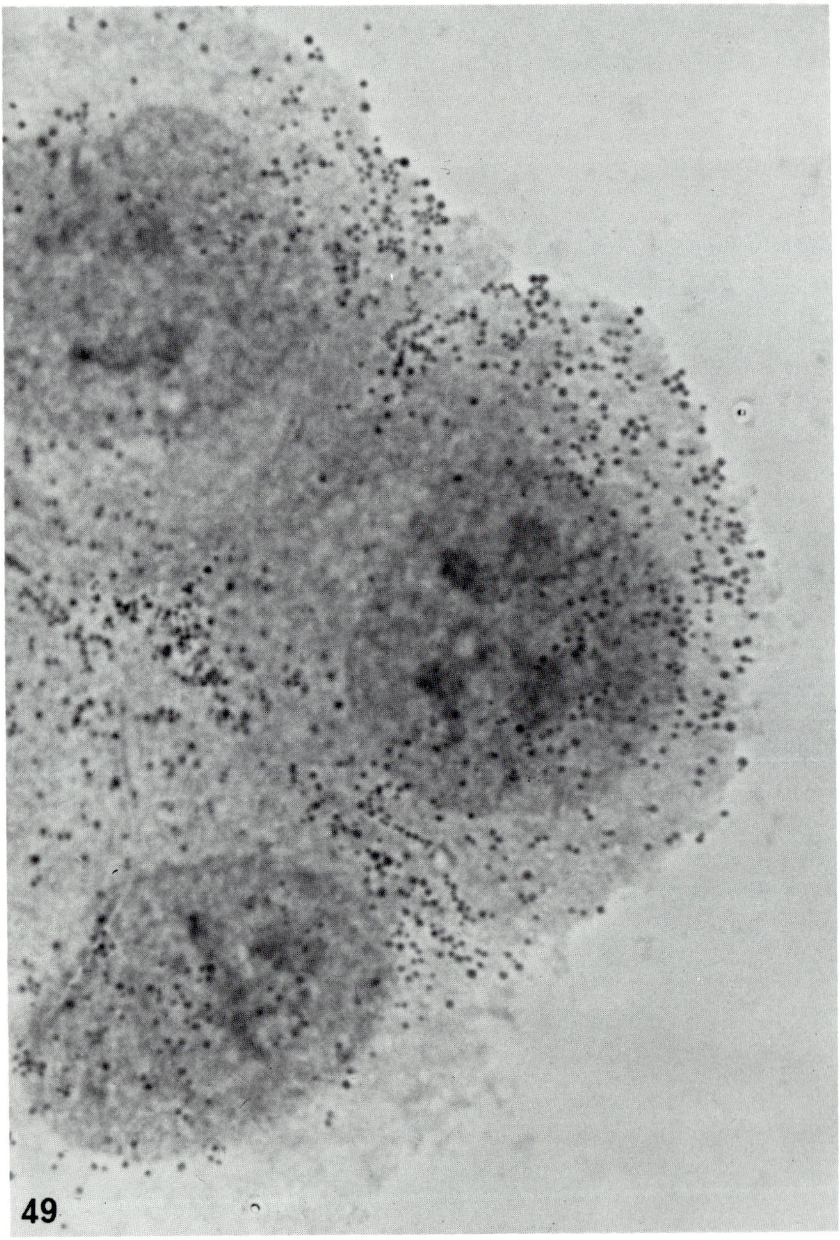

Fig. 49. FL cells infected with HT mycoplasma. The number of mycoplasmas per cell can easily be counted. Hypotonic treatment and Orcein staining. Phase optics. × 2400.

CHAPTER **5**

Contamination of Mammalian Cells in Culture by Mycoplasmata

GEORGE E. KENNY

I. Introduction 108
II. Biology of the Mycoplasmatales 108
A. Definition 108
B. Taxonomy 109
C. Ultrastructure 109
D. Physiology 110
III. Epidemiology of the Contamination of Cell Culture with Mycoplasmata 110
A. Incidence of Cell Culture Contamination 110
B. Source of Mycoplasmic Contaminants 112
C. Detection of Mycoplasmata in Cell Cultures 113
IV. Host–Parasite Relationships of Mycoplasmata and Mammalian Cells 117
A. Nature of the Mycoplasmata–Mammalian Cell Interaction 117
B. Cellular Habitat of the Mycoplasmatales 117
C. Interaction of Mycoplasmata with Mammalian Cell Surfaces 118
V. Effect of Mycoplasmata on Cell Cultures 119
A. The Arginine Effect 119
B. Effect of Mycoplasmata on Cell Morphology 121
C. Effects of Mycoplasmata on Cell Physiology and Metabolism 121
D. Effects of Mycoplasmata on Mammalian Chromosomes 122
E. Effects of Mycoplasmata on Virus Yields 123
VI. Elimination of Mycoplasmata from Cell Cultures 123
A. Elimination of Mycoplasmata with Antibiotics 123
B. Chemical and Physical Means of Elimination 124
C. Elimination with Specific Antiserum 125
D. Prevention of Mycoplasmic Contamination 125
VII. Conclusions 126
References 126

I. Introduction

When Robinson *et al.* (1956) reported that mammalian cell cultures were contaminated with pleuropneumonia-like organisms (PPLO) or *mycoplasmata** as they are now termed, this finding was treated with alarm by some and ignored by others. In any event, the new contaminants were poorly understood and extremely difficult to cultivate. In order to put this matter into perspective, it is important to recognize that 90% or more of the basic biological information that we have about mycoplasmata was elucidated in the 1960's. Similarly, as cell culture was a newly emerging technique of great promise for cellular and biological research, most laboratories had significant difficulties in producing sufficient cell cultures for experimental use. Thus, both cell culture and mycoplasmic technology were more in the form of an art than a science. Because few workers were skilled in both techniques, those who studied mycoplasmic contamination of cell cultures were largely drawn from the ranks of cell culturists. In fact, I became a mycoplasmologist quite by accident when I found that the cell cultures I was about to use for physiological experiments were contaminated with mycoplasmata. Thus, it will be helpful to the reader to consider the history of the study of mycoplasmata as cell culture contaminants in terms of the scientific knowledge available about mycoplasmata and cell cultures at the time when the work was done. Many concepts that appeared perfectly reasonable in their own time were later found to be groundless.

II. Biology of the Mycoplasmatales

A. Definition

Mycoplasmata are small, prokaryotic cells (0.3–0.5 μm in diameter). They form minute colonies (10–100 μm in diameter) on appropriate media with the morphology of a fried egg. This colony morphology resembles that of the agent of bovine pleuropneumonia (*Mycoplasma mycoides*); hence the earlier term *pleuropneumonia-like organism* (PPLO). Whereas the above criteria are the only positive statements that can be made about the definition of mycoplasmata, several negative factors are of considerable importance. The organisms do not possess a cell wall nor do they have the capability of making cell wall components. Thus, they do not revert to bacteria as is sometimes possible with the L form of bacteria, which they closely resemble in colonial morphology.

*The term *mycoplasmata* is the trivial name for organisms in the genus *Mycoplasma*. The term *mycoplasmas* is the common name in the English language and is equivalent to mycoplasmata.

B. Taxonomy

All the species (some 45) are classified in one order, Mycoplasmatales, and in one class, Mollicutes (Edward and Freundt, 1969). The order Mycoplasmatales is presently divided into two families: Mycoplasmataceae and Acholeplasmataceae. Most of the classified species are in the family Mycoplasmataceae, which has one genus, *Mycoplasma* [a list of species can be found in Edward and Freundt (1969) and in Chapter 6 by Barile]. Recently it has been proposed that those species that do not require cholesterol be classified in the family Acholeplasmataceae with one genus, *Acholeplasma*. The three species presently included are *A. Laidlawii* (formerly *Mycoplasma laidlawii*), *A. granularum* (formerly *Mycoplasma granularum*), and *A. axanthum* (Edward and Freundt, 1970). Because the organisms in the family Mycoplasmataceae are strikingly heterogeneous in DNA base ratio (Williams *et al.*, 1969; Neimark, 1970), protein patterns on polyacrylamide gel electrophoresis (Razin, 1968), physiological properties (Aluotto *et al.*, 1970), and in comparative analytic serology (Kenny, 1969), it is quite likely that additional families and genera will be created to encompass this diversity. For example, a survey of ten glycolytic species showed that four clusters could be recognized by double immunodiffusion (Kenny, 1969). One of these clusters already has been reclassified as the family Acholeplasmataceae. Although reclassification will have substantial biological utility for mycoplasmologists, it will make the literature much more difficult for those outside the field because of the complexity of the terminology.

Species are separated primarily by inhibition of their colonial growth on agar by specific antiserum (Clyde, 1964). The metabolic inhibition test (antibody-mediated inhibition of mycoplasmata growth in broth as measured by preventing the appearance of a metabolic product, acid production, or ammonia production from arginine or urea) also has been used to distinguish species (Purcell *et al.*, 1969).

C. Ultrastructure

The organisms are bounded by a unit membrane that encloses ribosomes and DNA. The morphology of the mycoplasmic cell has been controversial (Freundt, 1969; Anderson, 1969). What appears clear at present is that substantial heterogeneity in morphology is seen among the various species. *Mycoplasma gallisepticum* cells are ovoid with a specialized end structure (Maniloff and Morowitz, 1967). Cultures of *Mycoplasma felis* show star-shaped cells in log-phase growth (Boatman and Kenny, 1970). Star-shaped bodies are also seen in *Mycoplasma pneumoniae* spherules although the arms of the star are both longer and narrower than those of *M. felis* (Boatman and Kenny, 1971). The structure of *Mycoplasma pharyngis* (also termed *M. orale* type 1) appears to be markedly filamentous with frequent starlike shapes

(Clyde, 1969). Considerable caution must be employed in interpreting the literature concerning the ultrastructure of mycoplasmata because cells bounded only by a membrane can be distorted readily by high-speed centrifugation and by preparative techniques with resulting artifacts. As a generalization, it can be said that whereas individual species may have a reasonably consistent morphology, depending to some extent on the constitution of the growth medium (Rodwell, 1969), fundamental differences in morphology appear to exist between species and these differences may parallel the heterogeneity seen in other parameters.

D. Physiology

The growth requirements of mycoplasmata are complex; in addition to amino acids (Rodwell, 1969), the organisms require such nucleic acid precursors as nucleosides or even oligonucleotides. Serum protein is required for growth of most species in the genus *Mycoplasma* but not species in the genus *Acholeplasma*. Similarly, cholesterol is required by the genus *Mycoplasma* but not by *Acholeplasma*. Organisms in the order Mycoplasmatales can be divided into three major groups by fermentation reactions: (*1*) those which utilize arginine but not glucose or urea, (*2*) those which utilize glucose but neither arginine nor urea, and (*3*) those which utilize urea but not glucose or arginine (the T strains). Some of the glucose-utilizing species also utilize arginine. The glycolytic species can be divided into clusters on the basis of cross-reactions in double immunodiffusion (see Section II, B). More recent studies in this laboratory show that the species that utilize arginine but not glucose fall into a closely related cluster which shows little relationship to the glycolytic species (see Table I).

III. Epidemiology of the Contamination of Cell Cultures with Mycoplasmata

A. Incidence of Cell Culture Contamination

Following the report of Robinson *et al.* (1956), a number of studies confirmed the fact of mycoplasmic contamination of cell cultures and indicated that a large portion of the cell lines in common use were contaminated. A summation of three of the larger studies reported in 1959–1962 (Rothblat and Morton, 1959; Pollock *et al.*, 1960; Barile *et al.*, 1962) showed that 57% of the 295 continuous cell lines tested by isolation of the organisms on solid medium contained agents that formed colonies with morphologies resembling mycoplasmata. At that point in time, convenient methods for identifying the species had not yet been developed so most isolates were not further identified. Those isolates that were later identified from the study

TABLE I
Serotaxonomic Clusters of Mycoplasmata Isolated from Cell Cultures[a]

Arginine-positive, glucose-negative species
Serological Group 1[b]
M. pharyngis (*orale* type 1)
M. hominis
M. salivarium
M. arginini
M. arthritidis
M. gallinarum
Glucose-fermenting species
Serological Group 2
M. fermentans
M. hyorhinis
M. canis
M. pulmonis
Serological Group 3
A. laidlawii
A. granularum
Serological Group 4
M. gallisepticum

[a]Species listing from that compiled by Barile (Tables VI and VII, Chapter 6).

[b]Serotaxonomic data are provisional and numbers are assigned arbitrarily for discussion purposes and therefore should not be taken as definitive numbers for the groups of species. Group 1 organisms appear to be closely related when tested by double immunodiffusion (Taylor-Robinson *et al.*, 1963; Lemcke, 1965; Kenny, 1972). Group 2 corresponds to group with low guanine plus cytosine content in their DNA described in Kenny (1969). The two group 3 organisms are closely related to each other but not to other groups and *M. gallisepticum* (group 4) was found to be serologically unique when compared to groups 2 and 3 (Kenny, 1969) and to group 1 (Kenny, 1972).

by Pollock *et al.* (1960) turned out to be *M. hominis* (Pollock, 1972, Kenny, 1972). Similarly, most of the strains isolated in various laboratories at that time found to be *M. hominis* (Macpherson, 1966).

However, primary cell cultures have seldom been found to be contaminated (Rothblat and Morton, 1959; Barile *et al.*, 1962). As culture media have improved, particularly with the utilization of fresh yeast extract

(Hayflick, 1965) for isolating *M. pharyngis* (*M. orale* type 1), that species has also been encountered in cell culture. In fact, most spontaneous contaminants occurring in my laboratory in recent years have been *M. pharyngis*. Barile, in Chapter 6, has summarized the number of *Mycoplasma* species identified from cell cultures. Table I indicates the apparent serotaxonomic group of the most common contaminants in cell culture. Four different taxonomic groups of organisms have been found in cell cultures. Whereas the most frequent species identified in the 1960–1962 era was *M. hominis*, more recent data from this laboratory indicate that *M. pharyngis*, *M. arginini*, and *M. laidlawii* are the most frequent isolates. The experience with mycoplasmic contamination of cell cultures has varied widely from laboratory to laboratory (see Chapter 6 by Barile for a summary of his experience). The recent results from Hayflick's laboratory (Stanbridge, 1971) indicate that 12% of 3000 samples of cell cultures were contaminated with mycoplasmata.

B. Source of Mycoplasmic Contaminants

The initial reports that *M. hominis* was the species isolated resulted in some awkwardness in considering possible sources of contamination because *M. hominis* was considered to be a genital organism. However *M. hominis* has been isolated from human throats, eliminating some of the delicacy concerning possible sources of that species. Some *M. hominis* strains isolated from human throats were sufficiently resistant to streptomycin that cell cultures could readily become contaminated with these strains and show typical effects of naturally contaminated cultures (Kenny, 1965). Similarly, the *de novo* contamination of previously mycoplasma-free cell cultures with *M. pharyngis* is probably a result of direct contamination of cell cultures. The contamination of cultures with *M. laidlawii* and *M. arginini* can be explained by the occurrence of these organisms in animal serum (see Barile, Chapter 6). The source of *M. hyorhinis* as a cell culture contaminant is presently unknown.

The spread of cell culture contamination in various laboratories appears to be a result of the initial contamination of a cell line with a given *Mycoplasma* species, followed by dissemination of the contaminant by aerosols (O'Connell *et al.*, 1964). Thus, a given laboratory may have its cell cultures contaminated with a single *Mycoplasma* species. The original considerations of sources of mycoplasmic contamination were complicated because mycoplasmata were thought to be related to or were stable L forms of bacteria and because the mycoplasmata that were detected were thought to have arisen from the transformation of bacteria into stable L forms. This hypothesis was quite plausible because two protoplasting or L-for-inducing agents were commonly employed in cell culture medium: penicillin and serum. It is now generally felt that the Mycoplasmatales are not directly descended from bacteria and that they represent stable definable species (Edward and Freundt, 1969). Thus, contamination of cell cultures with mycoplasmata is no different than contamination with bacteria. The sterile techniques used for mass cell culture are quite casual and provide numerous opportunities for contamination with a variety of agents. Preparation of the numerous replicate cultures required in virology can rarely be accomplished without extensive antibiotic cover, thus providing ample opportunity for contamination by antibiotic-resistant organisms.

The inoculation of cell cultures with throat swab materials for virus isolation also provides substantial avenues for contaminating cell cultures because as many as half of these specimens may contain *M. pharyngis* and or *M. salivarium* and a small portion may contain *M. hominis*. The aerosilization of culture fluids during medium change of infected cultures may provide substantial opportunity for spread of these mycoplasmata into cell cultures and virus preparations.

C. Detection of Mycoplasmata in Cell Cultures

The detection of mycoplasmata in cell culture is ordinarily done by cultural means, although other means have been proposed (see Section III, C, 6).

1. *Media for Detection of Mycoplasmata in Cell Culture*

The formula for basic Soy Peptone–fresh yeast extract medium employed in this laboratory has been published elsewhere (Kenny, 1969; Grayston *et al.*, 1969) and is based on earlier formulations when commercial yeast extract was used as supplement (Pollock and Kenny, 1963). Alternate methods of isolation have been published by others (Hayflick, 1965; Barile *et al.*, 1962) and the medium most frequently used is that of Hayflick. This formulation is presented herein because I have substantial experience in detecting mycoplasmata in cell cultures and clinical materials using this preparation.

2. *Soy Peptone–Yeast Dialyzate Medium*

Basic broth	
Soy Peptone*	20 gm
NaCl	5 gm
H_2O	1000 ml

The pH is adjusted to 7.4, and 10 gm of Difco Noble Agar are added for basic agar. The medium is sterilized by autoclaving and the basic agar or broth may be stored at room temperature until used.

*Sheffield Chemical Company, Norwich, N.Y.

a. *Yeast Dialyzate.* This is prepared by adding 450 gm of Fleischmann's active dried yeast to 1250 ml of water. The suspension is heated in an autoclave for 5 minutes at 121°C. The suspension is poured into dialysis casings and permitted to dialyze against 1000 ml water for 2 days at 4°C. The dialysis casings are discarded and the dialyzate is sterilized by autoclaving. The sterile dialyzate is stored frozen until used. This method of preparing fresh yeast extract has been devised because it is less difficult than the method described by Hayflick (1965).

b. *Complete Medium*

Basic agar (or broth)	65 ml
Yeast dialyzate	10 ml
Horse serum	25 ml
Penicillin (10,000 units/ml)	2 ml
Thallium acetate (3.3%)	1 ml

A diphasic medium may be prepared by placing 2–3 ml of complete agar medium in a 16 × 125 mm test tube, permitting this to solidify, and overlaying the agar plug with complete fluid medium. For detecting glucose fermentation, 1% glucose and 0.4 mg-% Phenol Red may be included in the basic broth before autoclaving. Arginine utilization may be measured by the inclusion of 16 m*M* arginine and 0.4 mg-% Phenol Red in another culture tube. Glucose utilization will yield an acid (yellow) shift in color of the Phenol Red indicator, whereas arginine utilization will yield an alkaline (purple) shift.

3. Isolation of Mycoplasmata

Both cell culture supernatant and trypsinized cell suspension should be employed for inoculating plates and broth cultures (or diphasic cultures). Some plates should be incubated anaerobically (95% N_2, 5% CO_2) and others aerobically; broth cultures are incubated aerobically. Colonies will be recognizable in 1-10 days. Hayflick (1965) discusses the problems of isolation and some of the artifacts and shows pictures of colonial isolates. Certain gross cultural characteristics of the isolates are helpful. *Mycoplasma hominis* and *M. laidlawii* appear rapidly and most strains do not show any preference for anaerobic environment on plates. *Mycoplasma pharyngis* and most strains of the other species listed on Table I will form colonies much better under anaerobic conditions. Broth cultures should be transferred to solid medium on days 3 and 6, which have been chosen as the incubation times when most contaminating species should show their optimum growth in broth. Ordinarily, most contaminants will be detected in large abundance on the primary plates inoculated with culture supernatant. Because in a few instances isolates from cell cultures may only come from trypsinized or dispersed cells, that material should be used for specimen material if one plans to test only one specimen.

4. *Identification of Species*

Because the contaminating *Mycoplasma* species are heterogeneous and because various groups of species both have different effects on cell cultures and also come from different sources, it is important to identify the contaminants. Some preliminary identification may be made by determining whether the organisms utilize arginine or glucose. For final identification, the disk inhibition method of Clyde (1964) is the simplest method, particularly because antisera to most of the common types in cell culture are available commercially. Difficulties may be encountered in identifying some species by this method. Hayflick and Stanbridge (1967) have shown that some strains of *M. hyorhinis* are very difficult to type by this method. Similarly, I have found that some strains of the *M. laidlawii–M. granularum* group cannot be identified by disk inhibition. For these species, identification of the group to which they belong is readily possible by double immunodiffusion against potent rabbit antiserum (Kenny, 1969). This latter method may be restricted in utility because the antisera employed must be broadly reactive (form a number of double immunodiffusion lines in order to properly group species; Kenny, 1971). Similarly, identification of species by immunofluorescence is possible (Barile *et al.*, 1962) but requires considerable expertise and antisera of known specificity. Metabolic inhibition methods are reasonably simple (Purcell *et al.*, 1969) and species specific; thus, this method may serve as a backup for identifying species not typeable by disk inhibition.

5. *Significance of Cultural Results*

The significance of a positive culture result is clear; however, a negative result has a meaning dependent upon the technical competence of the laboratory attempting to isolate mycoplasmata. Culture media, particularly the lots of horse serum, vary greatly in their ability to detect mycoplasmata. Unless the laboratory attempting to test for mycoplasmata has a reasonable variety of control organisms (organisms isolated from cell cultures or, better, positive cell culture material), the failure to isolate mycoplasmata may have little significance. The fact that the medium prepared grows prototype strains well is of little utility, because the prototype strains have been selected for growth upon the particular medium, whereas the cell culture isolates have not been so selected. Furthermore, it is quite clear to this author that strains of mycoplasmata exist that cannot be grown on this or any other recommended medium. For example, if *M. hyopneumoniae* (also termed *M. suipneumoniae*) were a cell culture contaminant, it could not be detected upon this medium with any certainty because even strains selected for growth on this medium only form colonies in 4–6 weeks. Thus, the significance of a negative culture depends entirely upon the sensitivity of the culture medium used for detection. It is my opinion that considerable work needs to be done on culture media before any particular medium is considered standard. The Soy Peptone medium recommended here gives better plating efficiency than do similar media prepared from the lots of commercial mycoplasma medium that the author has tested. House and Waddell (1967) have compared various media and have shown variations in plating efficiency between different medium formulations. It should be noted that various lots of Soy Peptone may give markedly different plating efficiency so it is useful to pretest lots before adopting the use of any one. Similarly, the choice of solidifying agent, agar or agarose, maybe important because DEAE dextran has been shown to enhance the colony formation of certain mycoplasmic strains (Tauraso, 1967), possibly because of the presence of sulfated polysaccharides in ordinary agar.

6. Detection of Mycoplasmata by Noncultural Means

The facts that isolation of mycoplasmata is difficult and that noncultivatable strains may exist have stimulated the search for detection methods that do not rely upon growth of the organism. It is true, as Hayflick (1965) has stated, that the demonstration of typical colonies is presently the *sine qua non* for the definition of the Mycoplasmatales. Nevertheless, from the viewpoint of the cell culturist, the concern is more with the demonstration that the cell culture in question is free from contamination with extraneous organisms than with specific identification of the contaminant. Furthermore, as information increases about the Mycoplasmatales, it may be possible to add more information of an analytical nature to the definition of the order.

The simplest method will be the direct visualization of mycoplasmata in cell cultures. Fogh and Fogh (1964) have described an intensified staining method by which organisms may be directly observed with the light microscope. This method is reported to correlate well with cultural results; however it is not known whether mycoplasmic contamination from all taxonomic groups can be equally well detected. Barile *et al.* (1962) have demonstrated that mycoplasmata can be directly visualized with the fluorescent antibody technique. As the number of serotypes does appear limited, it would seem to be relatively simple to prepare a suitable polyvalent antiserum that will detect nearly all contaminants. Similarly, appropriately fixed material may be useful for detecting mycoplasmata by the electron microscope because organisms can be seen associated with the membrane (Anderson 1969).

Barile and Schimke (1963) proposed an enzymic method for detecting mycoplasmata based on the presence of the enzyme arginine deiminase. Although this method appeared reasonably sensitive, not all species contain arginine deiminase (Barile *et al.*, 1966; see also Table I). The thymidine cleavage method of detecting mycoplasmata was compared with isolation of the organism by House and Waddell (1967). In their hands, the former method appeared less sensitive than did direct isolation. Todaro *et al.* (1971) have recently proposed that mycoplasmata might be detected in cell cultures by the fact that the strains that they tested banded at a unique density in sucrose (1.22–1.24 gm/cm^3). Thus, labeling the cell culture with either an RNA or a DNA precursor permitted the detection of this unique band in material tested from contaminated cultures but not from mycoplasma-free cultures. Accordingly, a number of noncultural methods appear promising for the detection of mycoplasmata in cell cultures, although it appears likely that no single method will suffice to identify the various species and groups of species. Clearly, these methods cannot supplant cultural methods but will be most important in detecting contamination by noncultivatable mycoplasmata and other organisms.

IV. Host–Parasite Relationships of Mycoplasmata and Mammalian Cells

A. Nature of the Mycoplasmata–Mammalian Cell Interaction

The general nature of the interaction of mycoplasmata with mammalian cells in culture is that of a continuous infection of the cell culture with continuous propagation of both mammalian and mycoplasmic cells. As is usually the case, generalizations are not without hazards: mycoplasmata are quite diverse and species in various taxonomic groups may have different effects and cell lines may vary in susceptibility to the effects. Nevertheless, coexistence is the most likely form in which contamination of cell cultures is likely to be detected, because highly destructive contaminants will cause the cell culture to be lost and the contaminant will be lost in those systems where all its growth requirements cannot be met continuously. Examples of perpetual contaminants are the organisms shown in Table I. Attempts to establish T strains in culture have failed; the organisms persist for several days and then disappear (Mazzali and Taylor-Robinson, 1971).

B. Cellular Habitat of the Mycoplasmatales

Whether mycoplasmata are intracellular parasites of cultured mammalian cells is controversial. This question is very important to any consideration of this host–parasite system because the role an extracellular parasite plays must be radically different from that of an intracellular parasite. It is quite clear in the case of phagocytic cells that both monocytes and lymphocytes can phagocytize mycoplasmata (Zucker-Franklin *et al.*, 1966b). Similarly, phagocytic cells appear to take up mycoplasmata *in vivo* (Organic *et al.*, 1966). However, little clear evidence exists to demonstrate a clear intracellular site for mycoplasmic growth. The claims for an intracellular habitat of some *Mycoplasma* species in cultures of nonphagocytic cells by light microscopy (Hayflick and Stinebring, 1960; Shedden and Cole, 1966) are subject to the objection that organisms in close association with the cell membrane might well appear to be intracellular. However, ultrastructural studies of cells contaminated or deliberately infected with mycoplasmata have shown that the organisms are very strongly associated with the external membrane of the mammalian cell (Anderson, 1969; Boatman and Kenny, 1971; Edwards and Fogh, 1960). Only a few organisms are seen in locations that could be conceived of as intracellular and in these cases Zucker-Franklin *et al.* (1966a) have pointed out that the organisms could as well be at the bottom of a crypt as to be intracellular. The fact that mycoplasmata (*M hominis*) can be eliminated from cell cultures with specific antiserum (Pollock and Kenny, 1963; Herderschee *et al.*, 1963) argues for an extracellular site.

Finally, Jones and Hirsch (1971) have shown that *M. pulmonis* readily attaches to and grows on mouse peritoneal macrophages and mouse heteroploid cell cultures (strain L). Phagocytosis occurs rapidly with the macrophages when antibody is added to the medium but the organisms remain on the cell surface in the absence of antibody. Phagocytosis is not observed with the mouse heteroploid cell cultures with or without antibody.

Thus, the overall impression given by our present information is that mycoplasmata are surface parasites of mammalian cells in culture. The fact that a few organisms may be occasionally found intracellularly should not permit us to ignore the fact that the great majority of the mycoplasmata are associated with the external membrane. It should not be surprising if an occasional organism were associated with a piece of cell membrane that happened to be undergoing pinocytosis. However, it is also possible that the generalization of a preferred surface site for mycoplasmata may not hold for all species. The taxonomic heterogeneity of the Mycoplasmatales appears quite great (see Section II, B) and certain species or strains may be intracellular parasites, whereas those presently studied by reasonable methods appear to be surface parasites.

C. Interaction of Mycoplasmata with Mammalian Cell Surfaces

Mycoplasma gallisepticum and *M. pneumoniae* attach to mammalian cells by means of receptor sites on the mammalian cell membrane that are labile to neuraminidase; *M. hominis* and *M. salivarium* attach to cells with receptor sites that are not neuraminidase sensitive but are sensitive to proteolytic enzymes (Manchee and Taylor-Robinson, 1969; Sobeslavsky *et al.*, 1968). Most *Mycoplasma* species tested attach to glass and plastic surfaces in much the same manner that mammalian cells attach to glass (Taylor-Robinson and Manchee, 1967; Purcell *et al.*, 1971). Thus, the very small mycoplasmic cell behaves in much the same manner concerning glass surfaces as does the much larger animal cell. Therefore the interaction of the mycoplasmic cell with the mammalian cell in culture will be governed by the complex and poorly understood interactions of membrane systems, which characterize the interaction of mammalian cells with each other.

V. Effect of Mycoplasmata on Cell Cultures

Their small size, the difficulty of detecting them, and the apparent lack of any gross effect on cell cultures led many to believe that mycoplasmata had little effect on cell cultures. Moreover, because the preconceptions of researchers at any given time greatly influence the interpretation of results, it is important to note that many workers felt that mycoplasmata were intracellular parasites or at least had an intracellular phase in cultured cells. Finally, the possible relationship of mycoplasmata to bacterial L forms as discussed in Section III, B and the lack of knowledge about the properties of the Mycoplasmatales also confused the emerging picture of the interrelationships of mycoplasmata and mammalian cells. Because many of the cell lines in common use both now and in the past can be propagated indefinitely in spite of the presence of contaminating mycoplasmata, the presence of the contaminant is clearly not necessarily fatal to the culture nor does it appear to have much effect on the growth of the cell culture. However, it will be seen that much of the technology of cell culture, particularly that used in the late fifties and early sixties, was in fact arranged to compensate for the effects of mycoplasmic contamination.

A. The Arginine Effect

Human heteroploid cells in culture were found to grow both more slowly and to provide lower final yields in culture if they were contaminated with mycoplasmata, in contrast to the results obtained with cultures free of mycoplasmata (Kenny and Pollock, 1963). Addition of yeast extract to the cell culture medium gave an even greater differential effect between control and contaminated cultures, apparently because populations of mycoplasmata were increased, whereas addition of calf serum to the medium (ordinarily supplemented with human serum) decreased the effect, presumably because populations of mycoplasmata were reduced. Finally, the growth retardation effect could be eliminated by simply replacing the medium at 2 day intervals, whereas control cultures grew at the same rate whether or not the medium was changed. The growth retardation effect was found to be caused by depletion of arginine from the medium by the contaminating mycoplasmata (subsequently identified as *M. hominis*). Increasing the arginine content to 1 m*M* from the usual 0.1 m*M* permitted the contaminated cells to grow as fast as the control cells. No other component appeared to be involved in retarding growth in that system [human heteroploid cells grown in Basal Medium (Eagle, 1959) supplemented with 10% human serum and contaminated with strains of *M. hominis*]. When mycoplasmata were grown in cell culture medium without cells in the presence of 0.1% yeast extract, growth was equivalent to that in cell cultures. This medium was deficient for cell growth (Kenny and Pollock, 1963) and produced striking cytopathic effects (Pollock *et al.*, 1963); Because it was only necessary to add 0.1 m*M* arginine to this depleted medium to restore its growth potential, it appeared that the only component significantly depleted was arginine.

The fact that mycoplasmata cannot utilize citrulline (Smith, 1960), whereas certain cell lines can utilize citrulline *in lieu* of arginine, can be exploited. The arginine effect can be abolished in some human cell lines contaminated with *M. hominis* by the inclusion of 0.1 m*M* citrulline (Kenny, 1972; Pollock, 1972).

-Because glutamine was known to be metabolized by mycoplasmata (Smith, 1960), the concentration of this component was reduced to 0.2 m*M* from the usual 2.0 m*M* to test whether the requirement for such a large amount was related to mycoplasmic contamination. No difference in rate of growth between contaminated and free cell cultures was detected when the arginine level was maintained at 1.0 m*M*.

Thus, the most direct effect on growth of cells in culture that is produced by mycoplasmic contamination is the cessation of growth caused by the depletion of arginine in the culture medium. The rate of arginine depletion will be determined by the concentration of the contaminating organism, the ability of the particular strain to utilize arginine, and the concentration of arginine.

A more spectacular version of the arginine effect was the system studied by Kraemer *et al.* (1963). A line of murine lymphoma cells was found to undergo lysis in the presence of a contaminating mycoplasmic strain. Only certain mycoplasmic strains had the lytic ability and lytic strains lyzed only the lymphoma cell line. The lytic principle was found to be separable from the mycoplasmata and to be neutralizable by arginine (Kraemer, 1964). Gill and Pan (1970) showed that cell division in the same line of murine lymphoma cells did not occur when the culture was exposed to arginine deiminase prepared from yeast or *M. pharyngis* (*orale*). This effect was neutralizable by arginine or specific antiserum. It should be noted that the lytic effect did not occur in their experiments. Thus, if the lytic effect were equivalent to the growth inhibition effect, the enzyme system itself would appear to prevent cell replication and cause lysis of susceptible cells under suitable conditions.

Those mycoplasmata that do not contain arginine deiminase obviously do not induce the arginine effect. Although it is clear that such organisms as *M. hyorhinis*, *M. pulmonis*, and *A. laidlawii* decrease the net amount of growth from a cell culture, few experiments have been carried out to determine the magnitude and mechanism of this effect.

B. Effect of Mycoplasmata on Cell Morphology

The arginine effect can also be observed as a morphological effect. Pollock *et al.* (1963) has observed a spindling of the human heteroploid cells infected with mycoplasmata. This effect occurs much more rapidly when medium depleted of arginine by mycoplasmata is substituted. This effect is nonspecific, as similar effects can be obtained by depriving the cells of glucose in the absence of mycoplasmata. Little cytopathic effect has been observed in experimentally contaminated cultures unless yeast extract is added to the cultures (Fogh *et al.*, 1965; Pollock *et al.*, 1963; Wittler *et al.*, 1956).

The species that did not contain arginine deminase also produced cytopathic effects, some of which were more striking than those produced by the arginine deiminase-containing species. The acid-inducing agent described by Butler and Leach (1964), subsequently identified as *M. hyorhinis* (Purcell *et al.*, 1966), caused the death of infected human heteroploid cell cultures. This destruction of the cell monolayer was accompanied by increased acid production in the culture in contrast to the increased alkalinity sometimes observed in cultures contaminated with arginine deminase-containing mycoplasmata. The mechanism for the cytopathic effects with the glycolytic species is presently unknown.

C. Effects of Mycoplasmata on Cell Physiology and Metabolism

Arginine depletion in cell culture medium can be demonstrated biologically, because arginine is required for mammalian cell growth (Kenny and Pollock, 1963), or directly, by chemical means. Stanbridge *et al.* (1971) have shown that only arginine is depleted in cell culture medium in which arginine-utilizing *Mycoplasma* species have been grown. Species that do not utilize arginine do not reduce the arginine levels. This effect appears to have been observed in the analysis of cell culture media from various cell lines. Arginine disappears from the medium of certain human heteroploid cell lines and is replaced by citrulline (Chung *et al.*, 1966). Ammonia is also generated, suggesting that the arginine deiminase enzyme system (most likely derived from mycoplasmata) is functioning in those cell lines.

Some more subtle changes in cell metabolism can also be observed. Cells grown in medium containing amethopterin require a source of DNA thymine. This requirement can be met by thymidine or thymidylic acid; however, certain cell lines contaminated with mycoplasmata cannot be grown on either of these substrates (Hakala *et al.*, 1963). This effect has been found to be a result of the presence of mycoplasmata. Although the species has not been identified, it shows ability to cleave deoxyribonucleosides as does *M. laidlawii* B. Thus, it can remove the deoxyribonucleosides from the medium. The synthesis patterns for cellular RNA are changed in cell cultures contaminated with mycoplasmata: uridine label is not incorporated into cellular RNA but it is found in mycoplasmic RNA (Harley *et al.*, 1970; Levine *et al.*, 1968). The spontaneous contaminant tested by Levine *et al.* (1968) is *M. hyorhinis*; although certain other species have been tested, only some of these give rise to perpetually contaminated cultures that show abnormal patterns of RNA synthesis. The mechanism for this phenomenon is unknown.

Certain *Mycoplasma* species show substantial deoxyribonuclease activity. In cell culture, this activity is sufficient to cause significant degradation of host cell DNA (Randall *et al.*, 1965; Stock and Gentry, 1969). Stock and Gentry (1971) have suggested that the extracellular deoxyribonuclease produced by *M. hominis* may provide a scavenger route for the supply of thymidine nucleotides. This idea is of substantial interest with regard to the general findings that mycoplasmata do not grow in cell culture media, such as minimum essential medium (Eagle, 1959), unless yeast extract is added (Kenny and Pollock, 1963).

However, mycoplasmata readily grow and persist in medium with cells. Stanbridge *et al.* (1971) indicate that most of the *Mycoplasma* species that are commonly found in cell cultures will grow in BME (basal medium, Eagle) supplemented with 10% calf serum if a mixture of 5′-nucleoside diphosphates is added to the medium. This result suggests that the ability of the organisms to grow in cell culture may be a result of continuous supply of thymidine precursors resulting from the breakdown of host mammalian cell DNA by mycoplasmic deoxyribonuclease.

D. Effects of Mycoplasmata on Mammalian Chromosomes

Contaminated human heteroploid cell cultures show chromosome changes (Fogh and Fogh, 1965) that are apparently permanent. Contaminated human diploid cell cultures show increases in chromosome breakage (Paton *et al.*, 1965; Stanbridge *et al.*, 1969). Mycoplasmata (*M. salivarium,*

M. arthritidis, and *M. fermentans*) have been found to cause mitotic inhibition in human leukocyte cultures (Aula and Nichols, 1967). This effect is related to arginine concentration. This effect must be different in some respects from the chromosome breakage shown by some *Mycoplasma* species in continuous cell cultures because some of the glycolytic strains do not contain arginine deiminase and this amino acid is not depleted. The arginine neutralization effect may be similar to the effect of arginine deiminase on lymphoma cells (see Section III, A). Stanbridge *et al.* (1971) have suggested that the chromosome breakages observed in these systems may well be caused by the deoxyribonuclease activity described by Stock and Gentry (1969).

E. Effects of Mycoplasmata on Virus Yields

The initial study on mycoplasmal effects on virus yields went directly to the heart of the matter. Mycoplasmata were found to interfere with plaque assay of adenoviruses by depleting arginine (Rouse *et al.*, 1963). Both measles (Romano and Brancato, 1970) and vaccinia virus (Singer *et al.*, 1970) showed inhibition of virus growth that was restorable by arginine. The effects of mycoplasmata (*M. orale* type 1, *M. hominis*, and *M. hyorhinis*) on assay of a number of viruses (adenoviruses, herpes simplex, poliovirus, parainfluenza virus, respiratory syncytial virus, and vaccinia virus) in tube cultures of human heteroploid cell line were small, within tenfold (Hargreaves and Leach, 1970). Both enhancement and inhibition were observed; adenovirus titers were reduced tenfold and vaccinia titers were increased tenfold, in apparent contradiction to the results of Singer *et al.* (1970), cited above, who showed a decrease in yield of vaccinia virus.

IV. Elimination of Mycoplasmata from Cell Cultures

Two general approaches may be used for avoiding the problem of contamination of mammalian cell cultures with mycoplasmata: treating already contaminated cultures and preventing contamination of cultures. Little alternative to treatment exists for important cell lines that are known to be contaminated unless the cell line in question can be restarted from source material.

A. Elimination of Mycoplasmata with Antibiotics

Kanamycin was described as a useful antibiotic for eliminating mycoplasmata from cell cultures (Fogh and Hacker, 1960; Pollock *et al.*, 1960). Although the strains originally studied by these authors were kanamycin sensitive, strains were soon recovered that were kanamycin resistant (Pollock and Kenny, 1963; Perlman *et al.*, 1967).

Aureomycin was employed for the elimination of mycoplasmata by Hearn *et al.* (1959).

Tetracycline-resistant strains of mycoplasmata have been encountered (Perlman *et al.*, 1967). Gori and Lee (1964) observed that prolonged treatment with tetracycline reduced the severity of the infection but did not eliminate the organisms, whereas Carski and Shepard (1961) found that small doses (2.5 μg) were sufficient for elimination. Apparently, mycoplasmata exhibit a spectrum of sensitivity to tetracycline.

Tylosin has been used to eliminate mycoplasmata from cell cultures (Friend *et al.*, 1966; Cross *et al.*, 1967), and mixtures of antibiotics, cocktails, have been employed (Burskirk, 1967).

Gori and Lee (1964) used hypotonic shock coupled with antibiotic treatment for the elimination of mycoplasmata from cell cultures. Because the strains involved were not identified and because *Mycoplasma* species vary markedly in susceptibility to osmotic lysis, *M. laidlawii* being the most susceptible (Razin, 1964), this method should be most effective on that species.

A summary of present experience is somewhat hazardous because of the fragmentary data available and the fact that many of the species studied were not identified. It appears that strains and species of mycoplasmata vary widely in sensitivity to antibiotics. It also appears that the mycoplasmacidal antibiotics, such as kanamycin, are best suited for eliminating mycoplasmata, whereas the mycoplasmastatic antibiotics, such as tetracyline, are good for suppressing growth but appear to have relatively little efficacy as general measures for eliminating mycoplasmata. Unfortunately, strains resistant to most of the mycoplasmacidal drugs have developed.

B. Chemical and Physical Methods of Elimination

Hayflick (1960) showed that certain strains of mycoplasmata were more sensitive to heat than were mammalian cells and that simply heating the contaminating cell culture to 41°C resulted in differential killing of the mycoplasmata and survival of the mammalian cells. This method was combined with antibiotic therapy (Johnson and Orlando, 1967). Heat did not work for all systems, some cell cultures being unable to withstand that degree of heat and certain strains of mycoplasmata not being eliminated by heat (Herderschee *et al.*, 1963; Vogelzang and Compeer-Dekker, 1969). Sodium aurothiomalate was employed successfully for eliminating mycoplasmata (Shedden and Cole, 1966; Cross *et al.*, 1967), but this method did not work in all cases (Shedden and Cole, 1966). Recently, it was reported that the buffer *N*-tris-(hydroxymethyl) methylglycine (Tricine) when incorporated into cell culture medium led to the suppression of mycoplasmic growth (Spendlove *et al.*, 1971). Other buffers, such as *N*-2-hydroxyethyl piperazine-*N'*-2-ethanesulfonic acid (HEPES) and *N*-trishydroxymethyl)methyl-2-aminoethane sulfonic acid (TES) did not inhibit mycoplasmata. As with antibiotics, the chemical and physical means appear to have limited utility as universal measures for eliminating mycoplasmata.

C. Elimination with Specific Antiserum

Mycoplasmata (identified in retrospect as *M. hominis*; Pollock, 1972) were eliminated from cell cultures with rabbit immune serum to mycoplasmata (Pollock and Kenny, 1963). One strain could be eliminated with heated serum, wherease another strain could only be eliminated with fresh serum. Herderschee *et al.* (1963) reported that antiserum against *M. hominis* was successful in eliminating that species and that eliminating that organism permitted the detection of another species in the particular cell culture, which was then eliminated with antiserum specific to it. Both *M. orale* and *M. hyorhinis* were eliminated from cell cultures by specific antiserum to the respective species by Vogelzang and Compeer-Dekker (1969). However, they noted that some of the antisera were cytotoxic to the cell lines in question. Thus, eliminating mycoplasmata from cell cultures with specific antiserum appears promising, particularly as substantial information concerning the nature of the killing of mycoplasmata by antiserum is now coming forth. Barker and Patt (1967) showed that the inhibition or, more likely, the killing of *M. gallisepticum* by antibody required a complement-like factor. Riggs *et al.* (1967) have shown that an unidentified *Mycoplasma* species from cell culture could be inhibited by antiserum alone but that serum titers were greatly enhanced by the addition of fresh serum. *Mycoplasma pneumoniae* was killed by antiserum only in the presence of complement-like factors from normal guinea pig serum (Gale and Kenny, 1970), a finding that has been confirmed by Brunner *et al.* (1971). *Mycoplasma canis* also was killed in a complement-mediated reaction (Tachibana *et al.*, 1970). Therefore, it appears that a number of *Mycoplasma* species can be killed by antibody and complement, a finding that may enhance the use of antiserum for the elimination of mycoplasmata from cell cultures.

D. Prevention of Mycoplasmic Contamination

In view of the rather sizeable difficulties in curing cell cultures of mycoplasmata, as discussed in the preceding section, it would appear most useful to attempt to prevent contamination. The most logical method is cultivating reference cell cultures without the use of antibiotics, because any breach in aseptic technique will be accompanied by bacterial contamination and subsequent loss of the cell culture. This approach has been used by Pollock *et al.* (1960) with good results. However, this method is not practical in all cases. The use of laminar flow hoods, as discussed by McGarrity and Coriell (1971), should provide a most valuable method for propagating mass cell cultures for virus assay and physiological studies. The use of antibiotics, particularly aureomycin, has been recommended by Hayflick (1965) as a promising means for preventing contamination.

V. Conclusions

The problem of mycoplasmic contamination of cell cultures promises to be an enduring one. It is likely that the results of any biological experiment that employs contaminated cells are compromised to a greater or lesser degree depending upon the parameters being measured. However, the very ability of mycoplasmata to persist in a cell culture with little obvious damage to the host provides a very interesting host–parasite system for study. The finding that mycoplasmata as surface parasites of cultured cells can cause chromosome breakage in the intact cell, possibly mediated by a deoxyribonuclease, is of considerable interest.

It is clear that no single method appears suitable for either detecting or eliminating mycoplasmata, a finding likely a result of the heterogeneity of the Mycoplasmatales. Similarly, no specific effect or group of effects can be ascribed to each and every species or strain that contaminates cell cultures.

Acknowledgment

Some of the studies presented in this chapter were supported in part by Public Health Service Research Grants AI-06720 and AI-09586.

References

Aluotto, B. B., Wittler, R. G., Williams, C. O., and Faber, J. E. (1970). *Int. J. Syst. Bacteriol.* **20**, 35.

Anderson, D. R. (1969). *In* "The *Mycoplasmatales* and L-phase of Bacteria" (L. Hayflick, ed.), pp. 354–404. Appleton, New York.

Aula, P., and Nichols, W. W. (1967). *J. Cell. Physiol.* **70**, 281.

Barile, M. F., and Schimke, R. T. (1963). *Proc. Soc. Exp. Biol. Med.* **114**, 676.

Barile, M. F., Malizia, W. F., and Riggs, D. B. (1962). *J. Bacteriol.* **84**, 130.

Barile, M. F., Schimke, R. T., and Riggs, D. B. (1966). *J. Bacteriol.* **91**, 189.

Barker, L. F., and Patt, J. (1967). *J. Bacteriol.* **94**, 403.

Boatman, E. S., and Kenny, G. E. (1970). *J. Bacteriol.* **101**, 262.

Boatman, E. S., and Kenny, G. E. (1971). *J. Bacteriol.* **106**, 1005.

Brunner, H., Razin, S., Kalica, A. R., and Chanock, R. M. (1971). *J. Immunol.* **106**, 907.

Burskirk, H. H. (1967). *Appl. Microbiol.* **15**, 1442.

Butler, M., and Leach, R. H. (1964). *J. Gen. Microbiol.* **34**, 285.

Carski, T. R., and Shepard, C. C. (1961). *J. Bacteriol.* **81**, 626.

Chung, R. A., Brown, R. W., Beacham, L. H., and Juang, I. (1966). *Can. J. Biochem.* **44**, 1145.

Clyde, W. A. (1964). *J. Immunol.* **92**, 958.

Clyde, W. A. (1969). *In* "The *Mycoplasmatales* and L-phase of Bacteria" (L. Hayflick, ed.), pp. 349–364. Appleton, New York.

Cross, G. F., Goodman, M. R., and Shaw, E. J. (1967). *Aust. J. Exp. Biol. Med. Sci.* **45**, 201.

Eagle, H. (1959). *Science* **130**, 432.

Edward, D. G., and Freundt, E. A. (1969). *In* "The *Mycoplasmatales* and the L-phase of Bacteria" (L. Hayflick, ed.), pp. 147–200 Appleton, New York.

Edward, D. G., and Freundt, E. A. (1970). *J. Gen. Microbiol.* **62**, 1.

Edwards, G. A., and Fogh, J. (1960). *J. Bacteriol.* **79**, 267.

Fogh, J., and Fogh, H. (1964). *Proc. Sco. Exp. Biol. Med.* **117**, 899.

Fogh, J., and Fogh, H. (1965). *Proc. Soc. Exp. Biol. Med.* **119**, 233.

Fogh, J., and Hacker, C. (1960). *Exp. Cell Res.* **21**, 242.

Fogh, J., Hahn, E., and Fogh, H. (1965). *Exp. Cell Res.* **39**, 554.

Freundt, E. A. (1969). *In* "The *Mycoplasmatales* and L-phase of Bacteria" (L. Hayflick, ed.), pp. 281–316. Appleton, New York.

Friend, C., Patuleia, M. C., and Nelson, J. B. (1966). *Proc. Soc. Exp. Biol. Med.* **121**, 1009.

Gale, J. L., and Kenny, G. E. (1970). *J. Immunol.* **104**, 1175.

Gill, P., and Pan, J. (1970). *Can. J. Microbiol.* **16**, 415.

Gori, G. B., and Lee, D. Y. (1964). *Proc. Soc. Exp. Biol. Med.* **117**, 918.

Grayston, J. T., Foy, H. M., and Kenny, G. E. (1969). *In* "The *Mycoplasmatales* and L-phase of Bacteria" (L. Hayflick, ed.), pp. 651–682. Appleton, New York.

Hakala, M. T., Holland, J. F., and Horoszewicz, J. S. (1963). *Biochem. Biophys. Res. Commun.* **11**, 446.

Hargreaves, F. D., and Leach, R. H. (1970). *J. Med. Microbiol.* **3**, 259.

Harley, E. H., Rees, K. R., and Cohen, A. (1970). *Biochim. Biophys. Acta* **213**, 171.

Hayflick, L. (1960). *Nature* (*London*) **185**, 783.

Hayflick, L. (1965). *Tex. Rep. Biol. Med.* **23**, 285.

Hayflick, L., and Stanbridge, E. (1967). *Ann. N. Y. Acad. Sci.* **143**, 608.

Hayflick, L., and Stinebring, W. R. (1960). *Ann. N. Y. Acad. Sci.* **79**, 433.

Hearn, H. J., Officer, J. E., Elsner, V., and Brown, A. (1959). *J. Bacteriol.* **78**, 575.

Herderschee, D., Ruys, A. C., and van Rhijin, G. R. (1963). *Antonie van Leeuwenhoek; J. Microbiol. Serol.* **29**, 368.

House, W., and Waddell, A. (1967). *J. Pathol. Bacteriol.* **93**, 125.

Johnson, R. W., and Orlando, M. D. (1967). *Appl. Microbiol.* **15**, 209.

Jones, T. C., and Hirsch, J. G. (1971). *J. Exp. Med.* **133**, 231.

Kenny, G. E. (1972). Unpublished data.

Kenny, G. E. (1965). *Bacteriol. Proc.* p. 116.

Kenny, G. E. (1969). *J. Bacteriol.* **98**, 1044.

Kenny, G. E. (1971). *Infect. Immunity* **3**, 510.

Kenny, G. E., and Pollock, M. E. (1963). *J. Infec. Dis.* **112**, 7.

Kraemer, P. M. (1964). *Proc. Soc. Exp. Biol. Med.* **115**, 206.

Kraemer, P. M., Defendi, V., Hayflick, L., and Manson, L. A. (1963). *Proc. Soc. Exp. Biol. Med.* **112**, 381.

Lemcke, R. M. (1965). *J. Gen. Microbiol.* **38**, 91.

Levine, E. M., Thomas, L., McGregor, D., Hayflick, L., and Eagle, H. (1968). *Proc. Nat. Acad. Sci. U. S.* **60**, 583.

McGarrity, G. J., and Coriell, L. L. (1971). *In Vitro* **6**, 257.
Macpherson, I. (1966). *J. Cell Sci.* **1**, 145.
Manchee, R. J., and Taylor-Robinson, D. (1969). *Brit. J. Exp. Pathol.* **50**, 66.
Maniloff, J., and Morowitz, H. J. (1967). *Ann. N. Y. Acad. Sci.* **143**, 59.
Mazzali, R., and Taylor-Robinson, D. (1971). *J. Med. Microbiol.* **4**, 125.
Neimark, H. C. (1970). *J. Gen. Microbiol.* **63**, 249.
O'Connell, R. C., Wittler, R. G., and Faber, J. E. (1964). *Appl. Microbiol.* **12**, 337.
Organick, A. B., Siegesmund, K. A., and Lutsky, I. I. (1966). *J. Bacteriol.* **92**, 1164.
Paton, G. R., Jacobs, J. P., and Perkins, F. T. (1965). *Nature* (*London*) **207**, 43.
Perlman, D., Rahman, S. B., and Semar, J. B. (1967). *Appl. Microbiol.* **15**, 82.
Pollock, M. E. (1972). Personal communication.
Pollock, M. E., and Kenny, G. E. (1963). *Proc. Soc. Exp. Biol. Med.* **112**, 176.
Pollock, M. E., Kenny, G. E., and Syverton, J. T. (1960). *Proc. Soc. Exp. Biol. Med.* **105**, 10.
Pollock, M. E., Treadwell, P. E., and Kenny, G. E. (1 963). *Exp. Cell Res.* **31**, 321.
Purcell, R. H., Somerson, N. L., Fox, H., Wong, D. C., and Chanock, R. M. (1966). *J. Nat. Cancer Inst.* **37**, 251.
Purcell, R. H., Chanock, R. M., and Taylor-Robinson, D. (1969). *In* "The *Mycoplasmatales* and the L-phase of Bacteria" (L. Hayflick, ed.), pp. 221–264. Appleton, New York.
Purcell, R. H., Valdesuso, J. R., Cline, W. L., James, W. D., and Chanock, R. M. (1971). *Appl. Microbiol.* **21**, 288.
Randall, C. C., Gafford, L. G., Gentry, G. A., and Lawson, L. A. (1965). *Science* **149**, 1098.
Razin, S. (1964). *J. Gen. Microbiol.* **36**, 451.
Razin, S. (1968). *J. Bacteriol.* **96**, 687.
Riggs, S., Sharp, J. T., and Carpenter, R. R. (1967). *Ann. N. Y. Acad. Sci.* **143**, 784.
Robinson, L. B., Wichelhausen, R. H., and Roizman, B. (1956). *Science* **124**, 1147.
Rodwell, A. (1969). *In* "The *Mycoplasmatales* and L-phase of Bacteria" (L. Hayflick, ed.), pp. 413–450. Appleton, New York.
Romano, N., and Brancato, P. (1970). *Arch. Gesamte Virusforsch.* **29**, 39.
Rothblat, G. H., and Morton, H. E., (1959). *Proc. Soc. Exp. Biol. Med.* **100**, 87.
Rouse, H. C., Bonifas, V. H., and Schlesinger, R. W. (1963). *Virology* **20**, 357.
Shedden, W. I. H., and Cole, B. C. (1966). *Nature* (*London*) **210**, 868.
Singer, S. H., Fitzgerald, E. A., Barile, M. F., and Kirschstein, R. L. (1970). *Proc. Soc. Exp. Biol. Med.* **133**, 1439.
Smith, P. F. (1960). *Ann. N. Y. Acad. Sci.* **79**, 543.
Sobeslavsky, O., Prescott, B., and Chanock, R. M. (1968). *J. Bacteriol.* **96**, 695.
Spendlove, R. S., Crosbie, R. B., Hayes, S. F., and Keeler, R. (1971). *Proc. Soc. Exp. Biol. Med.* **137**, 258.
Stanbridge, E. (1971). *Bacteriol. Rev.* **35**, 206.
Stanbridge, E., Onen, M., Perkins, F. T., and Hayflick L. (1969). *Exp. Cell Res.* **57**, 397.
Stanbridge, E. J., Hayflick, L., and Perkins, F. T. (1971). *Nature (London)* **232**, 242.
Stock, D. A., and Gentry, G. A. (1969). *J. Virol.* **3**, 313.
Stock, D. A., and Gentry, G. A. (1971). *J. Gen. Microbiol.* **65**, 105.
Tachibana, D. K., Hayflick, L., and Rosenberg, L. T. (1970). *J. Infec. Dis.* **121**, 541.
Tauraso, N. M. (1967). *J. Bacteriol.* **93**, 1559.
Taylor-Robinson, D., and Manchee, R. J. (1967). *J. Bacteriol.* **94**, 1781.
Taylor-Robinson, D., Somerson, N.L., Turner, H.C., adn Chanock, R.M. (1963). *J. Bacteriol.* **85**, 1261.
Todaro, G. J ., Aaronson, S. A., and Rands, E. (1971). *Exp. Cell Res.* **65**, 256.

Vogelzang, A. A., and Compeer-Dekker, G. (1969). *Antonie van Leeuwenhoek; Microbiol. Serol.* **35**, 393.
Williams, C. O., Wittler, R. G., and Burris, C. (1969). *J. Bacteriol.* **99**, 341.
Wittler, R. G., Cary, S. G., and Lindberg, R. B. (1956). *J. Gen. Microbiol.* **14**, 763.
Zucker-Franklin, D., Davidson, M., and Thomas, L. (1966a). *J. Exp. Med.* **124**, 521.
Zucker-Franklin, D., Davidson, M., and Thomas, L. (1966b). *J. Exp. Med.* **124**, 533.

CHAPTER **6**

Mycoplasmal Contamination of Cell Cultures: Mycoplasma–Virus–Cell Culture Interactions

MICHAEL F. BARILE

I. Introduction 132
A. Historical Aspects 132
II. Classification 133
A. Properties of Mycoplasmas 133
B. Family and Species Designation 134
III. Detecting Cell Culture Contamination 135
A. Direct Culture Procedures 135
B. Indirect Procedures 140
IV. Mycoplasmal Contamination 140
A. Incidence 140
B. Sources 141
C. Prevalence of Mycoplasma Species Contaminants 147
D. Spread of Contamination 148
E. Preventing Mycoplasmal Contamination 148
F. Eliminating Mycoplasmal Contamination 149
V. Interactions between Mycoplasma and Animal Cells 150
A. Hemolysis 150
B. Hemadsorption 150
C. Hemagglutination 150
D. Mycoplasmal Attachment to Other Cell Membranes 151
E. Chromosomal Aberrations 152
F. Cytopathic Effects of Mycoplasmal Contamination 152
G. Inhibition of Lymphocyte Transformation 153
VI. Mycoplasma–Virus–Cell Interactions 154
A. Decreased Virus Yields 154
B. Increased Virus Yields 155
References 156

I. Introduction

In 1956, Robinson *et al.* reported the first isolation of a mycoplasma from a contaminated cell culture. Subsequently, mycoplasmas were established as common (Macpherson, 1966; Barile, 1968) and bothersome contaminants capable of altering cell activity and affecting results of study.

Some mycoplasmas (*a*) inhibit cell metabolism (Powelson, 1961; Hakala *et al.*, 1963), cell growth (Kenny and Pollock, 1963), and phytohemagglutinin (PHA) stimulation of lymphocyte cultures (Barile and Leventhal, 1968); (*b*) alter nucleic acid synthesis (Nardone *et al.*, 1965; Levine *et al.*, 1968; Markov *et al.*, 1969; Stanbridge *et al.*, 1969); (*c*) cause chromosomal aberrations (Fogh and Fogh, 1965; Paton *et al.*, 1965, 1967; see Chapter 7 by Fogh); and (*d*) alter the antigenicity of cell membranes (Schmidt *et al.*, 1965). Also, mycoplasmas are troublesome because they mimic viruses. In fact, mycoplasmas had been mistaken for viruses (Nocard and Roux, 1898; Borrel *et al.*, 1910; Chanock *et al.*, 1962; Negroni, 1964; Maré and Switzer, 1965; Goodwin *et al.*, 1965; Cross *et al.*, 1970; Stanbridge, 1971) because they are filterable, resistant to certain antibiotics and inhibited by antiserum alone and because they alter cell morphology (Pollock *et al.*, 1960; 1963), producing cytopathic effects (CPE) (Nelson, 1960) including plaque formation (O'Malley *et al.*, 1961; 1966; Kagan *et al.*, 1969) and destruction of the cell culture (Butler and Leach, 1964; Sabin, 1967). Some mycoplasmas hemagglutinate (Van Herick and Eaton, 1945) and hemadsorb (Berg and Frothingham, 1961; Manchee and Taylor-Robinson, 1968); for some mycoplasmas, this can be prevented by neuraminidase activity (Gesner and Thomas, 1966; Manchee and Taylor-Robinson, 1969a,b). Mycoplasmas affect virus growth, causing either an increase (Singer *et al.*, 1969a,b) or a decrease (Rouse *et al.*, 1963) in virus yields. Thus, mycoplasmas can alter cell function. The investigator must remain cognizant of mycoplasmas and their properties and must maintain a constant surveillance for mycoplasmal contamination in order to properly interpret the results of his study.

A. Historical Aspects

The first mycoplasma (later designated as *M. mycoides* var. *mycoides*) was isolated by Nocard and Roux (1898) and was shown to cause contagious pleuropneumonia in cattle. The organism was filterable, was considered a virus, and represented the first example of mistaken identity (see Section I). Subsequently, the pleuropneumonia organism (PPO) was grown in broth (Dujardin-Beaumetz, 1900) and on agar medium, producing a small characteristic colony (Borrel *et al.*, 1910). Years later, many microorganisms were isolated that produced small PPO-like colonies (see Sabin, 1941) and this group of organisms was called the *pleuropneumonia-like group of organisms* or *PPLO*.

II. Classification

In 1956, Edward and Freundt proposed a system of classification that was universally accepted and PPLO became *Mycoplasma* species. In 1967, mycoplasmas were given class designation, Mollicutes, with one order, Mycoplasmatales, and two families, Mycoplasmataceae and Acholeplasmataceae, each with one genus, i.e., *Mycoplasma* and *Acholeplasma* species, respectively (Edward and Freundt, 1969). Mycoplasma(s) is used as the trivial term for all *Mycoplasma* and *Acholeplasma* species belonging to the class Mollicutes. The class designation separated mycoplasmas (see Razin, 1972) taxonomically from bacteria, rickettsia, chlamydiae, viruses, fungi (yeasts), and protozoans on the basis of the following mycoplasma properties: growth in cell-free medium; absence of cell wall; smallest forms being 220 nm or less; containing both DNA and RNA; genome size; percent guanine plus cytosine (GC) base ratio of DNA; having metabolic systems; requiring sterols (cholesterol); growth being inhibited by antiserums alone; growth being inhibited by antibiotics; disease not being transmitted by vector; and cell not multiplying intracellularly. The salient properties and the procedures recommended for characterizing new species of mycoplasmas have been proposed by the Subcommittee on Taxonomy of Mycoplasmatales (1972).

A. Properties of Mycoplasmas

Identification of mycoplasmas is based on cellular (no cell wall, bounded by a three-layered unit cell membrane) (Figs. 7–18)* and colonial morphology (Figs. 1–4), filterability and size (< 220 nm), and the absence of reversion to bacteria. The mycoplasma cell is pleomorphic and varies in size and shape (Figs. 8–18), producing spherules (Figs. 13–18) or filaments (Figs. 9–11) with branching forms. Filtration (Tully *et al.*, 1970; DelGiudice *et al.*, 1971; Barile and DelGiudice, 1972; Barile *et al.*, 1972) and electron microscopy (Anderson, 1969) studies indicate that their size ranges from 0.22 to 2 μm (Figs. 9–18), and that a small portion (0.1–1.0%) of each culture is filterable (< 220–300 nm). Mycoplasmas have both DNA and RNA, have genome sizes from 0.45×10^9 to 1×10^9 daltons, can code for from 600 to 1000 proteins or enzymes, have GC base ratios of DNA from 24 to 40% (McGee *et al.*, 1967; Neimark, 1967, 1970; Williams *et al.*, 1969), and repre-

*See pages 163 through 172 for all figures.

sent the smallest free-living organism (Pirie, 1969; Morowitz, 1969). Mode (s) of reproduction are not well established (Freundt, 1969; Anderson, 1969) but there is good evidence that some mycoplasmas replicate by binary fission (Kelton, 1962; Morowitz and Maniloff, 1966; Maniloff and Morowitz, 1967). The evidence for replication by filament formation with condensation and constriction of blebs (budding) (Ørskov, 1942; Freundt, 1958; 1969; Anderson and Barile, 1966) (Figs. 9–11) or by condensation of internal granules or dense spherules (Anderson and Barile, 1966; Anderson, 1969; Freundt, 1969) (Figs. 14–16) is not universally accepted. Some investigators believe that budding, bleb, and spherule formation may be caused by artifacts. Also, Maniloff *et al.* (1965) have shown that blebs do not contain high concentrations of DNA.

Colonies are microscopic, ranging from 20 μm (T strains) to 500 μm in diameter (Figs. 1–4), grow beneath the surface of agar (Fig. 4), and have either a fried-egg (Fig. 1) or a granular (Fig. 2) appearance. Mycoplasmal colonies are readily distinguished from small, young bacterial colonies (Fig. 3) by scraping a bacteriological loop across the surface of the colony. Because mycoplasmal colonies grow embedded in agar (Fig. 4), they cannot be scraped away. However, some colonies of *M. pulmonis* and some freshly isolated colonies, e.g., *M. conjunctivae* (Barile *et al.*, 1972), do not grow deeply embedded in agar.

All mycoplasmas are resistant to penicillin (mycoplasmas do not have cell walls and penicillin inhibits cell wall synthesis), most are *resistant* to streptomycin, and most are *sensitive* to the broad-spectrum antibiotics (Newnham and Chu, 1965) when tested in cell-free, protein-free, *in vitro* systems. All mycoplasmas studied are sensitive to heat (Smith and Sasaki, 1958; Hayflick, 1960; Kim *et al.*, 1966), ether (Butler and Leach, 1964), chloroform (Barile, unpublished data), formaldehyde (Jensen *et al.*, 1965), β-propiolactone (Roberts, 1964b), ultraviolet irradiation (Holmgren and Smith, 1959; Wright and Bailey, 1969; Furness, 1969), and photodynamic inactivation (Grabowski *et al.*, 1969). Some of the biophysical characteristics of mycoplasmas were reviewed by Clyde (1969).

B. Family and Species Designation

The two recognized families are distinguished on the basis of their sterol (cholesterol) or serum requirements (Razin and Tully, 1970; Edward, 1971). The Mycoplasmataceae require sterols, whereas the Acholeplasmataceae do not. Species designation for *Mycoplasma* and *Acholeplasma* (see the Subcommittee's report, 1972) is based on the following properties: (*a*) antigenicity, using two or more of several serologic procedures, including growth inhibition (GI) (Clyde, 1964), metabolic inhibition (MI) of arginine

(Purcell *et al.*, 1966b) or dextrose (Purcell *et al.*, 1966a; Taylor-Robinson *et al.*, 1966), plate epi-immunofluorescence (FA) (DelGiudice *et al.*, 1967; Barile and DelGiudice, 1972), and complement fixation (CF) (Taylor-Robinson *et al.*, 1963; Lemcke, 1964b) tests; (*b*) biochemical activity, including glucose fermentation, arginine and urea hydrolysis (Barile *et al.*, 1968; Aluotto *et al.*, 1970), and carotenoid pigment production (Smith, 1963) for the *Acholeplasma* species only; (*c*) electrophoretic patterns of cell proteins (Razin, 1968; Theodore *et al.*, 1970; 1971); (*d*) genetic relatedness using nucleic acid homology (Reich *et al.*, 1966; Somerson *et al.*, 1967; Somerson and Weissman, 1969) and percent GC base ratio of DNA (Neimark, 1967, 1970; McGee *et al.*, 1967); (*e*) genome size (Morowitz *et al.*, 1967; Morowitz, 1969; Bak *et al.*, 1969); (*f*) mycoplasma-cell interactions (Section V), including hemolysis (Fig. 5) (Somerson *et al.*, 1963, 1965; Clyde 1963a), hemadsorption (Fig. 6) (Berg and Frothingham, 1961; DelGiudice and Pavia, 1964; Manchee and Taylor-Robinson, 1968), adsorption to epithelial (Collier and Clyde, 1971) and various tracheal cells (Collier *et al.*, 1969; 1971; Cherry and Taylor-Robinson, 1970a,b; Butler and Ellaway, 1971), and cytopathic effects (Sabin, 1967); and (*g*) pathogenicity. One species, *Mycoplasma neurolyticum*, produces a potent neurotoxin that causes a characteristic neurologic disease (rolling disease) in mice (Findlay *et al.*, 1938; Tully and Ruchman, 1964). The 55 distinct species or serotypes of mycoplasmas recognized are listed in Table I.

III. Detecting Cell Culture Contamination

A. Direct Culture Procedures

1. *Isolation of Mycoplasmas*

The culture procedures we developed include (*a*) a standard culture procedure for isolating mycoplasmas from contaminated cell cultures; (*b*) a more sensitive, large specimen-broth culture procedure for isolating low orders of mycoplasmas from infected or contaminated specimens, e.g., commercial bovine serums; and (*c*) a semisolid broth culture procedure for screening cell cultures for microbial (mycoplasmas, bacteria, fungi) contamination.

a. *Standard Culture Procedure*. The procedure (Barile *et al.*, 1966a) used successfully for years for isolating mycoplasmas from contaminated cell cultures is as follows: the Edward (1947)-Hayflick (1965) formula consisting of 70 parts basal medium (3% brain–heart infusion or trypticase soy), 20 parts of horse serum (Baltimore Biologic Laboratories, Tissue Culture Select), and 10 parts of a 25% autoclaved extract of fresh yeast (Chanock

TABLE I
The Established Species and Distinct Serogroups of Mycoplasma and Acholeplasma

Species and strain
Primate origin (10)
M. pneumoniae, FH[b]
M. hominis, PG21[a]
M. fermentans, PG18[a,b]
M. salivarium, PG20[a]
M. orale type 1, CH19299[a]
M. orale type 2, CH20247[a]
M. orale type 3, DC333[a]
M. primatum, HRC292 (Navel)[a]
M. lipophilum, MaBy[a]
Mycoplasma sp., Simian HRC 291[b]
Bovine origin (12)
M. mycoides var. *mycoides*, PG1[b]
M. bovigenitalium, PG11[c]
M. bovirhinis, PG43[b]
M. agalactiae var. *bovis* (*M. bovimastitidis*), Donetta[c]
Bovine group 6 (*Acholeplasma* sp.) Squire (PG49)[b]
Bovine group 7, N29 (PG50)[b]
Bovine group 8, D12 (PG51)[a]
Bovine group 9 (*M. dispar*), 462/2[b]
Bovine group 11, M165/69[a,b]
Mycoplasma sp., DBS 188 (California calf)[b]
Mycoplasma sp. HRC 70–159[b]
Mycoplasma sp. HRC 213[b]
Ovine and caprine origin (7)
M. mycoides var. *capri*, PG3[b]
M. agalactiae, PG2[c]
M. arginini, G230[a]
M. conjunctivae, HRC 581[b]
Mycoplasma sp., DBS 189 (California goat)[b]
Mycoplasma sp., BBL-G145[a]
Mycoplasma sp., KS-1 (goat strain)[b]
Swine origin (3)
M. hyosynoviae[b]
M. hyorhinis, GDL[b]
M. suipneumoniae (*M. hyopneumoniae*)[b]
Murine origin (3)
M. neurolyticum type A[b]
M. pulmonis, PG-34[b]
M. arthritidis, PG6, PG27[a]
Canine origin (5)
M. spumans, PG13[a]
M. canis, PG14[b]
M. maculosum, PG15[a]

Table I (cont.)

Species and strain
M. edwardii, PG24[b]
Mycoplasma sp., HRC 689[b]
Feline origin (4)
M. felis, 27[b]
M. gatae, KDC[a]
M. feliminutum, Ben[b]
Mycoplasma sp., KDA[b]
Guinea pig origin (1)
M. caviae, G122[b]
Avian origin (6)
M. gallisepticum, PG31[b]
M. gallinarum, PG16[a]
M. iners, PG30[a]
M. anatis, 1340[b]
M. meleagridis, 17529[a]
M. synoviae, WVU1853[b]
Acholeplasma (3)
A. laidlawii, PG8, PG9[b]
A. granularum, BTS 39[b]
A. axanthum, S 743[b]
T Strains (1)
T 960[d]

[a]Arginine user.
[b]Dextrose fermenter.
[c]Negative for arginine, dextrose, and urea.
[d]Urea user.

et al., 1962) (Standard Brand Yeast No. 2040) was supplemented with energy sources (0.5% dextrose and/or 0.5% l-arginine hydrochloride), 0.002% DNA (thymic nucleic acid, Mann) or diphosphopyridine nucleotides (DPN), vitamins [as used in Eagle's (1959) tissue culture medium], antibacterial agents (1:2000 parts thallium acetate and 1000 units/ml of penicillin G) and a dye indicator (0.002% Phenol Red) buffered at pH 7.2–7.4. It should be noted that fresh horse serum is toxic for certain mycoplasmas, e.g., *M. hyorhinis*. When indicated, horse serum is heat-inactivated at 56°C for 30 minutes before it is added to medium. The medium can be modified as follows: add 0.05% Noble agar for semisolid broth medium (Section III, A, 1, d) or add 1.0% Noble agar for solid medium; or add 0.5% urea, delete dextrose and arginine, and adjust pH to 6.0 for growth of T strains (Shepard, 1969). For culture, 1 ml and 0.1 ml of cell culture suspension or of exhausted "pour off" medium are inoculated into 10 ml of standard broth and onto 10 ml of agar medium, respectively. The broth cultures are incubated aerobically at 36 ± 1°C and then subcultured to agar media when the culture becomes turbid or at weekly intervals for 4 weeks. Duplicate agar cultures are incubated aerobically or in a 5% carbon dioxide in nitrogen (5% CO_2 in N_2) atmosphere (Barile *et al.*, 1958). Agar cultures are examined for mycoplasma colonies (Figs. 1–4) for at least 3 weeks before considered negative. For subculture, blocks (1 cm^2) of agar culture are inoculated into broth or onto agar media by friction transfer; i.e., the agar culture is placed colony-face down and pushed across the surface of the agar medium. Because mycoplasmas lose viability on prolonged incubation, early (2- to 7-day-old) broth and agar cultures are used for subculture. Agar plates are placed in jars and/or incubated in humidified incubators to prevent desiccation.

b. *Large Specimen Broth Culture Procedure.* Twenty-five ml of liquid specimen (e.g., commercial bovine serums) or pieces of unminced tissue specimen are inoculated into 100 ml of standard broth medium (Barile and Kern, 1971). Tissues are not homogenized or minced, which would release mycoplasmacidal substances (Tully and Rask-Nielson, 1967; Barker and Patt, 1967; Kaklamanis *et al.*, 1969; DelGiudice *et al.*, 1969; Barile *et al.*, 1970). Larger inoculums may be used to increase sensitivity; i.e., 50 or 100 ml of specimen are inoculated into 400 ml or more of broth medium. The broth medium is more sensitive than solid medium for isolating small numbers of mycoplasmas because larger quantities of specimens can be cultured. The broth culture is incubated at 36 ± 1°C and subcultured to agar media when the broth becomes turbid or at weekly intervals for 4 weeks. Duplicate agar media are inoculated; one plate is incubated aerobically and one in a 5% CO_2 in N_2 atmosphere. Agar cultures are examined for 3 weeks before the culture is considered negative. Each primary isolation-broth culture is stored frozen at −70°C and maintained for mycoplasma identification studies (Section II).

c. *Superiority of a 5% CO_2 Atmosphere.* The 5% CO_2 in N_2 atmosphere was first used for the growth of mycoplasmas by Barile *et al.* (1958), who subsequently showed that the 5% CO_2 in N_2 atmosphere can provide a superior environment for primary isolation of mycoplasmas from contaminated cell cultures (Barile, 1962a; Barile and Schimke, 1963) and from other animal tissues (Barile *et al.*, 1972). As shown in Table II, mycoplasmas were isolated more frequently when agar media were incubated in a 5% CO_2 in N_2 atmosphere (Barile and Schimke, 1963). However, to obtain the maximum number of isolations, the cell culture specimen must be inoculated into broth, and also onto agar media, in duplicate; one plate is incubated aerobically and one in a 5% CO_2 in N_2 atmosphere. The large specimen-broth culture procedure (Section III, 1, b) is also used when a more sensitive culture procedure is required for isolating small numbers of mycoplasmas.

TABLE II
Primary Isolation of Mycoplasmas from Contaminated Cell Cultures[a]

Growth atmosphere	Mycoplasmal agar media: BYE media[b] containing: Human blood	Mycoplasmal agar media: BYE media[b] containing: Horse serum	Mycoplasmal agar media: Hayflick[c] medium
Aerobic	22/73 (30)[d]	8/73 (11)	18/73 (25)
5% CO_2 in N_2 (anaerobic)[b]	27/73 (37)	25/73 (34)	27/73 (37)

[a]From Barile and Schimke (1963).
[b]Barile *et al.* (1958).
[c]Chanock *et al.* (1962).
[d]Number positive/number tested (percent).

d. *Semisolid Broth Procedure.* We have developed a semisolid broth culture procedure for screening cell cultures for microbial (mycoplasmas, bacteria, and fungi) contamination. To prepare semisolid broth medium (Beveridge, 1943; Edward, 1950), 0.05% of Noble agar is added to the broth medium (Section III, A, 1,a) which is dispensed in screw-capped tubes. The antibacterial agents (penicillin and thallium acetate) are deleted from the medium. For culture, 1 ml of tissue culture specimen is inoculated into 10 ml of medium. For increased sensitivity, larger quantities of specimen may be used. The cultures are incubated at 36 ± 1°C and examined periodically for at least 4 weeks for the presence of microbial growth. Uninoculated broth tubes are included in each test and serve as negative controls. The additional agar provides an oxygen gradient and permits mycoplasmas to produce dense layers of luxuriant turbid growth with microcolony formation (Fig. 26). The arginine-using mycoplasmas cause an alkaline shift and the fermenters an acid shift in pH (Table I), providing information on the biochemical activity of the contaminant. The changes in the pH and color of the culture also facilitate detection of growth by visual examination. Thus, the procedure has several advantages: it requires very few basic materials (media and tubes); it provides an oxygen gradient; it grows most mycoplasmas, bacteria, and fungi that contaminate cell cultures; and growth can be detected readily by visual examination. Also, broth medium provides a very sensitive procedure because large quantities of specimen can be cultured. Semisolid broth is also a good stabilizing menstruum for storing mycoplasma cultures at either refrigerating (2–5°C) or freezing (−70°C) temperatures and, in fact, some cultures maintain their viability at 37°C for many weeks. However, the semisolid medium has some disadvantages; e.g., it cannot be used for identifying or classifying mycoplasmas. Mycoplasmas are identified on the basis of colony morphology and agar cultures must be used. Also, it will not grow all species of mycoplasmas but no one medium can grow all mycoplasmas. In brief, the semisolid broth culture procedure grows most microbial contaminants and provides an improved broth culture procedure for surveillance of microbial contamination of cell cultures.

B. Indirect Procedures

The indirect procedures for detecting mycoplasmas include: immunofluorescence (Barile *et al.*, 1962), histologic staining (Fogh and Fogh, 1964; Figs. 27–38), arginine deiminase activity (Barile and Schimke, 1963), thymidinĕ degradation (Horoszewicz and Grace, 1964), electron microscopy (Dmochowski *et al.*, 1967; see Chapter 9 by de Harven), and genome size (Todaro *et al.*, 1971). Although these procedures are extremely useful, each of them has, in my opinion, limitations. (*a*) Multiple immunofluorescence tests using polyvalent antiserums must be used because there are at least 16 established mycoplasma contaminants (Table VII). (*b*) The microscopic CPE observed in stained preparations (Figs. 27–38) is characteristic but not necessarily pathognomonic for mycoplasmas (presence of mycoplasma must be established by culture procedures), and the gross, macroscopic CPE frequently resembles nutritional or toxic effects (Section V, F). (*c*) Not all of the mycoplasma contaminants have the arginine deiminase pathway (Barile and Schimke, 1963; Barile *et al.*, 1966b). Nonetheless, these indirect procedures have merit and should be considered under appropriate conditions, e.g., when mycoplasmal contamination is suspected but culture procedures fail to recover mycoplasmas (see discussion in Barile and DelGiudice, 1972). Also, the indirect procedure frequently provides the first evidence of contamination; e.g., an electron micrograph may show ultrastructures resembling mycoplasmas in contaminated cell cultures (Figs. 17 and 18). However, these observations must be confirmed by direct culture procedures (Section III, A).

IV. Mycoplasmal Contamination

A. Incidence

1. Primary versus Continuous Cell Cultures

The incidence of mycoplasmal contamination of uninoculated cell cultures is summarized in Table III. Because mycoplasmas are frequently present in oral and genital tissues of man (Hayflick and Chanock, 1965) and animals and in neoplastic tissues (Barile, 1967), cell cultures prepared from

TABLE III
Incidence of Mycoplasmal Contamination of Uninoculated Cell Cultures

Cell cultures	Number positive/ Number tested,	Percent	Investigators
Primary cells	0/37	(0)	Rothblat and Morton, 1959
	12/ > 1500	(0.8)	Barile, 1962, 1968; Barile and Schimke, 1963; Barile and DelGiudice, 1972; Barile *et al.*, 1962
	0/26	(0)	Herderscheê *et al.*, 1963
	?	(3–4)	Levashov and Tsilinskii, 1964
Continuous cells	22/37	(59)	Rothblatt and Morton, 1959
	94/166	(57)	Pollock *et al.*, 1960
	140/243	(57)	Barile, 1962, 1968; Barile and Schimke, 1963; Barile and DelGiudice, 1972; Barile *et al.*, 1962
	32/55	(58)	Herderscheê *et al.*, 1963
	45/49	(92)	Rakavskaya, 1965
	52/60	(87)	Ogata and Koshimizu, 1967

these tissues have a greater contamination risk. Nonetheless, primary cell cultures are rarely contaminated (1–3%) (Barile, 1968) and the original tissues are not a major source of contamination. However, continuous cell cultures are frequently contaminated (45–92%) (Barile, 1968) (Table III) and, consequently, most contamination occurs during cell propagation and comes from outside sources. The major sources of contamination are discussed elsewhere (Section IV, B) and are of human oral and genital, bovine, and swine origins.

2. *Cell Origin and Type of Culture*

Contamination occurs in cell cultures derived from all animal species tested, including human and nonhuman primates (rhesus, vervet, and grivet monkeys; chimpanzee), bovine, swine, canine, rabbit, hamster, mouse, rat, chicken, quail, duck, fish, and insect origins. Cell cultures derived from plant tissues and from all types of animal cells and tissue cultures are subject to mycoplasmal contamination, including primary, continuous, diploid, and heteroploid cell cultures; tissue cultures; fibroblastic, epithelial, lymphocytic, and macrophagocytic cultures; cells grown in monolayer and suspension; and cells derived from normal, infected, or neoplastic tissues. The mycoplasma titers for most contaminants range from 10^5 to 10^8 colony-forming units (CFU) per milliliter of medium fluids, regardless of cell type and origin. Occasionally, the cell culture may be resistant to mycoplasmal infection; Butler and Leach (1964) reported that *M. bovigenitalium* failed to grow in HEP-2 cells.

3. Role of Antibiotics and Serum

Contamination occured more frequently in cell cultures grown in media containing antibiotics and/or serum. In our surveys, cells grown in medium with antibiotics had 72% contamination, whereas cells grown in medium without antibiotics had 7% contamination (Barile *et al.*, 1962). Antibiotics might have provided a false sense of security and might have been used *in lieu* of rigid sterile procedures. Although only a small number of cell cultures grown in serum-free media were examined, none of these six cultures were contaminated, whereas 48 of 86 (56%) cell cultures grown in media with serums were contaminated (Barile *et al.*, 1962). Cell cultures grown in serum might have provided a better medium for the growth of mycoplasmas or medium free of serum might have eliminated a possible source of contamination.

B. Sources

1. Commercial Bovine Serums

Barile and Kern (1971) and Barile and DelGiudice (1972) have shown that commercial bovine serum is the major source of bovine mycoplasmal contamination of cell cultures. To date, of 677 lots examined, 235 mycoplasmas have been isolated, including 169 from raw unprocessed lots of calf (22) and fetal bovine (147) serums and 66 from final lots of serums (Table IV). Successful isolations have been made only when the large specimen broth procedure (Section II, A, 1, a) is used. The standard inoculum-culture procedures (Section III, A, 1, a) used successfully for isolating 1063 mycoplasmas from contaminated cell cultures (Table VII) have failed to isolate mycoplasmas from commercial serums (see Barile and Kern, 1971). However, most cell cultures are grossly contaminated, whereas most lots of bovine serums have low orders of contamination (Barile and DelGiudice, 1972).

The mycoplasmas isolated from commercial bovine serums are summarized in Table V. Of 151 mycoplasmas examined, 65 (43%) are identified as *A. laidlawii*, 56 (37%) are *M. arginini*, 23 (15%) are unclassified strains of *Acholeplasma*, and one is *A. axanthum*. Six (4%) are unclassified strains of *Mycoplasma*: three of these are related to Bovine group 8 (D12) (Leach, 1967; Cottew and Leach, 1969) (Table I), and three to Bovine group M165/69. At least six different species or serotypes of mycoplasmas have been isolated from contaminated commercial bovine serums. Six serum lots have had mixed contamination with *M. arginini* and *A. laidlawii*. Also, preliminary

TABLE IV
Isolation of Mycoplasmas from Commercial Bovine Serums

Serums[a]		
Source	Origin	Isolations of mycoplasmas[b]
Supplier	Calf	22/55
Supplier	Fetal bovine	147/384
Manufacturer	Fetal bovine	66/238
Total		235/677

[a] *Supplier* denotes unprocessed sublots of raw serums and *manufacturer* denotes final lots of serums produced for market.

[b] Number positive/number tested. These data do not necessarily reflect the incidence of contamination as many lots were selected because contamination was suspected.

evidence indicates that commercial bovine serums may be the source of swine mycoplasmal contamination; I have isolated one strain of *M. hyorhinis* from one raw lot of fetal bovine serum. Thus, commercial bovine serum

TABLE V
Identification of Mycoplasmas Isolated from Commercial Bovine Serums[a]

Species	Number of isolations	Percent
Acholeplasma laidlawii	65	(43)[b]
Mycoplasma arginini	56	(37)
Acholeplasma sp.[c]	23	(15)
Mycoplasma sp.[d]	6	(4)
Acholeplasma axanthum	1	(1)
Total	151	

[a] Species designation was established for 151 mycoplasmas isolated from 145 of the contaminated lots (Table IV). Two mycoplasmas were isolated from each of six serum lots with mixed contamination.

[b] Numbers in parentheses are percents.

[c] These 23 strains were unrelated to the three established species and may represent new species.

[d] These six strains were related to two bovine serogroups: three were related to bovine group 11 (M165/69) (see Table I) and three bovine group 8 (D12).

is a major source of cell culture contamination and serum should be treated prior to use to inactivate or to remove potential mycoplasmal (see Section IV, E) and viral contaminants (Molander *et al.*, 1972; see Chapter 10 by Kniazeff).

2. *Origin and Identification of Mycoplasma Species Contamination*

Table VI lists the species of mycoplasmas isolated from uninoculated, contaminated cell cultures by various investigators. Table VII summarizes our findings on the prevalence of mycoplasma species in contaminated cell cultures (Barile *et al.*, unpublished data). A total of 1063 mycoplasmas have

TABLE VI
Isolation and Identification of Mycoplasma and Acholeplasma Species from Uninoculated Cell Cultures: A Survey

Species	Normal habitat	Investigators
M. orale type 1	Human, oral	Herderscheê *et al.*, 1963; Lemcke, 1964a,b; Somerson and Cook, 1965; Hayflick and Stanbridge, 1967; Barile and DelGuidice, 1972
M. hominis	Human, genital (also primates, oral)	L. H. Collier, 1957; Edward, 1960, Coriell *et al.*, 1960; Bailey *et al.*, 1961; Barile *et al.*, 1962; Herderscheê *et al.*, 1963; Lemcke, 1964b; Hayflick and Stanbridge, 1967; Barile and DelGiudice, 1972
M. salivarium	Human, oral	Hayflick and Stanbridge, 1967
M. fermentans	Human, genital	Barile and DelGiudice, 1972
M. arginini	Bovine, oral, genital (also goat, sheep, cats, swine)	Barile *et al.*, 1968; Stewart and Young, 1971; Barile and DelGiudice, 1972
A. laidlawii	Bovine, genital (also chicken, duck, swine, murine, etc.)	Lemcke, 1964b; Hayflick and Stanbridge, 1967; Barile, 1968; Barile and DelGiudice, 1972
Acholeplasma sp.[a]	Bovine	Schwöbel and Leach, 1970; Barile and DelGiudice, 1972
A. granularum	Swine, nasopharynx	Tully, 1966
M. hyorhinis	Swine, nasal	Leach and Butler, 1966; Girardi *et al.*, 1966; Purcell *et al.*, 1966b; Barile, 1968; Barile and DelGiudice, 1972
M. arthritidis	Murine, nasopharynx	Bailey *et al.*, 1961; Edward and Freundt, 1965; Barile and DelGiudice 1972
M. gallisepticum	Chicken, nasopharynx	Edward, 1960; Barile and DelGiudice, 1972
M. gallinarum	Chicken, nasopharynx	Barile and DelGiudice, 1972
M. canis	Canine, oral, genital	Barile and DelGiudice, 1972

[a] Distinct, unclassified strains of *Acholeplasma* species.

TABLE VII
Prevalence of Mycoplasmas Isolated from Uninoculated Cell Cultures[a]

Species	Natural habitat	Isolations	
		Number	Totals[b]
	Human		
M. orale type 1	oral	370 (35%)	442 (42%)
M. hominis	genital	67 (6%)	
M. fermentans	genital	5 (0.5%)	
	Bovine		
M. arginini	oral, genital	239 (22%)	400 (38%)
Mycoplasma sp.[c]	oral, genital, conjunctival	80 (7.5%)	
A. laidlawii	genital	75 (7%)	
Acholeplasma sp.[d]	genital	6 (0.6%)	
	Swine		
M. hyorhinis	nasal	204 (19%)	204 (19%)
	Murine		
M. arthritidis	respiratory	5 (0.5%)	9 (0.9%)
M. pulmonis	respiratory	4 (0.4%)	
	Avian		
M. gallisepticum	respiratory	3 (0.3%)	5 (0.5%)
M. gallinarum	respiratory	2 (0.2%)	
	Canine		
M. canis	oral, genital	1 (0.1%)	3 (0.3%)
Mycoplasma sp.[e]	genital	2 (0.2%)	
Total		1063	

[a] Barile *et al.* (1973).

[b] Presents total number and percent of mycoplasmas from human, bovine, swine, murine avian, and canine origins.

[c] These 80 mycoplasmas were related to three distinct bovine serogroups: 67 were related to strain HRC 70–159, 12 to strain HRC 213, and one to strain M165/69.

[d] These six strains were unrelated to the three established species and may represent new species.

[e] These two strains were related to a distinct serogroup (canine strain HRC 689) (Barile *et al.*, 1970).

been isolated, using our standard culture procedure (Section III, A, 1, a), from over 6500 continuous and 3200 primary cell cultures examined from approximately 56 laboratories and commercial suppliers located throughout North America and Europe. These findings summarize the total experience from 1958 to 1972 of two laboratories: the author's and that of R. A. DelGiudice, Rockville, Md. Forty-two of the cell cultures have had mixed contamination with two mycoplasmas, and mixtures have been readily detected and identified by plate epi-immunofluorescence (Barile and DelGiudice, 1972). Mycoplasma contaminants fall into six groups, the human, bovine, swine, murine, avian, and canine species; thus, the major sources of contamination are of these origins. However, 99% of the contaminants have been of human, bovine, or swine origin (Table VII).

a. *Human Sources.* The human oral and genital species of mycoplasmas represent the largest group of contaminants and, therefore, the investigator and his environs are a major source of contamination. Of 442 (42%) human mycoplasmas isolated, 370 (35%) were identified as *M. orale*, type 1, 67 (6%) were *M. hominis*, and 5 were *M. fermentans*. The most common mycoplasma contaminant of cell cultures was *M. orale* type 1, which caused more than one third (35%) of the 1063 contaminations. The major vehicle of *M. orale* type 1 contamination is the mouth pipet. Saliva is introduced into the cell culture and the antibiotics present destroy the bacteria and allow the mycoplasmas to fluorish. In medium without antibiotics, the cell culture becomes grossly contaminated with bacteria and is destroyed and discarded. Contamination by the human genital mycoplasmas is probably caused by the use of faulty or inadequate sterile procedures.

b. *Bovine Sources.* The bovine mycoplasmas are the second major group of contaminants. Of 400 (38%) bovine mycoplasmas isolated, 239 (22%) have been identified as *M. arginini*; 75 (7%) are *A. laidlawii*, 80 are unclassified strains of *Mycoplasma* species [and fall into three distinct bovine (Table I) serogroups: 67 are related to strain HRC 70–159, 12 to strain HRC 213, and one to M165/69], and six are unclassified strains of *Acholeplasma* species. The unclassified strains of *Mycoplasma* and *Acholeplasma* may represent new bovine species. Commercial bovine serums (Section IV, B, 1) is the major source of bovine mycoplasmal contamination.

c. *Swine Sources.* The third major group of contaminants are the swine species of mycoplasmas, and especially *M. hyorhinis*: 204 (19%) of the 1063 contaminants have been identified as *M. hyorhinis* (Table VII). The source of swine contamination is unknown but trypsin has been incriminated (Barile, 1968; Edward, 1969) because it is prepared from swine tissues and is universally used in the preparation of cell cultures (Rous and Jones, 1916). Nonetheless, all attempts to isolate mycoplasmas from commercial trypsin have failed (see Barile and DelGiudice, 1972). Commercial bovine serums may also be a source of swine mycoplasma contamination. Because swine and cattle are frequently processed through the same slaughterhouses, bovine serum may become contaminated with swine mycoplasmas during collection and manufacture. I have isolated one strain of *M. hyorhinis* from one lot of commercial bovine serum.

d. *Original Tissue Source.* Occasionally, the original tissue used to prepare the primary cell culture is the source of contamination. The three canine mycoplasmas (one *M. canis* and two strains related to strain HRC 689) and one strain of *M. gallisepticum* have been isolated from primary canine kidney and from primary chick embryo cell culture lots, respectively (Table VII). In these few cases, the original tissues are the probable source of contamination. However, the original tissues are a minor source (1%) of contamination (Table VII).

C. Prevalence of Mycoplasma Species Contaminants

Human, bovine, and swine species of mycoplasmas have been responsible for 1046 (99%) of the contaminations observed (Table VII). The prevalence of mycoplasma contamination from these three sources have been determined for the periods 1960–1964, 1965–1968, and 1969–1972. The data presented graphically in Fig. 39 indicate that there have been three waves or "peaks" of mycoplasmal contamination.

The first peak of contamination occurred in the early 1960's and was caused by the human mycoplasmas; the major vehicle of contamination was the mouth pipet; The second peak occurred during the mid-1960's and was caused by the swine mycoplasmas (primarily *M. hyorhinis*). Of 204 *M. hyorhinis* isolated (Table VII), 192 (94%) were isolated from 1965 to 1968. By 1970 the incidence of swine contamination dropped drastically to 1% of the contaminants. The third peak of contamination began during the late 1960's and was caused by the bovine mycoplasmas.

Presently, the bovine mycoplasmas are the most frequent cause of cell culture contamination. Commercial bovine serum is the source of bovine mycoplasmal contamination (Tables IV and V). The relatively recent isolations of bovine mycoplasmas from contaminated cell cultures *vis-á-vis* the use of bovine serums as universal components of tissue culture media for more than a decade is unknown. However, a partial explanation may be found in changes made in the processing of commercial bovine serums during the past decade resulting from demands for large volumes of this product. These demands may have been responsible for changes in methods used for collection, harvest, and processing of serums at the abattoirs, by the supplier, and by the manufacturer of the final product (see Barile and Kern, 1971). Our surveys clearly indicate that commercial bovine serums that are still collected by sterile intracardiac punctures are rarely contaminated with mycoplasmas.

D. Spread of Contamination

Contamination can be spread from cell to cell either by contaminated virus pools (Carski and Shepard, 1961), or by antiserums or by other reagents commonly used in cell culture studies (see Chapter 2 by Coriell). Contamination can be spread also by aerosols, either of contaminated cell culture materials (O'Connell *et al.*, 1964) or of contaminated secretions (e.g., saliva, nasal discharges, and genital and intestinal droppings) of man and animals (Tully, 1966). For contamination of tissue cultures by cells of diverse origin, see Chapter 1 by Stulberg.

E. Preventing Mycoplasmal Contamination

The preventive measures used are designed to control the sources and the spread of contamination. In brief, the major sources of mycoplasmal contamination are of human, bovine, and swine origin (Section IV, B) and the major vehicles of spread are the aerosols of contaminated materials and the use of contaminated reagents (Section IV, D). Preventive measures used successfully for reducing the risk of mycoplasmal contamination (Coriell *et al.*, 1960; Coriell, 1962; see Chapter 2 by Coriell) include the following: (*a*) Eliminate the use of mouth pipeting; the mouth pipet is the major vehicle of human oral mycoplasmal contamination. (*b*) Use rigid sterile procedures, aseptic techniques, and good personal habits to prevent contamination by human genital mycoplasmas. (*c*) Treat commercial bovine serums, prior to use, to remove or inactivate potential contaminants; either heat serum at 56°C for 30 minutes or filter serum twice through a 220 nm filter. Commercial bovine serum is the major source of bovine mycoplasmal contamination; whenever possible, use serum substitutes instead. (*d*) Treat trypsin, prior to use, by filtering it through either a 100 nm filter once or a 220 nm filter twice. Whenever feasible, use trypsin substitutes instead. (*e*) Use absolute Cambridge-type filters to filter the affluent and effluent air. (*f*) Use laminar flow hoods (Coriell and McGarrity, 1968; McCarrity and Coriell, 1971). (*g*) Decontaminate the working area daily and the room weekly. (*h*) Discard contaminated cell cultures immediately and replace with mycoplasma-free cultures. (*i*) Monitor and treat contaminated virus pools, antiserums, and other reagents to inactivate or remove the mycoplasmal contaminants: the treatment of choice, when feasible, is either ether or chloroform inactivation. Mix 1 ml of ether with 1 ml of the contaminated virus pool, incubate the mixture at 22°–25°C overnight, remove ether by evaporation at 37°C, and test the virus pool for mycoplasmas by culture procedures (Section III, A). Another procedure used successfully is the combination of a pretested antibiotic at high concentrations with filtration, using either a 100 nm filter once (when feasible) or a 220 nm filter twice; e.g., add 10,000 μg/ml of kanamycin to the contaminated virus pool, incubate at 22°–25°C overnight, filter, and test the virus pool for mycoplasmas.

F. Eliminating Mycoplasmal Contamination

The procedures used to eliminate mycoplasmas include: (*a*) antibiotics [tetracycline (Carski and Shepard, 1961; Perlman *et al.*, 1967), chlorotetracycline (Hearn *et al.*, 1959), oxytetracycline (Robinson *et al.*, 1956; Nelson, 1960), kanamycin (Pollock *et al.*, 1960; Fogh and Hacker, 1960), tylosin (Friend *et al.*, 1966), novobiocin (Balduzzi and Charbonneau, 1964), and many others (Newnham and Chu, 1965)]; (*b*) prolonged heat at 41°C for 18 hours (Hayflick, 1960); (*c*) specific high-titered neutralizing antiserum (Barile, 1962; Pollock and Kenny, 1963; Heiderschee *et al.*, 1963); (*d*) aurothiomalate (Shedden and Cole, 1966), Tricine (Spendlove *et al.*, 1971); (*e*) Triton X (Reynolds and Hetrick, 1969); (*f*) photodynamic inactivation (Barile, 1962a); and (*g*) the combined use of an antiserum and an antibiotic (Barile, unpublished data). Although many of these reports conclude that mycoplasmas can be readily eliminated from contaminated cells, this widespread optimism is unfounded. To the best of my knowledge and experience, there is no simple, rapid, and universally effective procedure available for eliminating mycoplasmal contamination. Moreover, mycoplasmal contaminants develop resistance to antibiotics (Rahman *et al.*, 1967) and antibiotic resistance can be induced readily (Perlman *et al.*, 1967). Treatment is prolonged, is toxic to cells (Scriba, 1968), and may induce selection with serious consequences because the surviving cells may not retain the unique properties of the original culture. Thus, attempts to eliminate mycoplasmas from contaminated cells must be considered only as the last resort. Instead, the investigator must use preventive procedures (Section IV, E) to reduce the risk of contamination and must also routinely examine cell cultures in use for mycoplasmal contamination.

Contaminated cells must be discarded immediately and be replaced with mycoplasma-free cell cultures. When a cell culture cannot be replaced and must be treated, the treatment of choice, in my experience, is the combined use of a pretested and antiserum antibiotic. I would recommend either tetracycline or kanamycin. Pretested high-titered neutralizing antiserum must be used. The concentration of antibiotic is predetermined as that concentration which falls between the effective dose for the mycoplasma (Newnham and Chu, 1965) and the toxic dose for the cell culture. We have used either 10 μg/ml of tetracycline or 100 μg/ml kanamycin. The antibiotic and a 5% concentration of high-titered neutralizing antiserum are added to medium and the contaminated cell culture is grown in the presence of the antibiotic and antiserum for at least 4 weeks, with weekly subcultures and/or changes of medium. These pour-off medium fluids are examined periodically for the presence of mycoplasmas (Section III, A). In most cases, there is a rapid drop in the mycoplasma titer but low orders of mycoplasmal contamination can persist for extended periods of time (Friend *et al.*, 1966). Treatment must be continued even though the results of culture may appear to be negative. To establish an effective cure, treatment is stopped, the cell culture is grown without antibiotic and antiserum, and the cell culture must be negative for mycoplasmas after 4 weeks of cultivation. If the mycoplasmal contamination persists, the antibiotic sensitivity of the contaminant is retested. If resistance has developed, the antibiotic is replaced. The persisting mycoplasma contaminant must be identified because mixed contamination of cell cultures with two mycoplasma species is not uncommon (Table VII) (Section IV, B, 2). The procedure is repeated using an increased concentration of the antibiotic and antiserum(s).

V. Interactions between Mycoplasmas and Animal Cells

A. Hemolysis

Several pathogenic mycoplasmas produce hemolysis (Warren, 1942; Edward, 1950); these include *M. gallisepticum* (Adler, 1964; O'Malley *et al.*, 1966), *M. mycoides* var. *mycoides* (Adler, 1964), and *M. mycoides* var. *capri* (Cherry and Taylor-Robinson, 1970c). Hemolysin production has proved useful for the presumptive identification of *M. pneumoniae* (Somerson *et al.*, 1963; Clyde, 1963a) (Fig. 5). Hemolysins may also play an important role in the pathogensis of disease (Cohen and Somerson, 1967). The hemolysins of several mycoplasmas studied have been shown to be peroxides (Somerson *et al.*, 1965; Cole *et al.*, 1968; Cherry and Taylor-Robinson, 1970c).

B. Hemadsorption

Many of the pathogenic mycoplasmas adsorb erythrocytes (Manchee and Taylor-Robinson, 1968); these include *M. pneumoniae* (DelGiudice and Pavia, 1964), *M. gallisepticum* (Van Herick and Eaton, 1945), and *M. pulmonis* (Fig. 6). Mycoplasma-infected cell cultures can hemadsorb guinea pig erythrocytes (Berg and Frothingham, 1961).

C. Hemagglutination

Several pathogenic mycoplasmas can agglutinate erythrocytes and the hemagglutination inhibition test is used for the seroidentification of avian mycoplasmas (Van Herick and Eaton, 1945). The red cell binding site for some mycoplasmas is probably sialic acid because treatment of red cells with neuraminidase inhibits hemagglutination of *M. gallisepticum* (Gesner and Thomas, 1966), *M. pneumoniae*, and *M. synoviae* (Manchee and Taylor-Robinson, 1969a,b). However, it does not inhibit hemagglutination of all mycoplasmas, including *M. agalactiae*, *M. pulmonis*, and *M. bovigenitalium* (Manchee and Taylor-Robinson, 1969a,b). Thus, mycoplasmas can attach to cells in several ways (Zucker-Franklin *et al.*, 1966a,b; Manchee and Taylor-Robinson, 1969a,b; Biberfeld and Biberfeld, 1970; Collier and Clyde, 1971; Collier, 1972) (see Section, V, D).

D. Mycoplasmal Attachment to Other Cell Membranes

Mycoplasmas are frequently seen closely associated or attached to the plasma membranes of animal cells as determined by either ordinary, immunofluorescence, or electron microscopy, and in the following situations: (*a*) in *M. pneumoniae*-infected chick embryos (Liu, 1957), in infected cell cultures (Clyde, 1961, 1963b; Marmion and Goodburn, 1961; Goodburn and Marmion, 1962; Eaton *et al* 1962), and in infected hamsters (Collier, 1972); (*b*) in mycoplasma-contaminated cell cultures (Carski and Shepard, 1961; Barile *et al.*, 1962; Fogh and Fogh, 1964) (Figs. 21 and 27–38); (*c*) in electron photomicrographs of infected cell cultures (Edwards and Fogh, 1960; Anderson, 1965, 1969; Anderson and Manaker, 1966; Barile, 1965; Hummeler and Armstrong, 1967) (Figs. 17–18) and in *M. pulmonis*-infected macrophages (Jones and Hirsch, 1971; Jones *et al.*, 1972) and lung tissues of mice (Organick *et al.*, 1966; Kohn, 1971); (*d*) in infected HeLa cells, in neutrophils, eosinophils, monocytes, and lymphocytes (Zucker-Franklin *et al.*, 1966a,b); and (*e*) in *M. pneumoniae*-infected hamster and human tracheal organ cultures (Collier, 1972; Collier and Clyde, 1971; Collier *et al.*, 1969, 1971) and in *M. mycoides* var. *mycoides*-infected chicken tracheal organs (Cherry and Taylor-Robinson, 1970a,b; Butler and Ellaway, 1971).

Two of the *Mycoplasma* species examined have terminal apparatuses for attachment to mammalian cells. *Mycoplasma gallisepticum* attaches to cells by its terminal bleb (Zucker-Franklin *et al.*, 1966a,b) and *M. pneumoniae* attaches by a terminal spike (Biberfeld and Biberfeld, 1970; Collier and Clyde, 1971; Collier, 1972). Attachment apparatuses have not been noted for *M. pulmonis* (Anderson and Manaker, 1966; Jones and Hirsch, 1971; Jones *et al.*, 1972; Barile and DelGiudice, 1972) nor for other mycoplasmas.

Several mycoplasmas can stop ciliary activity of tracheal organ culture; these include *M. pneumoniae* (Collier and Clyde, 1971; Collier, 1972), *M. gallisepticum* and *M. mycoides* var. *mycoides*. Hemolysin (peroxide) was responsible for the rapid ciliary damage in the *M. mycoides* var *capri* infections but not in the *M. gallisepticum* infections (Cherry and Taylor-Robinson, 1970a,b,c). Cherry and Taylor-Robinson (1972) reported that preinfection with a nonpathogenic mycoplasma can inhibit the ciliary destructive effect of a pathogenic mycoplasma; i.e., preinfection of chicken tracheal organ cultures with *M. gallinarum* results in protection of the ciliated epithelium upon subsequent infection with *M. gallisepticum*. *Mycoplasma gallinarum* may either produce peroxidase and inhibit the destructive activity of liberated hemolysin (peroxides) on ciliary activity or it may destroy receptor sites and inhibit cell attachment by *M. gallisepticum*.

E. Chromosomal Aberrations

Chromosomal changes were observed in mycoplasma infected cell cultures, including human diploid-cell cultures (Paton *et al.*, 1965; 1967; Stanbridge *et al.*, 1969), in human amnion cells (Fogh and Fogh, 1965; 1967, see Chapter 7 by Fogh), in hamster cells (Macpherson and Russell, 1966), and in human leukocyte cultures (Aula and Nichols, 1967).

F. Cytopathic Effects of Mycoplasmal Contamination

To the untrained observer, mycoplasmal contamination of cell cultures may be insidious and may go undetected for the following reasons: (*a*) Mycoplasmas do not produce overt, turbid growth in contaminated cell cultures. (*b*) Early macroscopic cellular changes may be minimal or inapparent. (*c*) Mycoplasma contaminants compete for nutrients and frequently deplete the medium of an essential substrate (e.g., arginine; Schimke and Barile, 1963b) or they alter nucleic acid synthesis (Nardone *et al.*, 1965; Levine *et al.*, 1968; Markov *et al.*, 1969; Stanbridge *et al.*, 1969), causing toxic or nutritional cellular changes. Because these nutritional changes can be reversed by changing the medium, contamination may go unsuspected. To a large extent, the biochemical activity of the mycoplasma predetermines the nature of the cytopathology; i.e., the CPE produced depends on whether the mycoplasma utilizes arginine (Schimke and Barile, 1963b) or dextrose (VanDemark, 1969) for energy. Some mycoplasmas (Barile *et al.*, 1966b) can rapidly deplete arginine from the medium, depriving the cell culture of an essential amino acid (Morgan *et al.*, 1958; Eagle, 1959) and producing profound effects on cell morphology (Kenny and Pollock, 1963; Pollock *et al.*, 1963; Fogh *et al.*, 1971) and cell function (Sections V, G and VI). Miller *et al.* (1971) have reported that *M. arginini* (Barile *et al.*, 1968) produces cellular lysis in some but not all human lymphoblastoid cell cultures examined. They have shown that lysis can be arrested by adding arginine to the medium, suggesting that susceptibility of cell cultures to the arginine-using mycoplasmas is dependent on the arginine requirement of the contaminated culture.

Many of the mycoplasmal fermenters have been shown to produce severe CPE, characterized by development of small focuses of necrosis, lesions (Fig. 38), or plaque formations in agar overlay systems (Fig. 26). The rapid production of acidity is one of the factors responsible for the CPE produced. The cells detach from the glass and a destructive V-shaped CPE develops (Figs. 23–24) (Sabin, 1967). Certain bacterial fermenters (e.g., *Pseudomonas*) that contaminate cells can also cause the V-shaped CPE (Barile, unpublished data). The mycoplasmal fermenters known to produce severe CPE include *M. pulmonis* (Nelson, 1960), *M. hyorhinis* (Butler and Leach, 1964; Girardi *et al.*, 1966; Purcell *et al.*, 1966a; Sabin, 1967; Joncas *et al.*, 1969), *M. gallisepticum* (O'Malley *et al.*, 1961, 1966; Castrejon-Diez *et al.*, 1963; Grumbles *et al.*, 1964), *M. bovigenitalium* (Afshar, 1967), and *A. laidlawii* (Kagan *et al.*, 1969). The typical microscopic CPE produced by *M. hyorhinis*, *M. pulmonis*, *A. laidlawii*, and mycoplasma strain 14 are illustrated in Figs. 23, 24, and 27–38.

G. Inhibition of Lymphocyte Transformation

Mycoplasma pneumoniae stimulates transformation (blast formation) and mitosis in cultures of lymphocytes obtained from sensitized individuals (Leventhal *et al.*, 1969), as do other antigens (Oppenheim, 1968). In contrast, *M. hominis* (Copperman and Morton, 1966) and *M. arthritidis* (Cochrum *et al.*, 1969) inhibit lymphocyte transformation induced by phytohemmagglutinin (PHA). Barile and Leventhal (1968) have shown that the mechanism of mycoplasmal inhibition of PHA-induced transformation is a result of the depletion of arginine from the medium. Only those species that utilize arginine for energy are able to inhibit PHA stimulation. The mycoplasmal fermenters tested that do not utilize arginine do not inhibit PHA stimulation. Also, the inhibitory effect is blocked either by providing the cells with fresh medium or by adding arginine to the culture medium. The arginine-using mycoplasmas also inhibit thymidine uptake of nonstimulated lymphocyte culture, resulting in marked inhibition of cell function (Barile and Leventhal, 1968). The inhibitory effect on lymphocyte cultures is similar to the inhibitory effect of mycoplasmas on the growth of monolayer cell cultures (Kenny and Pollock, 1963) and is caused by the depletion of arginine from the medium, depriving the cells of an essential amino acid (Barile and Leventhal, 1968). These findings have been confirmed by others (Spitler *et al.*, 1968; Simberkoff *et al.*, 1969).

VI. Mycoplasma–Virus–Cell Interactions

Mycoplasmas can cause either an increase in virus yields (Singer *et al.*, 1969a,b) or a decrease in virus yields (Armstrong *et al.*, 1961; Brownstein and Graham, 1961; Rouse *et al.*, 1963; Butler and Leach, 1964; Somerson *et al.*, 1965; Kagan *et al.*, 1967; Afshar, 1967; Gafford *et al.*, 1969; Nakamura and Sakamoto, 1969; Singer *et al.*, 1970; Hargreaves and Leach, 1970; Romano and Brancata, 1970; Smirnova and Kagan, 1971) or can have no effect on virus yields (Schulze and Schlesinger, 1963; Herderscheê *et al.*, 1963; Armstrong and Paucker, 1966; Reed, 1971, 1972). In some mycoplasma–virus infections, the virus yield is unaffected but the mycoplasma yield is increased (Nakamura and Sakamoto, 1969; Reed, 1971). Nonpathogenic mycoplasmas can inhibit the pathogenicity of pathogenic mycoplasmas in tracheal organ cultures (Cherry and Taylor-Robinson, 1972). Mycoplasmas can alter the course of infection in animals; they may cause either a diminution of symptoms as seen in Marek's disease (Katzen *et al.*, 1969), or they may cause enhancement of disease, as seen in influenza A (Roberts, 1964a; Berry, 1969) and in swine adenovirus infections (Kasza *et al.*, 1969).

A. Decreased Virus Yields

1. Arginine Depletion

Mycoplasmas decrease virus yields by depleting arginine from the medium (Rouse *et al.*, 1963; Singer *et al.*, 1970) or by inhibiting cell metabolism and growth or by causing cell lysis and destruction (Butler and Leach, 1964). We have shown that mycoplasmas can deplete arginine from the medium and alter cell morphology and growth (Gill and Pan, 1970) (Sections V, F and G). These same arginine-using mycoplasmas have been shown to decrease the yield of certain arginine-requiring DNA viruses, including an intranuclear replicating virus (adenovirus type 2; Rouse *et al.*, 1963) and an intracytoplasmic (Archard and Williamson, 1971) replicating virus (vaccinia; Singer *et al.*, 1970). It would seem reasonable to suggest that the arginine-using mycoplasmas may also affect replication of other arginine-requiring DNA viruses, including adenovirus type 1 (Dubes *et al.*, 1969), adenovirus type 5 (Russell and Becker, 1968), herpesvirus (Tankersley, 1964; Becker *et al.*, 1967; Inglis, 1968), SV40 virus (Goldblum *et al.*, 1968), polyoma virus (Winters and Consigli, 1971), poliomyelitis virus (Ackermann and Wahl, 1966), and cytomegalovirus (Minamishima and Benyesh-Melnick, 1969). Arginine is not required for the early steps of virus replication, including viral DNA synthesis (Rouse and Schlesinger, 1967; Russell and Becker, 1968), but it is essential for the synthesis of viral coat protein and the production of complete infectious virion (viral progeny). Thus, the investigator examining the role of arginine in the synthesis of DNA virus must remain cognizant of mycoplasmal contamination and its effects on the results of his study.

B. Increased Virus Yields

Singer *et al.* (1969b) have shown that *M. arginini* and *M. hyorhinis* inhibits interferon production, interferon activity, and cell resistance, resulting in increased yields of Semliki-Forest virus (SFV). To illustrate this activity, the mycoplasma–virus–cell culture system must be standardized. Several factors can influence the results of study and must be considered: (*a*) Virus enhancement is only seen when low input of virus is used (Singer *et al.*, 1969b). With low virus input, a small number of cells are infected, leaving most of the cultured cells available to accept the interferon generated and to develop cell resistance to viral infection. With a high virus input, most of the cells are infected initially, and consequently, the effect of mycoplasmas on the inhibition of interferon activity and the development of cell resistance cannot be demonstrated. (*b*) The effect on cell resistance cannot be demonstrated in the agar overlay system (Gafford *et al.*, 1969; Singer *et al.*, 1969b, 1972) because the interferon generated is not able to disseminate to other cells in an agar overlay and induce cell resistance. (*c*) The number of cells in the mycoplasma-free and -infected cell cultures must be the same. (*d*) The same *Mycoplasma* species may have several properties, each of which can affect a cell culture in quite a different manner; e.g., *M. hyorhinis* can either produce CPE (Figs. 23–24, 31, 32, 36, and 38) and cause a decrease in virus yields (Butler and Leach, 1964) or can inhibit interferon production and cause an increase in virus yields (Singer *et al.*, 1969a,b, 1972). Thus, the virus–mycoplasma–cell culture system must be in proper balance in order to demonstrate the desired effect. Mycoplasmas do not produce interferon by themselves (Armstrong and Paucker, 1966; Yershov and Zhdanov, 1965; Singer *et al.*, 1969a,b) but they can inhibit interferon production and activity induced by either a virus (SFV) or by a synthetic complex RNA (polyinosinic and polycytidilic acid). In addition, mycoplasmal infection renders cells less sensitive to exogenously supplied interferon, and consequently, contamination can affect the results of interferon assay. Mycoplasmal infection of cell cultures can conceivably be used to advantage. Decreasing interferon production and activity with mycoplasmas may enhance virus yields and play a useful role in the detection of latent viruses.

In conclusion, mycoplasmas have been shown to affect cell cultures in many ways. The investigator using cell cultures must remain cognizant of mycoplasmas and their properties, must maintain a constant surveillance for mycoplasmal contamination, and must provide information in reports on procedures used and data obtained on the presence or absence of mycoplasmal contamination in order to properly interpret the results of his study.

References

Ackermann, W. W., and Wahl, D. (1966). *J. Bacteriol.* **92**, 1051.

Adler, H. E. (1964). *Amer. J. Vet. Res.* **25**, 243.

Afshar, A. (1967). *J. Gen. Microbiol.* **47**, 103.

Aluotto, B. B., Wittler, R. G., Williams, C. O., and Faber, J. E. (1970). *Int. J. Syst. Bacteriol.* **20**, 35.

Anderson, D. R. (1965). *In* "Methodological Approaches to the Study of Leukemias" (V. Defendi, ed.), pp. 113–146. Wistar Inst. Press, Philadelphia, Pennsylvania.

Anderson, D. R. (1969). *In* "The Mycoplasmatales and the L-phase of Bacteria" (L. Hayflick, ed.), pp. 365–402. Appleton, New York.

Anderson, D. R., and Barile, M. F. (1965). *J. Bacteriol.* **90**, 180.

Anderson, D. R., and Barile, M. F. (1966). *J. Nat. Cancer Inst.* **36**, 161.

Anderson, D. R., and Manaker, R. A. (1966). *J. Nat. Cancer Inst.* **36**, 139.

Archard, L. C., and Williamson, J. D. (1971). *J. Gen. Virol.* **12**, 249.

Armstrong, D., and Paucker, K. (1966). *J. Bacteriol.* **92**, 97.

Armstrong, J. A., Pereira, H. G., and Andrewes, C. H. (1961). *Virology* **14**, 276.

Aula, P., and Nichols, W. W. (1967). *J. Cell. Physiol.* **70**, 281.

Bailey, J. S., Clark, H. W., Felts, R. W., Fowler, R. C., and Brown, T. McP. (1961). *J. Bacteriol.* **82**, 542.

Bak, A. L., Black, F. T., Christiansen, C., and Freundt, E. A. (1969). *Nature* (*London*) **224**, 1209.

Balduzzi, P., and Charbonneau, J. R. (1964). *Experientia* **20**, 651.

Barile, M. F. (1962a). *Nat. Cancer Inst., Monog.* **7**, 50.

Barile, M. F. (1962b). *J. Bacteriol.* **83**, 430.

Barile, M. F. (1965). *In* "Methodological Approaches to the Study of Leukemias" (V. Defendi, ed.), pp. 171–181. Wistar Inst. Press, Philadelphia, Pennsylvania.

Barile, M. F. (1967). *Ann. N. Y. Acad. Sci.* **143**, 557.

Barile, M. F. (1968). *Nat. Cancer Inst., Monog.* **29**, 201.

Barile, M. F., and DelGiudice, R. A. (1972). "Pathogenic Mycoplasmas, Ciba Foundation Symposium," pp. 165–185. Elsevier, Amsterdam.

Barile, M. F., and Kern, J. (1971). *Proc. Soc. Exp. Biol. Med.* **138**, 432.

Barile, M. F., and Leventhal, B. G. (1968). *Nature* (*London*) **219**, 751.

Barile, M. F., and Schimke, R. T. (1963). *Proc. Soc. Exp. Biol. Med.* **114**, 676.

Barile, M. F., Yaguchi, R., and Eveland, W. C. (1958). *Amer. J. Clin. Pathol.* **30**, 171.

Barile, M. F., Malizia, W. F., and Riggs, D. B. (1962). *J. Bacteriol.* **84**, 130.

Barile, M. F., Bodey, G. P., Snyder, J., Riggs, D. B., and Grabowski, M. W. (1966a). *J. Nat. Cancer Inst.* **36**, 155.

Barile, M. F., Schimke, R. T., and Riggs, D. B. (1966b). *J. Bacteriol.* **91**, 189.

Barile, M. F., DelGiudice, R. A., Carski, T. R., Gibbs, C. J., and Morris, J. A. (1968). *Proc. Soc. Exp. Biol. Med.* **129**, 489.

Barile, M. F., DelGiudice, R. A., Carski, T. R., Yamashiroya, H. M., and Verna, J. E. (1970). *Proc. Soc. Exp. Biol. Med.* **134**, 146.

Barile, M. F., DelGiudice, R. A., and Tully, J. G. (1972). *Infect. Immunity* **5**, 70.
Barile, M. F., DelGiudice, R. A., Hopps, H. E., Grabowski, M. W., and Riggs, D. B. (1973) *N.Y. Acad. Sci.* (in press).
Barker, L. F., and Patt, J. K. (1967). *J. Bacteriol.* **94**, 403.
Becker, Y., Olshevsky, V., and Levitt, J. (1967). *J. Gen. Virol.* **1**, 471.
Berg, R. B., and Frothingham, T. E. (1961). *Proc. Soc. Exp. Biol. Med.* **108**, 616.
Berry, D. M. (1969). *Proc. Roy. Soc. Med.* **62**, 45.
Beveridge, W. I. B. (1943). *Med. J. Aust.* **2**, 479.
Biberfeld, G., and Biberfeld, P. (1970). *J. Bacteriol.* **102**, 855.
Borrel, A., Dujardin-Beaumetz, E., Geantet, P., and Jouan, C. (1910). *Ann. Inst. Pasteur, Paris* **24**, 168.
Brownstein, B., and Graham, A. F. (1961). *Virology* **14**, 303.
Butler, M., and Ellaway, W. J. (1971). *J. Comp. Pathol.* **81**, 359.
Butler, M., and Leach, R. H. (1964). *J. Gen. Microbiol.* **34**, 285.
Carski, T. R., and Shepard, M. C. (1961). *J. Bacteriol.* **81**, 626.
Castrejon-Diez, J., Fisher, T. N., and Fisher, E., Jr. (1963). *Proc. Soc. Exp. Biol. Med.* **112**, 643.
Chanock, R. M., Hayflick, L., and Barile, M. F. (1962). *Proc. Nat. Acad. Sci. U.S.* **48**, 41.
Cherry, J. D., and Taylor-Robinson, D. (1970a). *Appl. Microbiol.* **19**, 658.
Cherry, J. D., and Taylor-Robinson, D. (1970b). *Infect. Immunity* **2**, 431.
Cherry, J. D., and Taylor-Robinson, D. (1970c). *Nature (London)* **228**, 1099.
Cherry, J. D., and Taylor-Robinson, D. (1972). *Bacteriol. Proc.* p. 54.
Clyde, W. A., Jr. (1961). *Proc. Soc. Exp. Biol. Med.* **107**, 715.
Clyde, W. A., Jr. (1963a). *Science* **139**, 55.
Clyde, W. A., Jr. (1963b). *Proc. Soc. Exp. Biol. Med.* **112**, 905.
Clyde, W. A., Jr. (1964). *J. Immunol.* **92**, 958.
Clyde, W. A., Jr. (1969). *In* "The Mycoplasmatales and the L-phase of Bacteria" (L. Hayflick, ed.), pp. 349–363. Appleton, New York.
Cochrum, K. C., Dykman, L., Najarian, J. S., and Fudenberg, H. H. (1969). *In* "Proceedings of the Third Annual Leukocyte Culture Conference" (W. O. Rieke, ed.), pp. 169–176. Appleton, New York.
Cohen, G., and Somerson, N. L. (1967). *Ann. N. Y. Acad. Sci.* **143**, 85.
Cole, B. C., Ward, J. R., and Martin, C. H. (1968). *J. Bacteriol.* **95**, 2022.
Collier, A. M. (1972). "Pathogenic Mycoplasmas, Ciba Foundation Symposium," pp. 307–327. Elsevier, Amsterdam.
Collier, A. M., and Clyde, W. A., Jr. (1971). *Infect. Immunity* **3**, 694.
Collier, A. M., Clyde, W. A., Jr., and Denny, F. W. (1969). *Proc. Soc. Exp. Biol. Med.* **132**, 1153.
Collier, A. M., Clyde, W. A., Jr., and Denny, F. W. (1971). *Proc. Soc. Exp. Biol. Med.* **136**, 569.
Collier, L. H. (1957). *Nature (London)* **180**, 757.
Copperman, R., and Morton, H. E. (1966). *Proc. Soc. Exp. Biol. Med.* **123**, 790.
Coriell, L. L. (1960). *Ann. N. Y. Acad. Sci.* **79**, 610.
Coriell, L. L. (1962). *Nat. Cancer Inst.*, Monogr. **7**, 33.
Coriell, L. L., and McGarrity, G. J. (1968). *Appl. Microbiol.* **16**, 1895.
Coriell, L. L., Fabrizio, D. P., and Wilson, S. R. (1960). *Ann. N. Y. Acad. Sci.* **79**, 574.
Cottew, G. S., and Leach, R. H. (1969). *In* "The Mycoplasmatales and the L-phase of Bacteria" (L. Hayflick, ed.), pp. 527–570. Appleton, New York.
Cross, G. F., Goodman, M. R., Chatterji, J., Beswisk, T. S. L., and Chapman, J. A. (1970). *J. Gen. Microbiol.* **8**, 77.

DelGiudice, R. A., and Pavia, R. (1964). *Bacteriol. Proc.* p. 71.
DelGiudice, R. A., Robillard, N. F., and Carski, T. R. (1967). *J. Bacteriol.* **93**, 1205.
DelGiudice, R. A., Carski, T. R. Barile, M. F., Yamashiroya, H. M., and Verna, J. E. (1969). *Nature* (*London*) **222**, 1088.
DelGiudice, R. A., Carski, T. R., Barile, M. F., Lemcke, R. M., and Tully, J. G. (1971). *J. Bacteriol.* **108**, 439.
Dmochowski, L., Dreyer, D. A., Grey, C. E., Hales, R., Langford, P. L., Pipes, F., Recher, L., Seman, G., Shively, J. A., Shullenberger, C. C., Sinkovics, J. G., Taylor, H. G., Tessmer, C. F., and Yumoto, T. (1967). *Ann. N.Y. Acad. Sci.* **143**, 578.
Dubes, G. R., Moyer, K. B., Halliburton, B. L., and Van deBogart, R. K. (1969). *Acta Virol.* (*Prague*) English Ed. **13**, 8.
Dujardin-Beaumetz, E. (1900). "Le microbe de la péripneumonie et sa culture, étude bacteriologique d'un microorganisme á la limite de la visibilite (Oc. Doin, ed.). Paris.
Eagle, H. (1959). *Science* **130**, 432.
Eaton, M. D., Farnham, A. E., Levinthal, J. D., and Scala, A. R. (1962). *J. Bacteriol.* **84**, 1330.
Edward, D. G. ff. (1947). *J. Gen. Microbiol.* **1**, 238.
Edward, D. G. ff. (1950). *J. Gen. Microbiol.* **4**, 311.
Edward, D. G. ff. (1960). *Ann. N.Y. Acad. Sci.* **79**, 608.
Edward, D. G. ff. (1969). *Progr. Immunobiol. Stand.* **3**, 17.
Edward, D. G. ff. (1971). *J. Gen. Microbiol.* **69**, 205.
Edward, D. G. ff., and Freundt, E. A. (1956). *J. Gen. Microbiol.* **14**, 197.
Edward, D. G. ff., and Freundt, E. A. (1965). *J. Gen. Microbiol.* **41**, 263.
Edward, D. G. ff., and Freundt, E. A. (1969). *In* "The Mycoplasmatales and the L-phase of Bacteria" (L. Hayflick, ed.), pp. 147–200. Appleton, New York.
Edwards, G. A., and Fogh, J. (1960). *J. Bacteriol.* **79**, 267.
Findlay, G. M., Kleineberger, E., MacCallum, R. O., and MacKenzie, R. D. (1938). *Lancet* **235**, 1511.
Fogh, J., and Fogh, H. (1964). *Proc. Soc. Exp. Biol. Med.* **117**, 899.
Fogh, J., and Fogh, H. (1965). *Proc. Soc. Exp. Biol. Med.* **119**, 233.
Fogh, J., and Fogh, H. (1967). *Proc. Soc. Exp. Biol. Med.* **126**, 67.
Fogh, J., and Fogh, H. (1969). *Ann. N. Y. Acad. Sci.* **172**, 15.
Fogh, J., and Hacker, C. (1960). *Exp. Cell. Res.* **21**, 242.
Fogh, J., Holmgren, N. B., and Ludovici, P. P. (1971). *In Vitro* **7**, 26.
Freundt, E. A. (1958). "The Mycoplasmataceae (The Pleuropneumonia Group of Organisms), Morphology, Biology and Taxonomy." Munksgaard, Copenhagen.
Freundt, E. A. (1969). *In* "The Mycoplasmatales and the L-phase of Bacteria" (L. Hayflick, ed.), pp. 281–316. Appleton, New York.
Friend, C., Patuleia, M. C., and Nelson, J. B. (1966). *Proc. Soc. Exp. Biol. Med.* **121**, 1009.
Furness, G. (1969). *Appl. Microbiol.* **18**, 360.
Gafford, L. G., Sinclair, F., and Randall, C. C. (1969). *Virology* **37**, 464.
Gesner, B., and Thomas, L. (1966). *Science* **151**, 590.
Gill, P., and Pan, J. (1970). *Can. J. Microbiol.* **16**, 415.
Girardi, A. J., Hamparian, V. V., Somerson, N. L., and Hayflick, L. (1966). *Proc. Soc. Exp. Biol. Med.* **120**, 760.
Goldblum, N., Ravid, Z., and Becker, Y. (1968). *J. Gen. Microbiol.* **3**, 143.
Goodburn, G. M., and Marmion, B. P. (1962). *J. Gen. Microbiol.* **29**, 271.
Goodwin, R. F. W., Pomeroy, A. P., and Whittlestone, P. (1965). *Vet. Rec.* **77**, 1247.
Grabowski, M. W., Riggs, D. B., Moore, D. E., and Barile, M. F. (1969). *Bacteriol. Proc.* p. 33.

Grumbles, L. C., Hall, C. F., and Cummings, G. (1964). *Avian Dis.* **8**, 274.
Hakala, M. T., Holland, J. F., and Horoszewicz, J. S. (1963). *Biochem. Biophys. Res. Commun.* **11**, 466.
Hargreaves, F. D., and Leach, R. H. (1970). *J. Med. Microbiol.* **3**, 259.
Hayflick, L. (1960). *Nature (London)* **185**, 783.
Hayflick, L. (1965). *Tex. Rep. Biol. Med.* **23**, Suppl. 1, 285.
Hayflick, L., and Chanock, R. M. (1965). *Bacteriol. Rev.* **29**, 185.
Hayflick, L., and Stanbridge, E. (1967). *Ann. N. Y. Acad. Sci.* **143**, 608.
Hearn, H. J., Officer, J. E., Elsner, V., and Brown, A. (1959). *J. Bacteriol.* **78**. 575.
Herderscheê, D., Ruys, A. C., and van Rhijn, G. R. (1963), *Antonie van Leeuwenhoek; J. Microbiol. Serol.* **29**, 368.
Holmgren, N. B., and Smith, N. H. (1959). *Bacteriol. Proc.* p. 58.
Horoszewicz, J. S., and Grace, J. T., Jr. (1964). *Bacteriol. Proc.* p. 131.
Hummeler, K., and Armstrong, D. (1967). *Ann. N. Y. Acad. Sci.* **143**, 622.
Inglis, V. B. M. (1968). *J. Gen. Virol.* **3**, 9.
Jensen, K. E., Senterfit, L. B., Chanock, R. M., Smith, C. B., and Purcell, R. H. (1965). *J. Amer. Med. Ass.* **194**, 248.
Joncas, J., Chagnon, A., Lussier, G., and Pavilanis, V. (1969). *Can. J. Microbiol.* **15**, 451.
Jones, T. C., and Hirsch, J. G. (1971). *J. Exp. Med.* **133**, 231.
Jones, T. C., Yeh, S., and Hirsch, J. G. (1972). *Proc. Soc. Exp. Biol. Med.* **139**, 464.
Kagan, G. Y., Ershov, F. I., Rakovskaya, I. V., Tsareva, A. T., and Zhdanov, V. M. (1967). *Vop. Virusol.* **4**, 478.
Kagan, G. Y., Fleer, G. P., Rakovskaya, I. V., Smirnova. T. D., and Chumakov, M. P. (1969). *Zh. Mikrobiol., Epidemiol. Immunobiol.* **46**, 63.
Kaklamanis, E., Thomas, L., Stavropoulos, K., Borman, I., and Boshivitz, C. (1969). *Nature (London)* **221**, 860.
Kasza, L., Hodges, R. T., Betts, A. O., and Trexler, P. C. (1969). *Vet. Rec.* **84**, 262.
Katzen, S., Matsuda, K., and Reid, B. L. (1969). *Poultry Sci.* **48**, 1504.
Kelton, W. H. (1962). *J. Bacteriol.* **83**, 948.
Kenny, G. E., and Pollock, M. E. (1963). *J. Infec. Dis.* **112**, 7.
Kim, K. S., Clyde, W. A., and Denny, F. W. (1966). *J. Bacteriol.* **92**, 214.
Kohn, D. F. (1971). *Lab. Anim. Sci.* **21**, 856.
Leach, R. H. (1967). *Ann. N. Y. Acad. Sci.* **143**, 305.
Leach, R. H., and Butler, M. (1966). *J. Bacteriol.* **91**, 934.
Lemcke, R. (1964a). *J. Hyg.* **62**, 351.
Lemcke, R. (1964b). *J. Hyg.* **62**, 199.
Levashov, V. S., and Tsilinskii, Y. Y. (1964). *Tr. Inst. Poliomielita Verusn Entsefalitov* **5**, 150; *Ref. Zh., Biol.* No. 20B70 (1965) (translation).
Leventhal, B. G., Smith, C. B., Carbone, P. P., and Hersch, E. M. (1969). *In* "Proceedings of the Third Annual Leucocyte Culture Conference" (W. O. Rieke, ed.), pp. 519–532. Appleton, New York.
Levine, E. M., Thomas, L., McGregor, D., Hayflick, L., and Eagle, H. (1968). *Proc. Nat. Acad. Sci. U.S.* **60**, 583.
Liu, C. (1957). *J. Exp. Med.* **106**, 455.
McGarrity, G. J., and Coriell, L. L. (1971). *In Vitro* **6**, 257.
McGee, Z. A., Rogul, M., and Wittler, R. G. (1967). *Ann. N. Y. Acad. Sci.* **143**, 21.
Macpherson, I. (1966). *J. Cell Sci.* **1**, 145.
Macpherson, I., and Russell, W. (1966). *Nature (London)* **210**, 1343.
Manchee, R. J., and Taylor-Robinson, D. (1968). *J. Gen. Microbiol.* **50**, 465.
Manchee, R. J., and Taylor-Robinson, D. (1969a). *J. Bacteriol.* **98**, 914.

Manchee, R. J., and Taylor-Robinson, D. (1969b). *Brit. J. Exp. Pathol.* **50**, 66.
Maniloff, J., and Morowitz, H. J. (1967). *Ann. N. Y. Acad. Sci.* **143**, 59.
Maniloff, J., Morowitz, H. J., and Barrnett, R. J. (1965). *J. Bacteriol.* **90**, 193.
Maré, C. J., and Switzer, W. P. (1965). *Vet. Med.* **60**, 841.
Markov, G. G., Bradvarova, I., Mintcheva, A., Petrov, P., Shishkov, N., and Tsanev, R. G. (1969). *Exp. Cell. Res.* **57**, 374.
Marmion, B. P., and Goodburn, G. M. (1961). *Nature (London)* **189**, 247.
Miller, G., Emmons, J., and Stitt, D. (1971). *J. Infec. Dis.* **124**, 322.
Minamishima, Y., and Benyesh-Melnick, M. (1969). *Bacteriol. Proc.* p. 170.
Molander, C. W., Kniazeff, A. J., Boone, C. W., Paley, A., and Imagawa, D. T. (1972). *In Vitro* **7**, 168.
Morgan, J. F., Morton, H. E., and Pasieka, A. E. (1958). *J. Biol. Chem.* **233**, 664.
Morowitz, H. J. (1969). *In* "The Mycoplasmatales and the L-phase of Bacteria" (L. Hayflick, ed.), pp. 405–412. Appleton, New York.
Morowitz, H. J., and Maniloff, J. (1966). *J. Bacteriol.* **91**, 1638.
Morowitz, H. J., Bode, H. R., and Kirk, R. G. (1967). *Ann. N. Y. Acad. Sci.* **143**, 110.
Nakamura, M., and Sakamoto, H. (1969). *Proc. Soc. Exp. Biol. Med.* **131**, 343.
Nardone, R. M., Todd, J., Gonzales, P., and Gaffney, E. V. (1965). *Science* **149**, 1100.
Negroni, G. (1964). *Brit. Med. J.* **1**, 927.
Neimark, H (1967). *Ann. N. Y. Acad. Sci.* **143**, 31.
Neimark, H. C. (1970). *J. Gen. Microbiol.* **63**, 249.
Nelson, J. B. (1960). *Ann. N. Y. Acad. Sci.* **79**, 450.
Newnham, A. G., and Chu, H. P. (1965). *J. Hyg.* **63**, 1.
Nocard, E., and Roux, E. (1898). *Ann. Inst. Pasteur. Paris* **12**, 240.
O'Connell, R. C., Wittler, R. G., and Faber, J. E. (1964). *Appl. Microbiol.* **12**, 337.
Ogata, M., and Koshimizu, K. (1967). *Jap. J. Microbiol.* **11**, 289.
O'Malley, J. P., Meyer, H. M., Jr., and Smadel, J. E. (1961). *Proc. Soc. Exp. Biol. Med.* **108**, 200.
O'Malley, J. P., McGee, Z. A., Barile, M. F., and Barker, L. F. (1966). *Proc. Nat. Acad. Sci. U.S.* **56**, 895.
Oppenheim, J. J. (1968). *Fed. Proc., Fed. Amer. Sci. Exp. Biol.* **27**, 21.
Organick, A. B., Siegesmund, K. A., and Lutsky, I. I. (1966). *J. Bacteriol.* **92**, 1164.
Ørskov, J. (1942). *Acta Pathol. Microbiol Scand.* **19**, 586.
Paton, G. R., Jacobs, J. P., and Perkins, F. T. (1965). *Nature (London)* **207**, 43.
Paton, G. R., Jacobs, J. P., and Perkins, F. T. (1967). *Ann. N. Y. Acad. Sci.* **143**, 626.
Perlman, D., Rahman, S. B., and Semar, J. B. (1967). *Appl. Microbiol* **15**, 82.
Pirie, N. W. (1969). *In* "The Mycoplasmatales and the L-phase of Bacteria" (L. Hayflick, ed.), pp. 3–14. Appleton, New York.
Pollock, M., E. and Kenny, G. (1963). *Proc. Soc. Exp. Biol. Med.* **112**, 176.
Pollock, M. E., Kenny, G. E., and Syverton, J. T. (1960). *Proc. Soc. Exp. Biol. Med.* **105**, 10.
Pollock, M. E., Treadwell, P. E., and Kenny, G. E. (1963). *Exp. Cell Res.* **31**, 321.
Powelson, D. M. (1961). *J. Bacteriol.* **82**, 288.
Purcell, R. H., Somerson, N. L., Fox, H. H., Wong, D. C., Turner, H. C., and Chanock, R. M. (1966a). *J. Nat. Cancer Inst.* **37**, 251.
Purcell, R. H., Taylor-Robinson, D., Wong, D. C., and Chanock, R. M. (1966b). *J. Bacteriol.* **92**, 6.
Rahman, S. B., Semar, J. B., and Perlman, D. (1967). *Appl. Microbiol.* **15**, 970.
Rakovskaya, I. V. (1965). *Vop. Virusol.* **10**, 233.
Razin, S. (1968). *J. Bacteriol.* **96**, 687.
Razin, S. (1972). *In* "Handbook of Microbiology," Vol. 1, pp. 105–129. Chem. Rubber Publ. Co., Cleveland, Ohio.

Razin, S., and Tully, J. G. (1970). *J. Bacteriol.* **102**, 306.
Reed, S. E. (1971). *J. Infec. Dis.* **124**, 18.
Reed, S. E. (1972). "Pathogenic Mycoplasmas, Ciba Foundation Symposium," pp. 329–348. Elsevier, Amsterdam.
Reich, P. R., Somerson, N. L., Rose, J. A., and Weissman, S. M. (1966). *J. Bacteriol.* **91**, 153.
Reynolds, R. K., and Hetrick, F. M. (1969). *Appl. Microbiol.* **17**, 405.
Roberts, D. H. (1964a). *Vet. Rec.* **76**, 470.
Roberts, D. H. (1964b). *Brit. Vet. J.* **120**, 479.
Robinson, L. B., Wichelhausen, R. H., and Roizman, B. (1956). *Science* **124**, 1147.
Romano, N., and Brancato, P. (1970). *Arch. Gesamte Virusforsch.* **29**, 39.
Rothblat, G. H., and Morton, H. E. (1959). *Proc. Soc. Exp. Biol. Med.* **100**, 87.
Rous, P., and Jones, F. S. (1916). *J. Exp. Med.* **23**, 549.
Rouse, H. C., and Schlesinger, R. W. (1967). *Virology* **33**, 512.
Rouse, H. C., Bonifas, V. H., and Schlesinger, R. W. (1963). *Virology* **20**, 357.
Russell, W. C., and Becker, Y. (1968). *Virology* **35**, 18.
Sabin, A. B. (1941). *Bacteriol. Rev.* **5**, 1.
Sabin, A. B. (1967). *Ann. N. Y. Acad. Sci.* **143**, 628.
Schimke, R. T., and Barile, M. F. (1963a). *Exp. Cell Res.* **30**, 593.
Schimke, R. T., and Barile, M. F. (1963b). *J. Bacteriol.* **86**, 195.
Schmidt, P. J., Barile, M. F., and McGinniss, M. H. (1965). *Nature* (*London*) **205**, 371.
Schulze, I. T., and Schlesinger, R. W. (1963). *Virology* **19**, 40.
Schwöbel, W., and Leach, R. H. (1970). *Zentralbl. Bakteriol., Parasitenk., Infektionskr. Hyg., Abt. Orig.* **214**, 495.
Scriba, M. (1968). *Z. Med. Mikrobiol. Immunol.* **154**, 267.
Shedden, W. I. H., and Cole, B. C. (1966). *J. Pathol. Bacteriol.* **92**, 574.
Shepard, M. C. (1969). *In* "The Mycoplasmatales and the L-phase of Bacteria" (L. Hayflick, ed.), pp. 49–65. Appleton, New York.
Simberkoff, M. S., Thornbeck, J. G., and Thomas, L. (1969). *J. Exp. Med.* **129**, 1163.
Singer, S. H., Kirschstein, R. L., and Barile, M. F. (1969a). *Nature* (*London*) **222**, 1087.
Singer, S. H., Barile, M. F., and Kirschstein, R. L. (1969b). *Proc. Soc. Exp. Biol. Med.* **131**, 1129.
Singer, S. H., Fitzgerald, E. A., Barile, M. F., and Kirschstein, R. L. (1970). *Proc. Soc. Exp. Biol. Med.* **133**, 1439.
Singer, S. H., Ford, M., Barile, M. F., and Kirschstein, R. L. (1972). *Proc. Soc. Exp. Biol. Med.* **139**, 56.
Smirnova, T. D., and Kagan, G. Y. (1971). *Zh. Mikrobiol., Epidemiol. Immunobiol.* **48**, 54.
Smith, P. F. (1963). *J. Gen. Microbiol.* **32**, 307.
Smith, P. F., and Sasaki, S. (1958). *Appl. Microbiol.* **6**, 184.
Somerson, N. L., and Cook, M. K. (1965). *J. Bacteriol.* **90**, 534.
Somerson, N. L., and Weissman, S. M. (1969). *In* "The Mycoplasmatales and the L-phase of Bacteria" (L. Hayflick, ed.), pp. 201–220. Appleton, New York.
Somerson, N. L., Taylor-Robinson, D., and Chanock, R. M. (1963). *Amer. J. Hyg.* **77**, 122.
Somerson, N. L., Purcell, R. H., Taylor-Robinson, D., and Chanock, R. M. (1965). *J. Bacteriol.* **89**, 813.
Somerson, N. L., Reich, P. R., Chanock, R. M., and Weissman, S. M. (1967). *Ann. N. Y. Acad. Sci.* **143**, 9.
Spendlove, R. S., Crosbie, R. B., Hayes, S. F., and Keeler, R. F. (1971). *Proc. Soc. Exp. Biol. Med.* **137**, 258.
Spitler, L., Cochrum, K., and Fudenberg, H. H. (1968). *Science* **161**, 1148.
Stanbridge, E. (1971). *Bacteriol. Rev.* **35**, 206.
Stanbridge, E., Onen, M., Perkins, F. T., and Hayflick, L. (1969). *Exp. Cell. Res.* **57**, 397.

Stewart, S. M., and Young, D. (1971). *Lancet* **1**, 347.
Subcommittee on Taxonomy of Mycoplasmatales (1972). *Int. J. Syst. Bacteriol.* **22**, 184.
Tankersley, R. W. (1964). *J. Bacteriol.* **87**, 609.
Taylor-Robinson, D., Somerson, N. L., Turner, H. C., and Chanock, R. M. (1963). *J. Bacteriol.* **85**, 1261.
Taylor-Robinson, D., Purcell, R. H., Wong, D. C., and Chanock, R. M. (1966). *J. Hyg.* **64**, 91.
Theodore, T. S., King, J. R., Tully, J. G., and Cole, R. M. (1970). *Develop. Ind. Microbiol.* **11**, 122.
Theodore, T. S., Tully, J. G., and Cole, R. M. (1971). *Appl. Microbiol.* **21**, 272.
Todaro, G. J., Aaronson, S. A., and Rands, E. (1971). *Exp. Cell Res.* **65**, 256.
Tully, J. G. (1966). *Proc. Soc. Exp. Biol. Med.* **122**, 565.
Tully, J. G., and Rask-Nielson, R. (1967). *Ann. N. Y. Acad. Sci.* **143**, 345.
Tully, J. G., and Ruchman, I. (1964). *Proc. Soc. Exp. Biol. Med.* **115**, 554.
Tully, J. G., Barile, M. F., DelGiudice, R. A., Carski, T. R., Armstrong, D., and Razin, S. (1970). *J. Bacteriol.* **101**, 346.
VanDemark, P. J. (1969). *In* "The Mycoplasmatales and the L-phase of Bacteria" (L. Hayflick, ed.), pp. 491–502. Appleton, New York.
Van Herick, W., and Eaton, M. D. (1945). *J. Bacteriol.* **50**, 47.
Warren, J. (1942). *J. Bacteriol.* **43**, 211.
Williams, C. O., Wittler, R. G., and Burris, C. (1969). *J. Bacteriol.* **99**, 341.
Winters, A. L., and Consigli, R. A. (1971). *J. Gen. Virol.* **10**, 53.
Wright, D. N., and Bailey, G. D. (1969). *Can. J. Microbiol.* **15**, 1449.
Yershov, F. I., and Zhdanov, V. M. (1965). *Virology* **27**, 451.
Zucker-Franklin, D., Davidson, M., and Thomas, L. (1966a). *J. Exp. Med.* **124**, 521.
Zucker-Franklin, D., Davidson, M., and Thomas, L. (1966b). *J. Exp. Med.* **124**, 533.

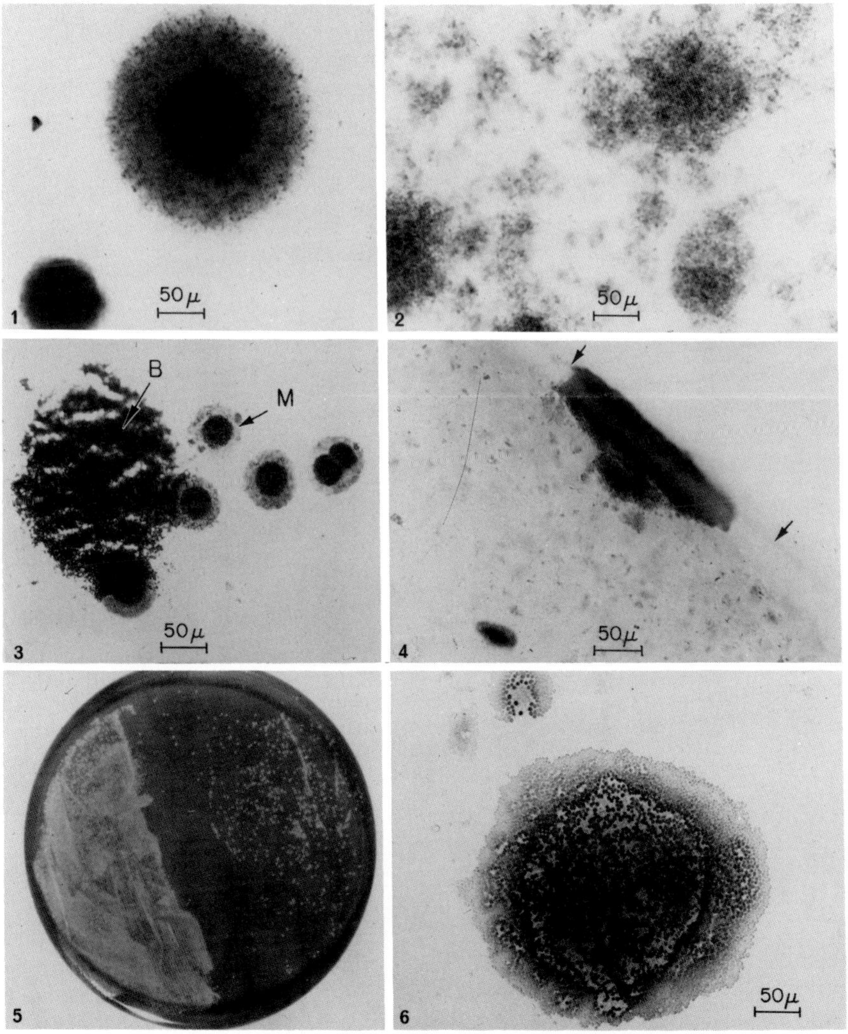

Figs. 1–4. Mycoplasma colonies are microscopic, have a fried-egg (*M. orale* type 1, Fig. 1) or granular (*M. pneumoniae*, Fig. 2) appearance, and grow beneath the surface of the agar (*M. hominis*, Figs. 3 and 4). In a sagittal section preparation, the central button of the colony grows more deeply embedded than does the peripheral growth (Fig. 4). Compare the general appearance of the young bacterial (*Corynebacterium* sp.) colony (B) with the well-defined morphology of the mycoplasma colonies (M) (Fig. 3). Note the surface growth of the bacterial colony and compare it to the subsurface growth of the mycoplasma colonies (Fig. 3). (1 μ = 1 μm.)

Figs. 5 and 6. A positive hemolysis reaction by colonies of *M. pneumoniae* to guinea pig erythrocytes (Fig. 5) and a positive hemadsorption reaction by a colony of *M. pulmonis* to sheep erythrocytes (Fig. 6). (1 μ = 1 μm.)

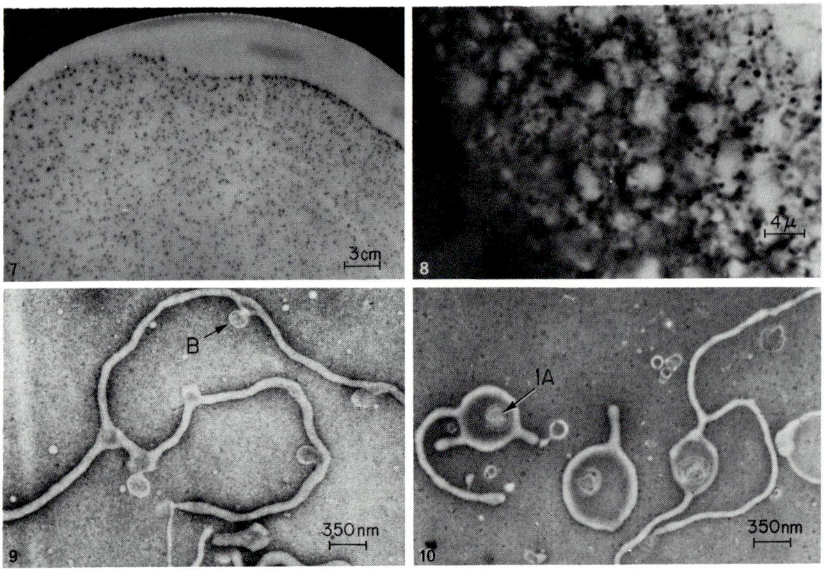

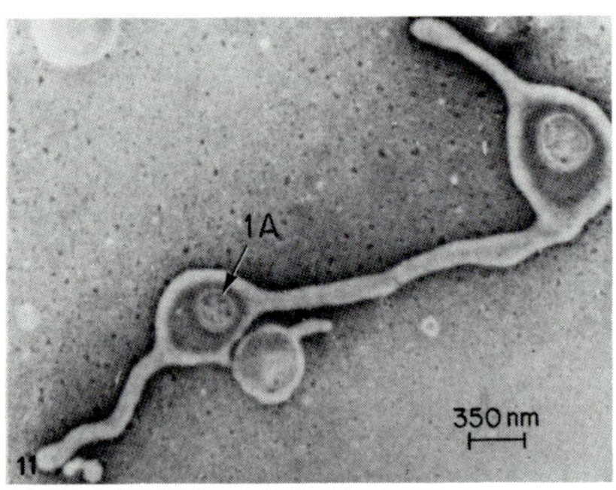

Figs. 7 and 8. Growth of *M. hominis* colonies on a 47 mm membrane filter disk. For growth, the filter disk is placed on agar medium and examined as described by Barile (1962b). Filtration was used to isolate low orders of mycoplasmas from large volumes of specimen, using 300 or 450 nm filters. For microscopic examination, the filter containing mycoplasma colonies was fixed with isopropyl alcohol, stained with one-tenth diluted Crystal Violet stain, dried thoroughly and observed using oil immersion at high magnification (Fig. 8). Note the pleomorphic mycoplasma structures at the periphery of the stained colony (*M. hominis*). Owing to its size, the morphology and ultrastructure of the mycoplasma cell must be examined by electron microscopy (Figs. 9–18). (1 μ = 1 μm.)

Figs. 9–11. Ultrastructures of *M. orale* type 1 in negatively stained preparations (Anderson and Barile, 1966). *Mycoplasma orale* produces large branching filaments that form small blebs (B, Fig. 9) and large swollen spherules containing internal apparatuses (IA, Figs. 10 and 11).

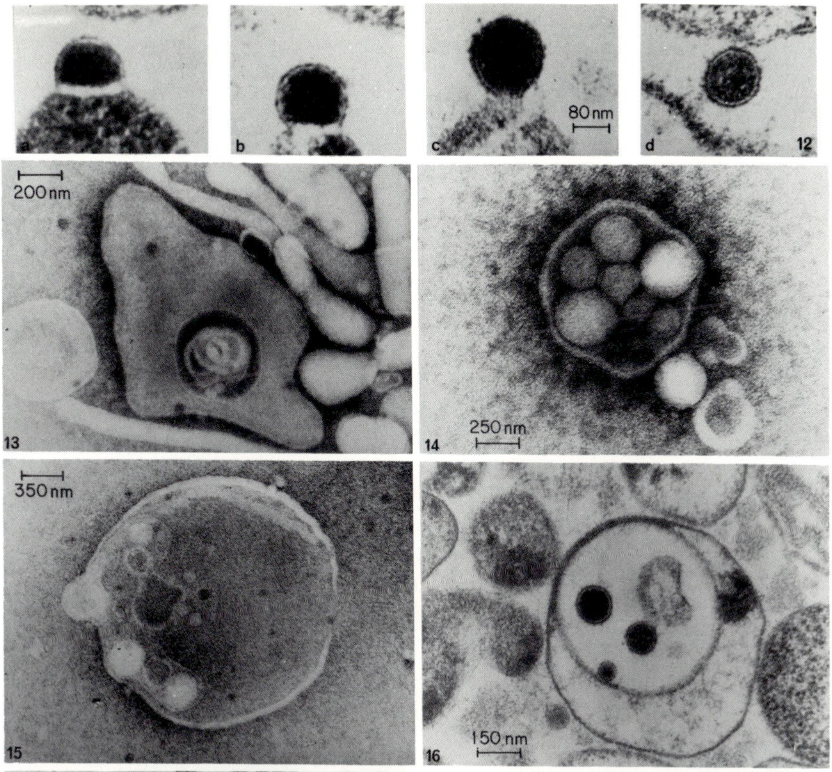

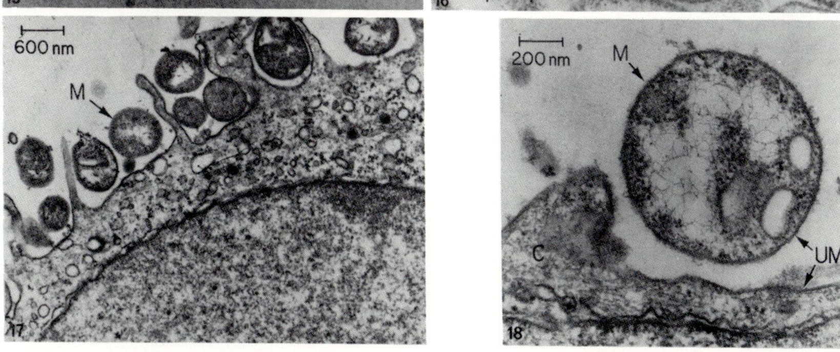

Figs. 12–16. The ultrastructures of *M. hominis* in negatively stained (Figs. 13–15) (Anderson and Barile, 1965) and in thin-sectioned preparations (Figs. 12 and 16) (Anderson, 1965). Note the pleomorphic size, shape, and structures of the *M. hominis* cell. Some cells have sharply defined internal apparatuses containing coiled strands (Fig. 13), whereas others contain either small internal lucent spheres (Fig. 14) or an envelop (vacuole) with small lucent spheres (Fig. 15). In thin-sectioned preparations, cells are seen producing an apparent budding of dense spherical forms at the cell periphery (Fig. 12,a–d) or containing a central vacuole with small dense spheres (Fig. 16).

Figs. 17 and 18. Thin-sectioned preparations of *M. pulmonia*-infected HeLa (Fig. 17)(Barile, 1965) and L cell (Fig. 18) (Anderson, 1965) cultures. The mycoplasmas (M) are seen closely associated with the plasma membranes of the cell cultures. Note the morphologic similarities of the three-layered unit membrane (UM) of the mycoplasma (M) and the L cell (C) (Fig. 18).

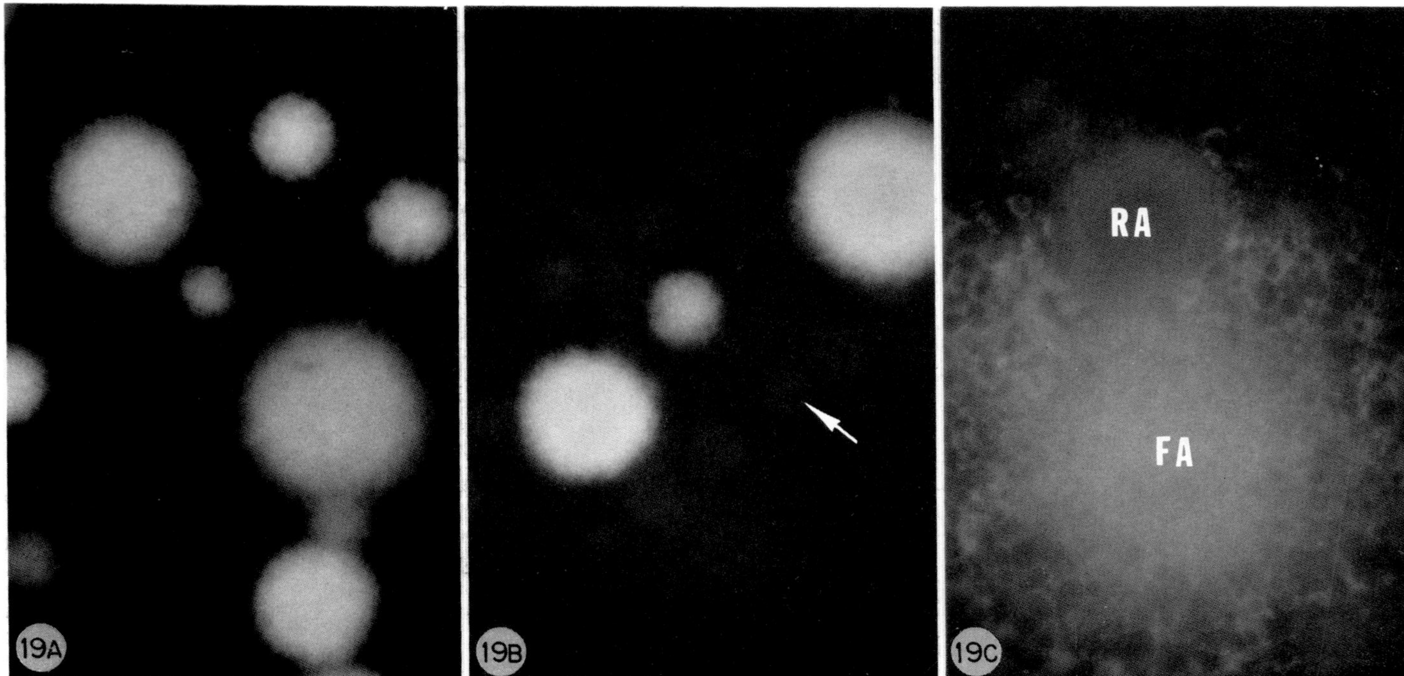

Fig. 19. Immunofluorescence reactions of mycoplasma colonies by the plate epi-immunofluorescence procedure (Barile and DelGiudice, 1972), including (A) a positive homologous reaction with colonies of *A. laidlawii*, (B) a negative (arrow) and a positive reaction (with colonies of *A. laidlawii*) in a mixed mycoplasma culture, and (C) two positive reactions with colonies to two different antiserums, including a fluorescein-conjugated antiserum (FA) to *A. laidlawii* and a rhodamine-conjugated (RA) antiserum to *M. orale* type 1 in a mixed mycoplasma culture.

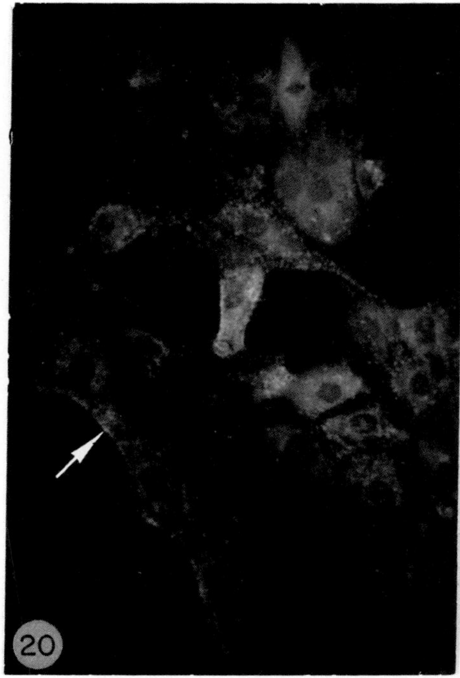

Figs. 20 and 21. A positive homologous immunofluorescence reaction with a *M. hominis*-infected cell culture (Fig. 20) and a negative reaction with a mycoplasma-free cell culture (Fig. 21).

Fig. 22. Growth in semisolid broth media of an arginine-utilizer, *M. arginini* (left), and a dextrose fermenter, *M. canis* (right). The center tube is the uninoculated negative broth control. Note the luxuriant turbid growth and the formation of microcolonies. *Mycoplasma arginini* utilizes arginine, producing an alkaline shift in pH, whereas *M. canis* ferments dextrose, producing an acid shift in pH, each with corresponding changes in the color of the culture, providing a useful index on growth and on the biochemical activity of the mycoplasma. The semisolid broth medium grows most mycoplasmas, bacteria, and fungi that contaminate cell cultures and the medium is recommended for testing cell cultures for microbial contamination (Section III, A, 1, d).

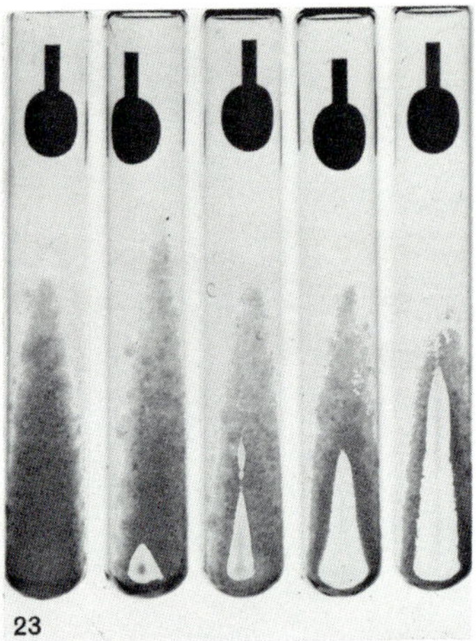

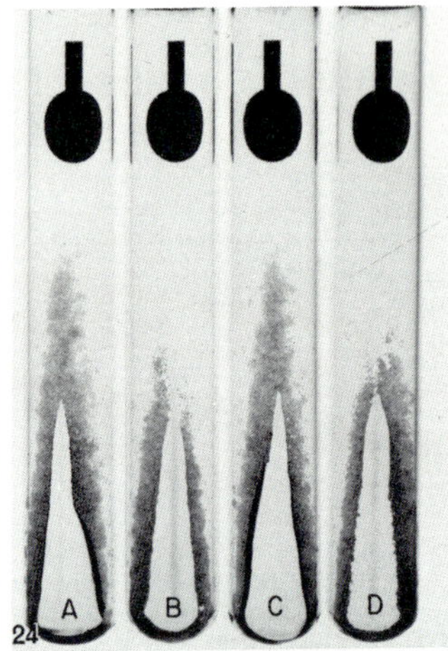

Figs. 23 and 24. The typical V-shaped cytopathic effect (CPE) produced by mycoplasmal fermenters, including *M. pulmonis* (Fig. 24a), *Mycoplasma* strain 14 (Fig. 24b), *A. laidlawii* (Fig. 24c), and *M. hyorhinis* (Fig. 24d) in L cells. Note the progressive stages of cell destruction in a *M. hyorhinis*-infected L cell (Fig. 23) stained with one-tenth diluted Gentian Violet.

Fig. 25. The effect on reduction of plaque size and on number of Semliki-Forest viruses (SFV) in *M. hyorhinis* (b) and (c) *M. arginini*-infected mouse embryo fibroblast cultures (Singer *et al.*, 1972). (a) SFV plaques in a mycoplasma-free culture.

Fig. 26. Plaque formation produced by *M. gallisepticum* in primary chick embryo cell cultures (O'Malley *et al.*, 1961, 1966). Certain mycoplasmas mimic properties heretofore associated only with viruses (Sections I and II). Fig. 26c illustrates a mycoplasma-free culture.

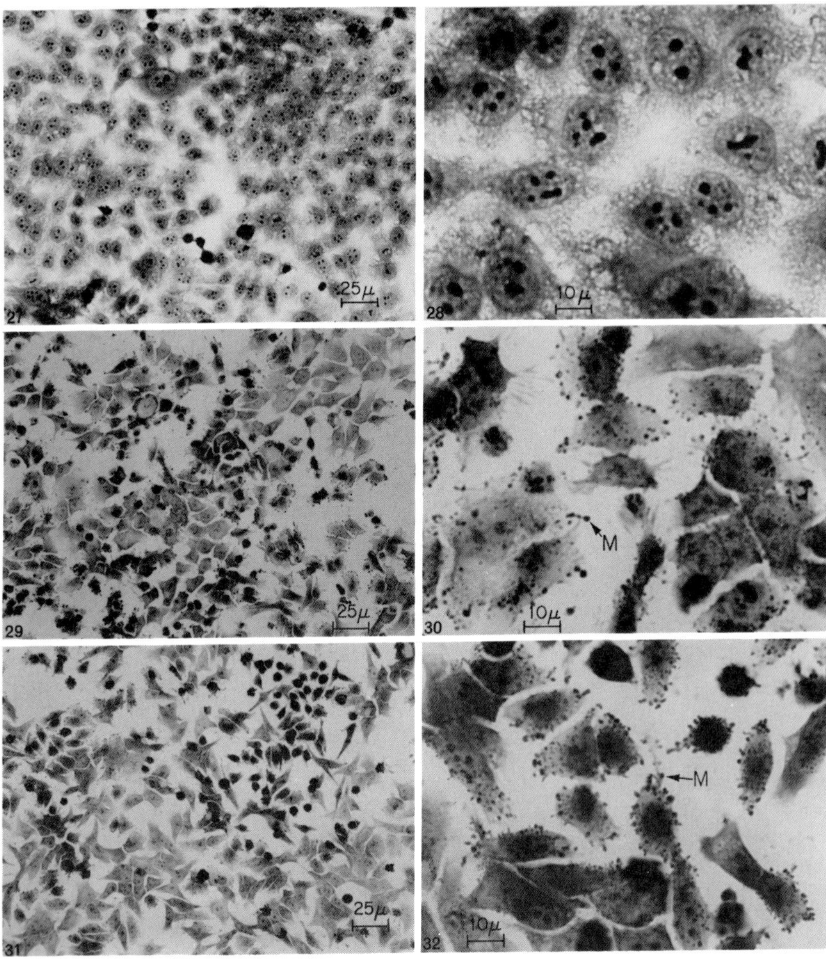

Figs. 27–32. The early, initial, cytopathic effects of *M. pulmonis* and *M. hyorhinis* infection on Hep-2 (human carcinoma of larynx) cell cultures at low (× 125) and high (× 500) magnifications in Hemotoxylin Eosin-stained preparations. Compare the general appearance of the mycoplasma-free cells (Figs. 27 and 28) with the cellular changes in the *M. pulmonis* (Figs. 29 and 30) and *M. hyorhinis* (Figs. 31 and 32) infected cells. On macroscopic examination, mycoplasma-contaminated or -infected cell cultures lose their transparency, become grainy or granular, have a dirty appearance, appear toxic, and develop changes in the orientation of cell growth. On microscopic examination, mycoplasmas (M) appear as pleomorphic coccal, oval, and filamentous bodies closely associated with the plasma membrane of the cell cultures (Figs. 29–32, 34, 36, and 38). Note also the multiple extensions of the plasma membrane in the mycoplasma-infected cells (Figs. 30 and 32). (1 μ = 1 μm.)

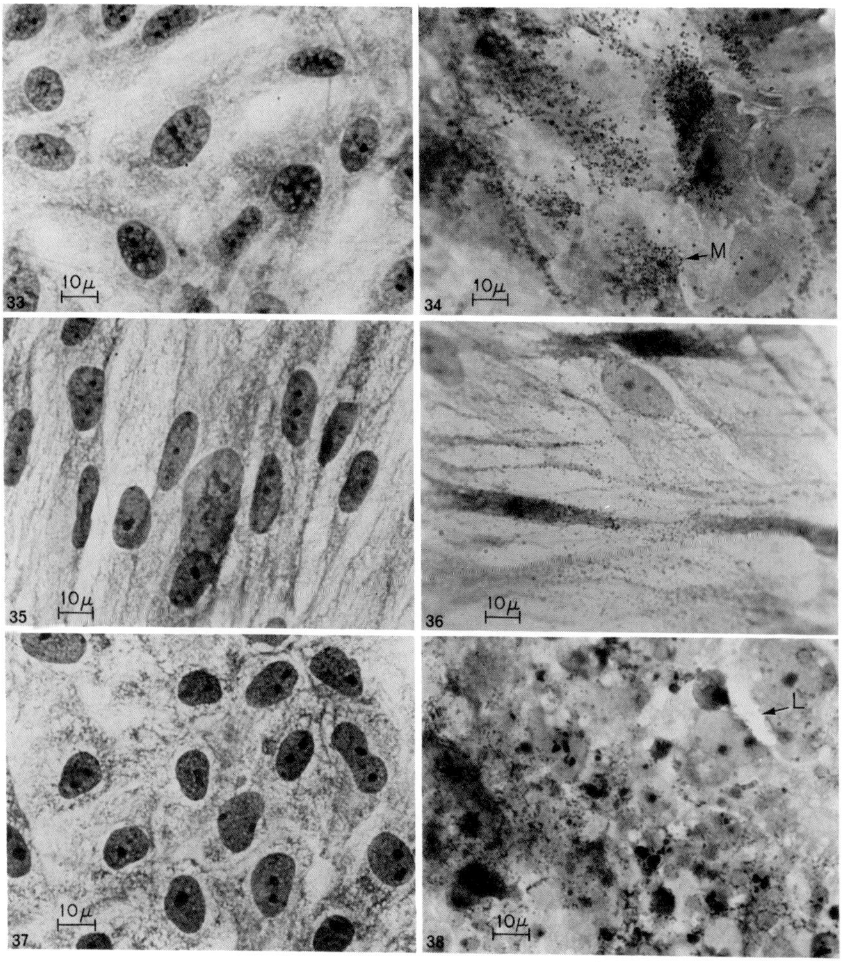

Figs. 33–38. Note the general appearance of the normal, mycoplasma-free cell cultures, including the primary rabbit kidney (PRK) (Fig. 33), human embryonic lung (WI-38) (Fig. 35), and primary African green monkey (PVK) (Fig. 37) cells and compare them to the initial cytopathic effects produced in mycoplasma-infected cells, including the *M. pulmonis*-infected PRK (Fig. 34), the *M. hyorhinis*-infected WI-38 (Fig. 36), and the *M. hyorhinis*-infected PVK (Fig. 38) cell cultures. Note the cellular changes produced by the same *Mycoplasma* species in different cell cultures, e.g., the CPE produced in Hep-2 (Figs. 29 and 30) and in PRK (Fig. 34) cell cultures infected with *M. pulmonis* and the CPE produced in Hep-2 (Figs. 31 and 32), WI-38 (Fig. 36) and PVK (Fig. 38) cell cultures infected with *M. hyorhinis*. *Mycoplasma pulmonis* produces microcolonies in PRK cells (Fig. 34) but attaches to the plasma membranes of Hep-2 cells (Figs. 29 and 30), whereas single cells of *M. hyorhinis* attach to the plasma membranes of Hep-2 (Figs. 31 and 32) and of WI-38 (Fig. 36) cells and produces microcolonies and small focuses of cell destruction with lesion (L) formation in PVK cells (Fig. 38). The CPE proceeds with development of V-shaped destruction of the cell cultures (Figs. 23 and 24). In most cases, the microscopic CPE varies with the *Mycoplasma* species and the type of cell culture (Figs. 27–38). Occasionally, different *Mycoplasma* species can produce similar microscopic CPE (Figs. 29–32). However, most mycoplasmal fermenters produce similar macroscopic changes, i.e., a V-shaped CPE (Figs. 23 and 24), resulting from cell destruction. Stained with Paragon multiple stain, Paragon C. § C. Co., Inc., New York, N.Y. (1 μ = 1 μm.)

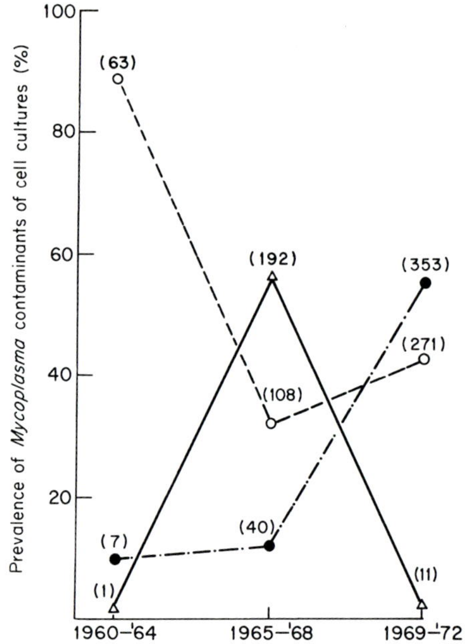

Fig. 39. The prevalence of mycoplasmal species contamination from human (○), bovine (●), and swine (△) sources during the period 1960–1972. Three peaks of mycoplasmal contamination occurred: human mycoplasmas were prevalent in the early 1960's, followed by swine mycoplasmas in the mid-1960's and bovine mycoplasmas by the late 1960's. The numbers in parentheses are the number of mycoplasmas isolated and identified from human, bovine, swine sources (see Table VII).

CHAPTER 7

Reversible and Irreversible Alterations of Cultured FL Human Amnion Cells after Experimental Mycoplasmal Infection

JØRGEN FOGH

I. Effects of Mycoplasmal Infection on the *in Vitro* Growth of FL Cells 175

II. *In Vitro* Growth and Susceptibility to Mycoplasmal Infection of Previously Mycoplasma-Infected Cells from which Mycoplasmas Have Been Eliminated 176

III. Chromosome Changes Following Mycoplasmal Infection of FL Cells 178

A. Reduction in Chromosome Number 178

B. Increase in Chromosome Aberrations 180

C. New Chromosome Varieties 181

D. Karyotypic Changes after Mycoplasmal Infection of FL Cells 181

E. Discussion of Chromosome Data 182

IV. Reduced Tumor-Producing Capacity of Mycoplasma-Modified FL Cells 183

V. Effects of Mycoplasmas on SV40 Transformation of Human Amnion Cells 188

VI. Decreased Resistance of Transformed Human Cells to the Destructive Effects of Mycoplasmal Infection 189

VII. Cell-Associated Mycoplasmas: Preferential Attachment to Transformed Cells and Cells of Malignant Origin 190

VIII. Discussion of Results 192

References 193

The frequent presence of mycoplasmas as tissue culture contaminants (Robinson *et al.*, 1956) is well documented in other chapters of this book. This contamination represents a most serious problem. During the last 10 to 15 years, it has been shown that mycoplasmas can cause severe effects on the host cells, including cytologic, biochemical, genetic and immunologic alterations, as well as changes in virus susceptibility. In addition, mycoplasmas may modify mammalian cell surfaces to make them capable of hemadsorption (Berg and Frothingham, 1961), and they can agglutinate several types of red blood cells (Tully, 1963). Attachment of mycoplasmas to the host cells and various host cell reactions have interfered with electron microscopic studies of cells and virus–cell interactions (Edwards and Fogh, 1960). Obviously, therefore, experiments performed with mycoplasma-contaminated cultures may give misleading information.

This chapter deals primarily with the effects of a strain of mycoplasma, strain HT, that was isolated as a contaminant in this laboratory many years ago (Fogh *et al.*, 1965). This strain has been identified as *Mycoplasma fermentans* (PG-18) by Dr. M. Barile, Division of Biologic Standards, N.I.H.; Dr. J. G. Tully, Laboratory of Bacterial Diseases, National Institute of Allergy and Infectious Diseases, N.I.H.; and Dr. Donald Armstrong, Sloan-Kettering Institute.

Many different types of cultured cells were destroyed when this mycoplasma was present. The observed changes often resembled virus cytopathic effects. The most striking effects were observed after experimental infection of cells of the FL human amnion line (Fogh and Lund, 1957). There were pronounced morphologic changes, and many cells were completely destroyed.

During the last several years, a number of alterations have been noted in cultures of FL cells infected with this mycoplasma. They included the growth *in vitro*, the chromosome constitution, and the tumor-producing capacity of the cells. Distinct effects have also been observed when primary human amnion cells were concurrently infected with SV40 and the HT mycoplasma. Some changes were noticeable during the early events of the SV40 transformation and several of the characteristics of the transformed amnion cell strains were also changed. More recently, the observation has been made that this mycoplasma has a specific affinity to transformed cells and to cells of malignant origin. The amounts of mycoplasma associated with such cells are much greater than those associated with the normal cells from which they are derived. Also, cell destructive effects are more pronounced for the transformed and malignant cells than for their normal counterparts.

I. Effects of Mycoplasmal Infection on the *in Vitro* Growth of FL Cells

When cultures of FL cells were inoculated with mycoplasma, within a couple of days many of the cells showed cytopathic effects and became rounded and detached from the glass. However, a few days later, an increase in cell number occurred. This surviving cell population could be transferred to subcultures but it continued to divide at a reduced rate (Fogh *et al.*, 1965). This initial response of FL cells to mycoplasmal infection has been studied many times. One such experiment is shown in Fig. 1. Shortly after infection, the growth rate was decreased, and from day 6 after infection there was also a loss of cells. During the following days, the combined increase in cell number and loss of cells resulted in a population at day 12 that was no more than that at day 2. In the third week, the number of cells increased steadily. The general appearance of the cells was healthy in spite of the constant presence of the mycoplasma at a concentration that exceeded 10^7 colony forming units per milliliter.

Morphologic changes in the mycoplasma-infected cultures of FL cells could be observed as early as 1–2 days after infection, depending on the infecting dose. The changes were accentuated by suboptimal nutritional conditions; for example, by reducing the serum concentration.

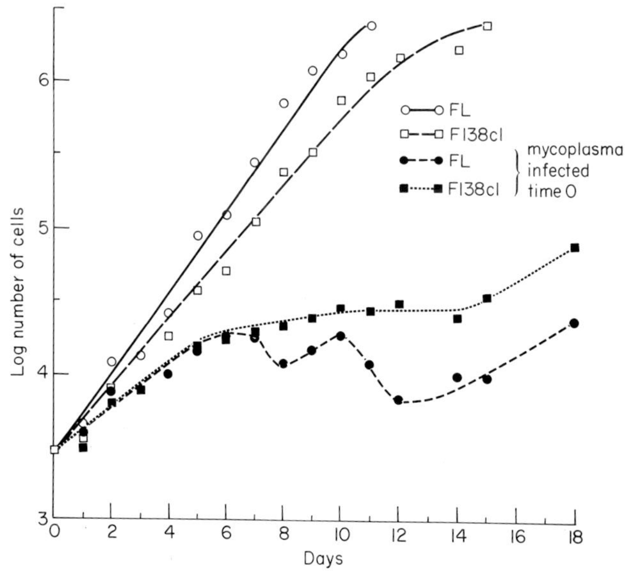

Fig. 1. Growth curves demonstrating the different response of FL and F138cl cells to mycoplasmal infection (Fogh *et al.*, 1971).

A number of uncontaminated cultures of other cells types were experimentally infected in order to study their reactions. FL cells were destroyed more by this mycoplasma and showed more morphological changes of individual cells than did cells of any other line we examined, including 11 epithelial human lines, six of normal, and five of malignant origin, and one fibroblastic rat embryo cell line (Fogh *et al.*, 1965). Primary amnion cells and diploid human fibroblasts showed considerable resistance to infection. When mycoplasma-containing supernatants from cultures of all these cell types were inoculated into cultures of FL cells, their destructive effects on FL cells were the same both qualitatively and quantitatively.

We have carried FL cells surviving the initial mycoplasmal infection in continuous cell cultures for more than 6 years. There is more variation in cell size and shape than in the corresponding uninfected cultures and occasional cells of odd morphology appear (Fig. 2). The following data have some bearing on the question as to whether this population represents a selection of a cell-type already present in the uninfected cultures or whether the mycoplasmal infection induces cell changes, making the cells suitable for survival in the presence of high titers of mycoplasmas.

II. *In Vitro* Growth and Susceptibility to Mycoplasmal Infection of Previously Mycoplasma-Infected Cells from which Mycoplasmas Have Been Eliminated

After the HT strain of mycoplasma was eliminated from infected FL cell lines (Fogh and Fogh, 1967; Fogh *et al.*, 1971), the growth *in vitro* differed from the parent FL cell line as illustrated by line F138cl in Fig. 1. As compared to the infected line, the doubling time had decreased but was still longer (21 hours) than for the uninfected parent cells (17.5 hours). When F138cl cells were reinfected with mycoplasma, their rate of increase in number was similar to infected FL cells for the first 6 days. From this time, however, the reduction in numbers was much less pronounced (approximately ten times). After 2 weeks the increase in number was similar to that of infected FL cells.

This increased resistance to mycoplasmal infection of the F138cl line (and other similar lines), as compared to the FL line, was also obvious from the different degrees of cytopathic effects and cell destruction (Fogh *et al.*, 1971). Infected cultures of FL cells showed considerable destruction of the cell sheet at a time when cell destruction in cultures of mycoplasma-reinfected F138cl cells was only minor. In infected cultures of FL cells, only a very light population of mostly spindle-shaped cells remained on the glass some days after infection. In the reinfected F138cl cell cultures,

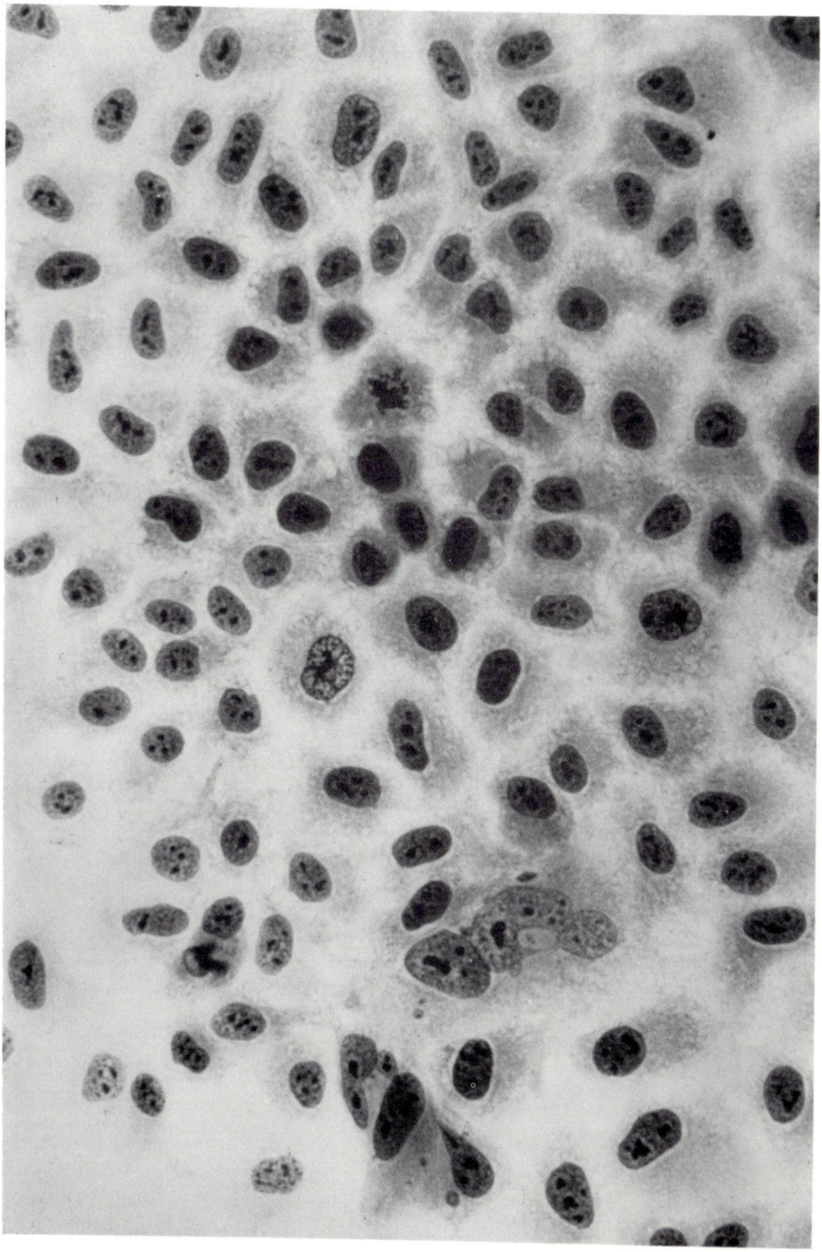

Fig. 2. Hematoxylin and Eosin stain of mycoplasma-carrying culture (F138) of FL cells, 2 years after infection. ×800.

most cells maintained epithelial morphology and cell destruction was quite minor.

The increased resistance, the reduced *in vitro* rate of cell division, and certain morphologic changes have been observed to be maintained several years after the elimination of mycoplasmas. Thus, these changes must be considered irreversible. They may be interpreted as being the results of selection, adaptation, or genetic changes directly or indirectly induced by the mycoplasmal infection. Other observations, including those to be reported later in this chapter, have supported the hypothesis that this mycoplasmal infection may indeed affect the cell genome.

These results have some bearing on the practical problems of tissue culture contamination. The frequent lack of obvious signs of morphologic effects in cultures of cell lines, even in the presence of mycoplasmas at high titers, may well be the result of inapparent, previous infections that may have been cured by the application of various routinely used antibiotics. Such infections may have selected for or changed the original mycoplasma-sensitive cell population to a population of relatively resistant cells that do not respond to reinfection with early cell destruction or morphological changes (Fogh *et al.*, 1971).

III. Chromosome Changes Following Mycoplasmal Infection of FL Cells

Following infection of cultures of FL cells with the HT strain of mycoplasma, characteristic chromosome changes were observed. They could be classified as three types: reduction in chromosome numbers, increase in chromosome aberrations, and the appearance of new chromosome varieties. The major chromosome changes developed very slowly (Fogh and Fogh, 1965).

A. Reduction in Chromosome Number

The chromosome numbers, which are predominantly between 70 and 76 for uninfected FL cells, continued to decrease in the infected cultures, as shown in Fig. 3. A temporary reduction in numbers occurred within the first days after infection. After 1 month, however, the numbers were similar to those of uninfected cells. Further reduction was seen in the following months and, after 10 months, more than three quarters of the cells had numbers of less than 68. As indicated by the growth of the cultures, the majority of the cells was obviously proliferative. Even beyond this time, the chromosome numbers continued to decrease and when counts were made as late as 3 years after the initial infection (Fogh and Fogh, 1967),

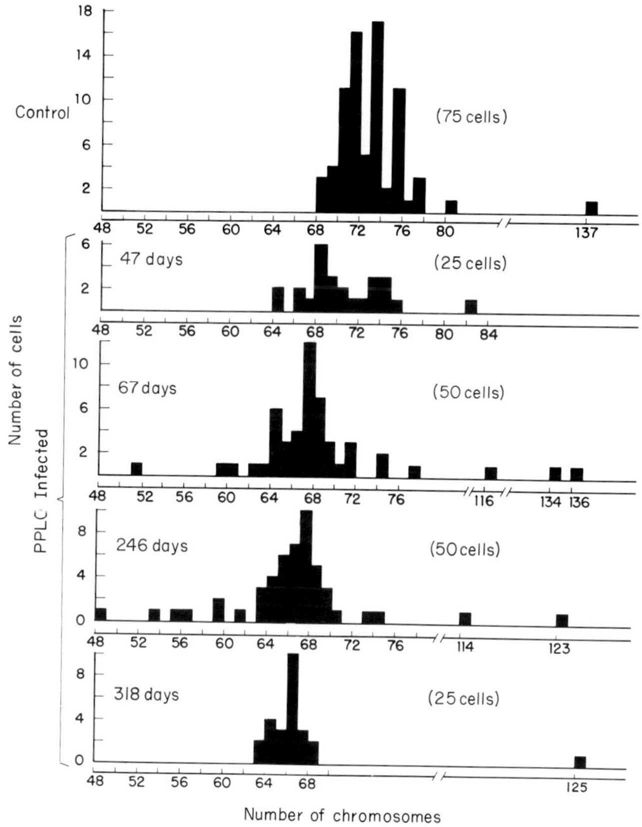

Fig. 3. Chromosome numbers for uninfected and mycoplasma mycoplasm-infected FL cells (F138) at various times after infection (Fogh and Fogh, 1965).

the predominant numbers were 62–64 chromosomes per cell (Fig. 4).

After mycoplasma elimination, there was no reversion to higher chromosome numbers (Fig. 4). At 26 days, 155 days, and 348 days after elimination, the numbers closely resembled those counted for the corresponding mycoplasma-infected lines examined at the same times (Fogh and Fogh, 1967).

We have determined that the DNA contents for FL cells and cells of the F138cl line were similar (26.7 $\pm$0.57 for FL, 27.1 $\pm$0.68 for F138cl) on the basis of analysis of 18 samples of each cell type (Gaffney and Fogh, 1970). This similarity indicates that the mycoplasma-effected reduction of approximately 15% of the chromosome number cannot be explained by a loss of DNA but may be interpreted as being a rearrangement of the genetic material.

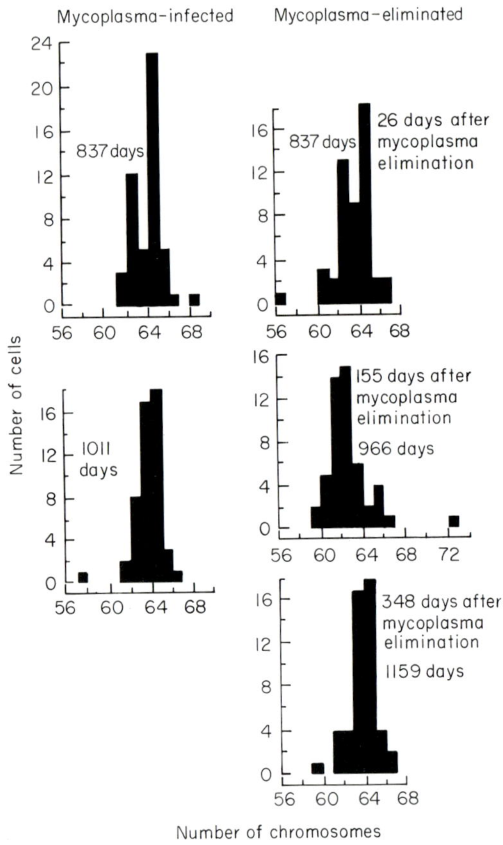

Fig. 4. Chromosome numbers for mycoplasma-infected FL cells (F138) at 837 and 1011 days after infection and for these cells at various times after mycoplasma elimination (F138cl) (Fogh and Fogh, 1967).

B. Increase in Chromosome Aberrations

Chromosome aberrations observed as results of the mycoplasmal infection represented mainly accentuations of aberrations present to a minor degree in uninfected FL cells (Fogh and Fogh, 1965). Whereas chromatid breaks were seen consistently within the first weeks of infection, they were rare or absent after a longer period of infection. The incidence of cells in endoreduplication increased with time of cultivation in the presence of mycoplasmas. Dicentric chromosomes and acentric fragments were observed more frequently. Translocational exchanges, observed extremely rarely in uninfected FL cells, were quite frequent in the infected cultures. Ring chromosomes, never observed in uninfected FL cells, were seen after long periods of infection. Secondary constrictions appeared to be more frequent and the frequency of destruction of chromosomes was increased. Cells showing fragmentation of the entire chromosome complement were seen. Most of these changes developed also very slowly.

After mycoplasma elimination, many of the chromosome aberrations decreased in frequency (Fogh and Fogh, 1967). Dicentrics were rare and cells with minute chromosomes, translocational exchanges, ring chromosomes, and chromosome destruction became less frequent. In general, the aberration frequencies some time after mycoplasma elimination resembled those of uninfected FL cells.

C. New Chromosome Varieties

The new chromosome varieties observed after mycoplasmal infection included a very large telocentric chromosome, which increased in frequency with time of culture. In addition, a large metacentric chromosome and a large subtelocentric chromosome were observed, both with less frequency (Fogh and Fogh, 1965). None of these chromosomes have ever been seen in uninfected FL cells. The same new varieties appeared in several independent experiments after infection with mycoplasma of FL cells. Three years after the initial mycoplasmal infection, the new large telocentric chromosome was present in practically all metaphase plates (98.3%) (Fogh and Fogh, 1967). The large subtelocentric chromosome was seen in 16.5% of metaphase plates at this time. The large metacentric chromosome varied in frequency and occasionally was absent.

Following elimination of mycoplasmas, the large telocentric chromosome persisted at very high frequencies. In one line it was observed in 99.1% of metaphase plates at 155 days, and in 95.7% of cells at 348 days after mycoplasma elimination. The large metacentric chromosome was present at low frequencies and the large subtelocentric chromosome was very infrequent (Fogh and Fogh, 1967).

D. Karyotypic Changes after Mycoplasmal Infection of FL Cells

Examination of the karyotypes of cells of the F138 and F138cl line revealed a considerable variation among the metaphase plates within the lines (Fogh and Fogh, 1968). This was also true for cells of the uninfected FL cell line. Mean values of chromosome distribution per group for ten karyotypes of each of these three cell types are shown in Fig. 5. For comparison, the number per group in a diploid human cell is shown. The X chromosome is included in the C group, the Y chromosome in the G group. There is an

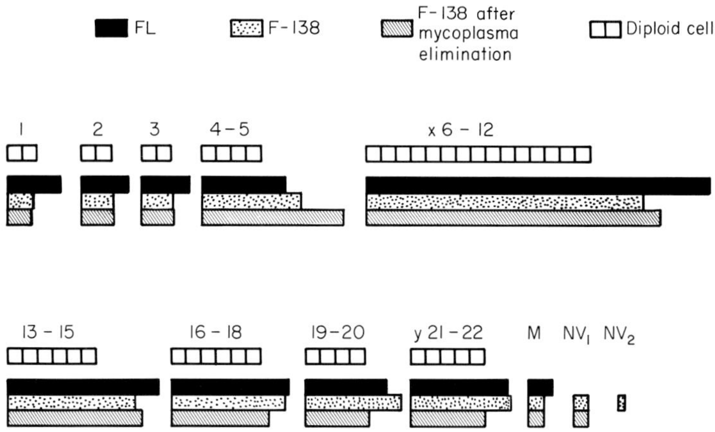

Fig. 5. Mean values of chromosomes per group from ten karyotypes of FL cells, F138 cells, F138cl cells, and diploid cells for comparison (Fogh and Fogh, 1968).

increase in the average number of chromosomes for all chromosome groups in FL cells, most for chromosome number 1, group D, chromosome number 3, and group C as compared to the number per group in the diploid cell.

The mean values for the mycoplasma-infected F138 cells show that all chromosome groups contain fewer chromosomes than FL cells, except for an increase in group B and group F. Numbers 1, 2, and 3 chromosome are reduced in numbers to the level of the diploid cell. The remaining groups are increased in numbers from the diploid level. F138cl karyotypes are similar to F138 karyotypes in groups A, C, and D. In group A, E, F, and G, the numbers are close to the diploid level. There is a definite increase in group B. The mean value for this group is greater than the value for FL and for F138 cells.

E. Discussion of Chromosome Data

Other studies have described chromosome changes after mycoplasmal infection of mammalian cell cultures. In human diploid cell cultures infected with mycoplasma, polyploid counts and break frequencies are high and occasional cells with fragmented chromosomes are seen (Paton *et al.*, 1965). Some of the infected cells have been suspected of containing structurally abnormal chromosomes and translocational transfigurations have been observed. Human leukocyte cultures also show increased chromosome breakage after infection with *M. salivarium* (Aula and Nichols, 1967). There is evidence that this early effect is caused by arginine deficiency. With various types of mycoplasmas, including human *Mycoplasma orale* type 1, murine *Mycoplasma pulmonis*, porcine *Mycoplasma hyorhinis*, and a saprophytic *Mycoplasma laidlawii*, a significant increase in the frequency of chromosome aberrations has been seen in the infected WI-38 cells (Stanbridge *et al.*, 1969). There have also been fluctuations in the breakage frequencies.

In view of the frequent contamination of cell and tissue cultures with mycoplasma, it is pertinent to point out that the chromosome changes observed by us and by others as consequences of mycoplasmal infection in many respects resemble those reported as consequences of virus infection of mammalian cells. For example, extensive changes in chromosome numbers with many of the cells showing low counts have been seen after SV40 infection of diploid human fibroblasts (Koprowski *et al.*, 1962). Dicentrics and acentric fragments are increased in frequency following infection with herpes simplex virus (Hampar and Ellison, 1961a) and SV40 (Koprowski *et al.*, 1962), and dicentrics are also seen in patients with chicken pox (Aula, 1963). Minute chromosomes are present in SV40-transformed cells (Koprowski *et al.*, 1962). Rearrangements of chromosomes with exchanges and translocations are effects of these viruses, of human adenovirus type 12 (Stitch *et al.*, 1964), and to a minor degree of measles virus (Nichols *et al.*, 1962). Chromosome breaks have been observed as an early effect of most of the viruses studied (Hampar and Ellison, 1961a; Nichols *et al.*, 1962; Aula, 1963; Stitch *et al.*, 1964). Fragmented chromosomes are a characteristic effect of adenovirus type 12-infected Chinese hamster cells (Stitch *et al.*, 1964) and were seen in diploid human embryonic cells infected with herpes zoster virus (Benyesh-Melnich *et al.*, 1964). New chromosome varieties have been reported in Chinese hamster cells infected with herpes simplex virus (Hampar and Ellison, 1961b), or with adenovirus type 12 (Stitch *et al.*, 1964). Several abnormal new chromosomes have been observed in SV40-transformed human cells (Koprowski *et al.*, 1962).

IV. Reduced Tumor-Producing Capacity of Mycoplasma-Modified FL Cells

Long-term mycoplasmal infection of FL cells reduced their tumor-producing capacity (Fogh, 1969). In the cheek pouch of cortisonized weanling hamsters, the number of cells necessary for tumor production was increased as compared to uninfected FL cells; also after elimination of mycoplasma from the cells. Tumors resulting from a certain inoculum of mycoplasma-modified cells were smaller and regression occurred earlier.

Detailed descriptions of tumor growth in X-irradiated and cortisone-treated weanling rats after inoculation of FL cells and of a tumor developing after subcutaneous inoculation in a late cancer patient have been reported (Fogh and Hok, 1958; Friedman and Fogh, 1958). In the human host, the FL cell tumor appears to satisfy the criteria of Warren (1957) for malignancy. Toolan's suggested criteria (Toolan, 1957) for malignancy in the conditioned heterologous host are also fulfilled. Intramuscular inoculation of FL cells into the thigh muscles of young female mice, which were given cortisone subcutaneously, has resulted in the production of tumors containing bone and cartilage (Anderson *et al.*, 1964). Because the hamster technique provides the opportunity for continuous and quantitative direct observations of developing tumors, it was chosen for all experiments involving mycoplasmal modification.

The cell types include uninfected FL cells as controls, cells of three mycoplasma-infected and mycoplasma-modified FL lines (F138, F159, HTP8), and cells of the corresponding lines after mycoplasmas have been eliminated (F138cl, F159cl, HTP8cl), as shown in Table I. This table also indicates the range of chromosome numbers and the frequency of metaphase plates with approximately double stem line (2S) for each of the seven cell lines. The time of mycoplasmal infection and elimination and the interval during which the tumor assays were done is indicated for each of the modified lines.

The incidence of animals developing tumors at various cell inoculation levels, as well as the average number of cells needed for tumor production in 50% of inoculated animals (TPD_{50}), is shown in Table II. For the mycoplasma-modified lines, the TPD_{50} is increased in the order of five to ten times, as compared to uninfected FL cells. The differences in TPD_{50} between the lines are unrelated to whether they still carried the mycoplasmal infection or to whether mycoplasmas have been eliminated. In order to obtain 100% tumor incidence with cells of the mycoplasma-modified lines, 600,000 to 1,000,000 cells must be inoculated per animal. For FL cells, 100% tumor incidence is obtained with 300,000 cells.

Mean values for tumor volumes (in mm^3) at various cell inocula, calculated from the largest tumor measured in each animal at any time, are shown in Table III for the cell types FL, F138cl, F159cl, and HTP8cl. It is obvious that tumor volumes for the three mycoplasma-modified lines are reduced.

An apparent increase in tumor volume observed after inoculating mycoplasma-modified cells still carrying the infection was not an expression of a high rate of tumor cell growth. Most of the volume was a large mass consisting of central necrosis. The concurrent infection with extensive infiltration of neutrophils, perhaps combined with a reduction of cellular debris removal, explained the increased volume. In contrast, the tumors resulting from inoculating the uninfected cells contained predominantly solid masses of viable tumor cells and showed only a minor degree of necrotic changes.

Although tumor-producing capacity cannot be considered an absolute measure of the malignancy of the cells, these tumor results, when cor-

TABLE I

Range of Chromosome Numbers and Percentage of Metaphase Plates with Double Stem Line (2S) during the Tumor-Testing Period for FL Cells and the Six Lines of Mycoplasma-Modified FL Cells[a]

	Type of cell						
	FL	F138	F138cl	F159	F159cl	HTP8	HTP8cl
Months after mycoplasma inoculation	—	22–47	27–47	13–38	22–38	15–40	24–40
Months after mycoplasma elimination	—	—	0–20	—	0–16	—	0–16
Number of chromosomes	68–75	62–66	59–66	64–66	61–67	66–68	63–68
2S	3.3	6.0	5.0	3.3	12.0	1.7	5.0

[a] Fogh (1969).

TABLE II
Incidence of Animals Developing Tumors at Various Cell Inoculation Levels for the Seven Different Cell Types[a]

Cell type	Number of cells inoculated per animal									
	6×10^6	3×10^6	1×10^6	6×10^5	3×10^5	1×10^5	3×10^4	1×10^4	1×10^3	TPD_{50}[b] $\times 10^4$
FL	20/20	28/28	22/22	51/51	62/62	46/63	10/50	1/17	0/10	5.9
F138	23/23	10/10	6/6	8/8	4/16	0/16	0/16	—	—	38
F138cl	26/26	—	—	26/29	26/39	0/39	0/38	—	—	24.5
F159	10/10	28/28	20/22	5/6	6/14	2/18	0/8	0/10	0/10	35
F159cl	16/16	—	—	7/8	7/8	0/8	0/8	—	—	20
HTP8	10/10	10/10	6/6	10/10	5/10	2/10	0/10	—	—	25
HTP8cl	16/16	—	—	9/10	5/10	0/10	0/10	—	—	32

[a]Fogh (1969).
[b]TPD_{50} is the number of cells needed for tumor production in 50% of inoculated animals.

TABLE III
Mean Value for Tumor Volumes (in mm³) at Various Cell Inoculation Levels for Four Cell Types[a]

	Number of cells inoculated per animal								
Cell type	1×10^7	6×10^6	3×10^6	1×10^6	6×10^5	3×10^5	1×10^5	3×10^4	1×10^3
FL	767	133	181	114	54	33	3.6	0.7	0
F138cl	—	29.5	—	—	2.9	0.7	0	0	—
F159cl	—	20	—	—	6	4	0	0	—
HTP8cl	—	34	—	—	3.5	0.2	0	0	—

[a] Fogh (1969).

related with the changes in *in vitro* behavior and in chromosome constitution, add further support to the conclusion that the malignant state of FL human amnion cells has become degraded after a long-term mycoplasmal infection. As clearly demonstrated, the reduction in tumor-producing capacity is also irreversible, because the mycoplasma-modified cell lines from which mycoplasmas have been eliminated maintain the reduced tumor-producing capacity even when examined many months after the mycoplasma elimination. With reference to the changes characteristic of virus-induced *in vitro* transformations, the irreversible changes in mycoplasma-modified FL cells have, quite appropriately, been referred to as the effects of a reverse transformation, or, perhaps, retransformation (Fogh and Fogh, 1967).

V. Effects of Mycoplasmas on SV40 Transformation of Human Amnion Cells

Many changes from the normally observed picture were seen in cultures or primary human amnion cells exposed to and transformed by SV40 when such cultures were also infected with mycoplasmas of the HT strain (Fogh, 1970). The incidence of transformation by SV40 was reduced, and the time of appearance of transformed foci was delayed. The number of population doublings and the number of culture passages prior to crisis were reduced. The time before crisis was shortened. When SV40-transformed cell strains, already in serial cultivation, were infected with mycoplasmas, pronounced cell destruction was observed and the population doubling number was considerably reduced. Concurrent mycoplasmal infection changed the virus production pattern for the SV40-transformed cell strains. When mycoplasmas were eliminated from such cell strains, the number of culture passages, the number of cell divisions, and the period of time prior to crisis increased.

The reduced incidence of transformation, the late appearance of transformed foci, and the decreased growth potentials resulting from concurrent infection with SV40 and mycoplasmas may, in part, be explained by the cell destructive effects of the mycoplasmas. SV40-transformed amnion cells are more susceptible to the effects of mycoplasmas than are untransformed amnion cells. This increased susceptibility to destruction may have reduced the number of cells that underwent transformation in the primary culture, and the number of transformed foci observed, therefore, may be similar to that observed after a lesser multiplicity of virus exposure (Fogh *et al.*, 1970a). There is also some cell destruction during subcultivation. Depletion of the culture medium by the mycoplasmas may be a factor of importance. Other data from such experiments have shown that cell strains resulting from concurrent mycoplasmal and SV40 infection of primary amnion cell cultures demonstrate an increased resistance to later infection with mycoplasma. Although apparently not a transforming agent in this system by itself, the HT mycoplasma has modified the SV40-transformed human amnion cells to the extent that long-lasting and possibly permanent changes have occurred.

The chromosome picture in this system was also changed after concurrent mycoplasma–SV40 infection of cultures of primary amnion cells (Fogh *et al.*, 1970b). Chromosome numbers were maintained in the diploid range for a longer period of time. Most chromosome aberrations were present at lower frequencies. The chromosome numbers and aberration frequencies were not reversed when mycoplasmas were eliminated from the cultures. However, a large telocentric chromosome increased in frequency after mycoplasma elimination. The relative absence of hypertriploid–hypotetraploid cells in the presence of mycoplasmas could be the result of an adaptation or it could be the response to a selection pressure in the population toward cells with lower numbers. Another interpretation is a delay of the appearance of, or a specific elimination of, cells with higher numbers, conceivably by cell destruction.

VI. Decreased Resistance of Transformed Human Cells to the Destructive Effects of Mycoplasmal Infection

A number of different experiments have shown that primary human amnion cells and diploid human fibroblasts were quite resistant to infection with the HT strain of mycoplasma (Fogh *et al.*, 1965, 1971). For example, 6 weeks after infection with a low mycoplasma inoculum, there was no cell loss in primary amnion cell cultures. During the first month, morphologic changes were insignificant. A few weeks later, however, cells appeared slightly more elongated than in corresponding uninfected cultures. In mixed cultures of FL cells and primary amnion cells infected with this mycoplasma, FL cell colonies were reduced in size within 1 week after infection as compared to similar uninfected cultures, and selective and pronounced destructive effects of FL cells were apparent. The morphology of the primary amnion cells in these cultures remained unchanged (Fogh *et al.*, 1971).

There is evidence from other studies of human cell systems that cells with higher chromosome numbers are more susceptible to destruction by mycoplasmas than are the diploid cell from which they are derived. Such a difference has been demonstrated among SV40-transformed human amnion cells and primary amnion cells. Mycoplasmal effects on the growth of these cells may also be evaluated from the appearance and structure of cell colonies in infected and uninfected cultures (Fogh, 1971). In uninfected cultures, the colony centers consist of several to many layers resulting from pronounced vertical growth. Mycoplasma-infected cultures show a reduction in the number of cells per colony and a tendency for the cells to grow as a monolayer within the colonies. There is a more pronounced reduction in growth per colony for cells originating from postcrisis established cell lines (Gaffney *et al.*, 1970) than for cells of precrisis strains (Fogh *et al.*, 1969).

Both pre- and postcrisis SV40-transformed amnion cells were susceptible to the destructive effects of the mycoplasmas in mixed cultures of either of these cells with primary human amnion cells. Also, under these conditions, the postcrisis cells showed the earliest and most pronounced cell destruction. The appearance of the primary cells appeared unchanged for weeks. Morphological studies showed the precrisis cells grew less organized after infection, with nuclear changes including enlargement, irregular shapes, and pycnosis and prevalent phagocytosis. Post crisis cells showed much more pronounced destruction, and the appearance of the cell layer was changed from an epithelial-like growth to a completely unorganized growth of spindle-shaped cells with oblong nuclei of irregular shapes, mixed with dead cells and debris. Again, the morphology of the primary amnion cells appeared completely unchanged. Growth-curve experiments have confirmed the differences between these three types of cells.

As previously mentioned, FL cells are destroyed more and show more pronounced morphological changes than do cells of any of the other lines we have examined. However, all transformed lines, and all lines of malignant origin, respond with some degree of cell destruction, varying from cell type to cell type. Many of the examined lines include old laboratory lines, in most cases obtained from other laboratories. Of these lines, only FL cells developed in our laboratory have been carried continuously here since the establishment of the cell line many years ago. Because of regular and frequent tests, we have had good evidence that our line of FL cells has never been accidently contaminated with mycoplasmas. It is conceivable that many of the other lines at some time have been exposed to mycoplasmas. Hence, it has been proposed (Fogh *et al.*, 1971) that their lesser susceptibility to mycoplasmas could be related to a previous infection, which is responsible for a change to a more resistant cell population.

VII. Cell-Associated Mycoplasmas: Preferential Attachment to Transformed Cells and Cells of Malignant Origin

There is a distinct difference in the amounts of mycoplasmas of the HT strain that may be seen in association with FL cells and with primary normal amnion cells from which FL cells have been derived by spontaneous transformation (Fogh *et al.*, 1971). A very similar picture is obtained when primary amnion cells are compared with SV40-transformed amnion cells (Fogh, 1971). Other types of transformed cells and several types of cells derived from malignant human tumors have also shown an increased amount of cell-associated HT mycoplasma in comparison with the normal cells. This comparison has been performed by using the method for direct demonstration of mycoplasmas in infected mammalian cell cultures (Fogh and Fogh, 1964), by which the cultured cells are exposed to a hypotonic treatment, Carnoy's fixative followed by air drying, and staining with Orcein. The stained cultures are observed under phase optics.

Figures 6 and 7 show mixed cultures of FL cells and primary amnion cells after several days infection with the HT mycoplasma. In both micrographs, great amounts of mycoplasmas can be seen associated with the cell surfaces, in particular with the free borders of the FL cells. In contrast, the normal cells are mainly devoid of associated mycoplasmas. There are reasons to assume that the differences in the amounts of associated mycoplasmas, as well as in the susceptibility to the destructive effects of the mycoplasmas, are related to the different state of *in vitro* malignancy of the cultured cells.

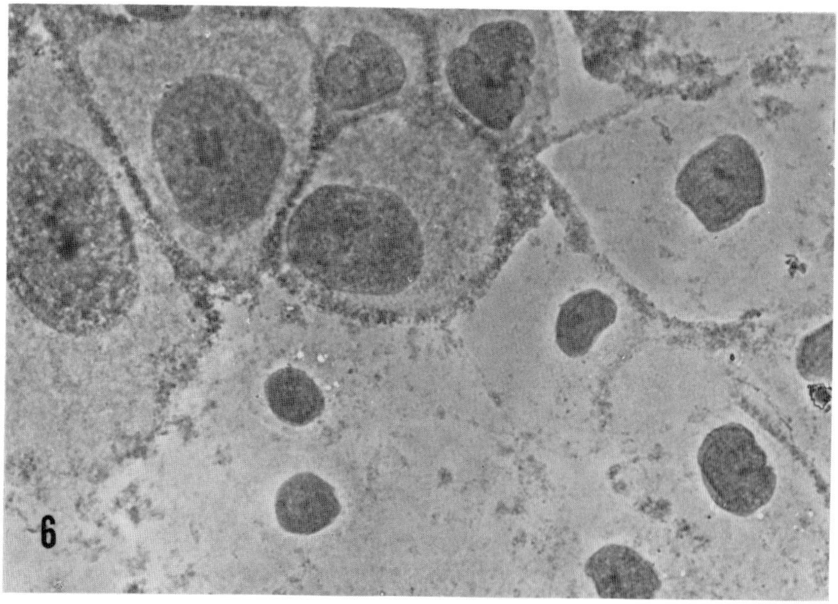

Fig. 6. Mycoplasma-infected culture of FL (top of photograph) and primary amnion cells (bottom) stained with Orcein. × 672.

VIII. Discussion of Results

The present data have demonstrated several very pronounced effects of infection of human cells in culture with the HT *Mycoplasma fermentans*. The ultimate answer to the question of which mechanism is responsible for these changes is still not available. It is yet to be finally determined whether the effects are due to an intracellular action by the mycoplasmas or to mycoplasma-associated products. The consistency of the appearance of new chromosome varieties in several independent experiments following mycoplasmal infection of FL cells indicates that mycoplasma-induced genetic changes are a more likely mechanism than an adaptation or selection process. It also suggests that specific loci on the FL cell chromosomes are affected by infection with our HT strain of *Mycoplasma fermentans*. It is conceivable that at least some of the chromosome differences are related to the other observed differences among uninfected cells, mycoplasma-infected cells, and mycoplasma-modified cells from which mycoplasmas have been eliminated. It has been suggested (Stanbridge *et al.*, 1969) that the close mycoplasma–host cell relationship, intracellular or membrane bound, allows the mycoplasmas access to the nucleic acid precursors manufactured by the cell. This may be accomplished by an alteration of the cell

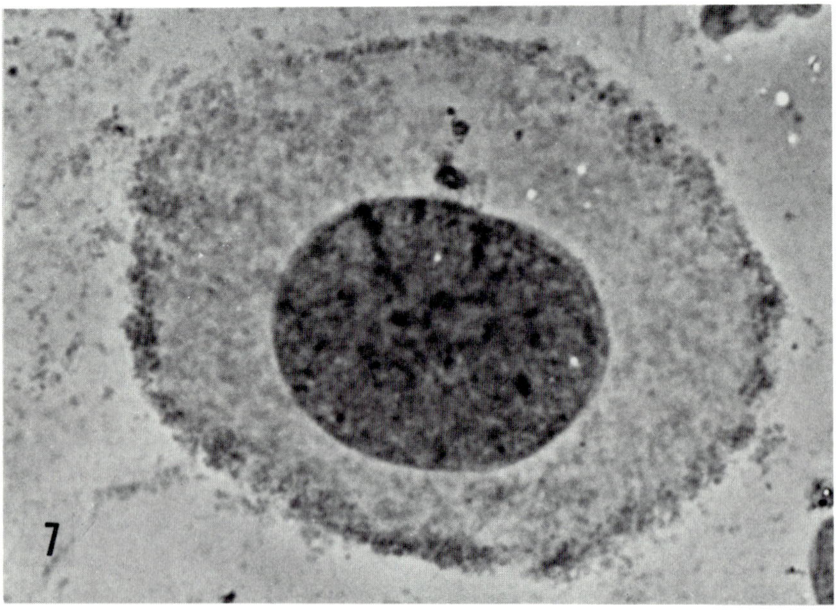

Fig. 7. High magnification of a single, very large mycoplasma-infected FL cell surrounded by primary amnion cells, stained with Orcein. × 1092.

permeability, and the induced chromosome aberrations may be the result of a competitive effect for nucleic acid precursors that may interfere with the host cell DNA synthesis.

Certain cell changes after mycoplasmal infection are reversible, for example, cytopathic effects and some chromosome changes. However, many of the effects of the HT mycoplasma on FL cells, and also on primary amnion cells when concurrently infected with SV40, are irreversible. These irreversible effects include changed *in vitro* growth and cell morphology, increased resistance to mycoplasmal infection, changes in chromosome number and karyotypes, and the presence of new chromosome varieties. Reduced tumor-producing capacity is also maintained after mycoplasma elimination. The observation that irreversible changes occur in characteristics of human cells after infection, both *in vitro* and *in vivo*, with this mycoplasma calls for great caution in considering mycoplasma contamination an innocent phenomenon, even when mycoplasma elimination has been accomplished. As illustrated by our studies, the cured cell population may indeed differ considerably from the population of the initial uninfected cells.

References

Anderson, H. C., Merker, P. C., and Fogh, J. (1964). *Amer. J. Pathol.* **44**, 507.

Aula, P. (1963). *Hereditas* **49**, 451.

Aula, P., and Nichols, W. W. (1967). *J. Cell. Physiol.* **70**, 281.

Benyesh-Melnick, M., Stitch, H. F., Rapp, F., and Hsu, T. C. (1964). *Proc. Soc. Exp. Biol. Med.* **117**, 546.

Berg, R. B., and Frothingham, T. E. (1961). *Proc. Soc. Exp. Biol. Med.* **108**, 616.

Edwards, G. A., and Fogh, J. (1960). *J. Bacteriol.* **79**, 267.

Fogh, J. (1969). *Cancer Res.* **29**, 1721.

Fogh, J. (1970). *Proc. Soc. Exp. Biol. Med.* **134**, 217.

Fogh, J. (1971). *Proc. Soc. Exp. Biol. Med.* **137**, 498.

Fogh, J., and Fogh, H. (1964). *Proc. Soc. Exp. Biol. Med.* **117**, 899.

Fogh, J., and Fogh, H. (1965). *Proc. Soc. Exp. Biol. Med.* **119**, 233.

Fogh, J., and Fogh, H. (1967). *Proc. Soc. Exp. Biol. Med.* **126**, 67.

Fogh, J., and Fogh, H. (1968). *Proc. Soc. Exp. Biol. Med.* **129**, 944.

Fogh, J., and Hok, K. (1958). *Cancer Res.* **18**, 692.

Fogh, J., and Lund, R. O. (1957). *Proc. Soc. Exp. Biol. Med.* **94**, 532.

Fogh, J., Hahn, E., and Fogh, H. (1965). *Exp. Cell Res.* **39**, 544.

Fogh, J., Ramos, L., and Fogh, H. (1969). *In* "Axenic Mammalian Cell Reactions" (G. L. Tritsch, ed.), p. 59. Dekker, New York.

Fogh, J., Gaffney, E. V., and Loveless, J. D. (1970a). *Proc. Soc. Exp. Biol. Med.* **134**, 174.

Fogh, J., Fogh, H., and Dowling, A. M. (1970b). *Proc. Soc. Exp. Biol. Med.* **135**, 206.

Fogh, J., Fogh, H., and Ramos, L. (1971). *Proc. Soc. Exp. Biol. Med.* **136**, 809.

Friedman, M., and Fogh, J. (1958). *Cancer Res.* **18**, 879.

Gaffney, E. V., and Fogh, J. (1970). *Proc. Soc. Exp. Biol. Med.* **133**, 607.

Gaffney, E. V., Fogh, J., Ramos, L., Loveless, J. D., Fogh, H., and Dowling, A. M. (1970). *Cancer Res.* **30**, 1668.

Hampar, B., and Ellison, S. A. (1961a). *Nature* (*London*) **192**, 145.
Hampar, B., and Ellison, S. A. (1961b). *Proc. Nat. Acad. Sci. U.S.* **49**, 474.
Koprowski, H., Pontén, J. A., Jensen, F., Ravdin, R. G., Moorhead, P., and Saksela, E. (1962). *J. Cell. Comp. Physiol.* **59**, 281.
Nichols, W. W., Levan, A., Hall, B., and Östergren, G. (1962). *Hereditas* **48**, 367.
Paton, G. R., Jacobs, J. P., and Perkins, F. T. (1965). *Nature* (*London*) **207**, 43.
Robinson, L. B., Wichelhausen, R. H., and Roizman, B. (1956). *Science* **124**, 1147.
Stanbridge, R., Onen, M., Perkins, F. T., and Hayflick, L. (1969). *Exp. Cell Res.* **57**, 397.
Stitch, H. F., Van Hoosier, G. L., and Trentin, J. J. (1964). *Exp. Cell Res.* **34**, 400.
Toolan, H. W. (1957). *Cancer Res.* **17**, 248.
Tully, J. G. (1963). *Proc. Soc. Exp. Biol. Med.* **114**, 704.
Warren, S. (1957). *In* "Pathology" (W. D. Anderson, ed.), 3rd ed., p. 417. Mosby, St. Louis, Missouri.

CHAPTER **8**

Contamination in Tissue Culture by Parasites

NELDA B. HOLMGREN

I. Introduction 195
II. Identification of the Amoebae 196
A. Cytopathology 196
B. Characteristics Simulating Virus Properties 197
III. Source of Contamination 197
IV. Methods of Detection, Prevention, and Cure 198
A. Methods of Detection 198
B. Prevention and Cure 200
V. Summary 202
References 203

I. Introduction

On several occasions, *Hartmannella* species of amoeba have been identified as contaminants of cultured tissue cells. Although this potential source of contamination is relatively infrequent, lack of awareness of it has caused considerable difficulty because of the similarity of the trophozoite form to damaged tissue cells and because of the similarity of cytopathology produced by amoebae in tissue cell culture to that produced by viruses.

Various species of the genus *Hartmannella* have been identified either as spontaneous or as coincidental contaminants of tissue culture cells. Spontaneous contamination has been reported in freshly trypsinized monkey kidney cells (Jahnes, *et al.*, 1957; Culbertson, *et al.*, 1958); Chang's liver cells (Chang, 1961); canine lymphosarcoma cells, LS30 (Moore and Hlinka, 1968); and HeLa cells (Casemore, 1969). Coincidental contamination with amoebae has been detected in cell cultures inoculated from pharyngeal swabs or other experimental or clinical material (Hull *et al.*, 1958; Chi *et al.*, 1959; Wang and Feldman, 1967). The ease of isolation from the human throat (Wang and Feldman, 1967) and successful cultivation in many types (primary, diploid, established cell lines) and several species (human, monkey, canine, bovine, rabbit, and chick embryo) of tissue culture cells emphasizes the efficacy of tissue culture as an excellent growth medium for these species of *Hartmannella*.

II. Identification of the Amoebae

A. Cytopathology

1. Light and Phase Microscopy of Living Cultures

Cells infected with *Hartmannella* amoeba become granular and gradually progress to complete disintegration, the time required depending on the size of the inoculum and whether cysts or trophozoites are the predominant form present in the inoculum. Culbertson (1971) has noted a general relationship between the pathogenicity of the amoeba for mice and the rate of growth in tissue culture. In an experiment reported by Jahnes *et al.* (1957), cysts predominate up to day 9 in monkey kidney cell culture. By day 11, only trophozoites are found and cysts again begin to appear on day 16 of culture. On day 20 only cysts are found.

In a culture containing many cysts, which vary in size from 12 to 31 μm in diameter (Kingston and Warhurst, 1969), the cysts have been mistaken on microscopic examination of the living culture for the cytopathic effect on tissue cells caused by some viruses. However, if a living culture is examined during the trophozoite phase, close microscopic examination shows the sluggishly motile amoeba protruding and retracting filiform pseudopodia. The *Hartmannella* trophozoites, varying in diameter from 10 to 30 μm (Kingston and Warhurst, 1969) when rounded, have a granular, vacuolated cytoplasm. Sometimes in such preparations, a small, spherical, well-defined nucleus (3 μm) containing a relatively large eosinophilic karyosome measuring approximately one-half the diameter of the nucleus can be clearly seen (Armstrong and Pereira, 1967). Another distinguishing feature that can be occasionally seen, especially in preparations containing bacteria or other particles, is a contractile vacuole with a periodicity of 1 to 2 minutes.

2. Electron Microscopy

The cytopathology of infected cultures was shown by electron microscopy (Armstrong and Pereira, 1967) to be of a degenerative nature that at the fine structural level was revealed as swelling or rupture of the mitochondria and endoplasmic reticulum with eventual fragmentation of the cells. In late stages, only trophozoites could be found among the cellular debris. The same authors reported nuclear changes caused by the amoebae that were suggestive of viral infection: altered nucleoli, eosinophilic inclusions, and clumped chromatin. The altered nucleoli showed the phenomenon of nucleolar capping, a segregation of nucleolar components. Similar changes were described as resulting from mycoplasmal infection (Jézéquel *et al.*, 1967).

B. Characteristics Simulating Virus Properties

As contaminants in tissue culture, the trophozoite forms of the parasite have been mistaken for damaged tissue cells. They bear some resemblance to the macrophage. Lack of awareness of the possibility of amoebic contamination has led some investigators to the mistaken conclusion that a virus is the responsible agent; some terms for this agent have been *virus* (Chavis) (Feldman and Wang, 1967), *Ryan virus* (Pereira *et al.*, 1966); and *lipovirus* (Dunnebacke and Williams, 1967; Armstrong and Pereira, 1967). Certain properties of both viruses and amoebae have characteristics in common and these can lead to confusion (Pereira *et al.*, 1966). They bear some resemblance to each other in cytopathology, in transmission serially in tissue culture, in sensitivity to ether, in resistance to acid at pH 4, in inhibition by 5-fluorodeoxyuridine, in failure to grow in bacteriological culture medium, in resistance to antibiotics, and in animal pathogenicity. The eosinophilic karyosome in the nucleus of the amoeba has been mistaken for a viral inclusion. The amoeba differ from viruses in filtrability, as they are retained by bacteriological filters.

III. Source of Contamination

The *Hartmannella* amoeba belong to a relatively large group found living free in the soil. They resemble species first discovered by Castellani (1930), described by Douglas (1930), and classified as *Hartmannella castellanii* Douglas (1930). A key for the identification of these species can be found in a paper by Kingston and Warhurst (1969). The taxonomic status of the free-living amoebae is somewhat confused and unsettled. Some authorities (Kudo, 1966; Chang, 1971) list two genera, *Hartmannella* and *Acanthamoeba*. Kudo places them in the family Amoebidae Bronn, whereas Chang classifies them in the family Hartmannellidae Singh (1952). Kingston and Warhurst (1969) also classify the species in the family Hartmannellidae but list some in the genus *Hartmannella*, and others in the genus *Hartmannella*, subgenus *Acanthamoeba*. Most of the isolates from tissue culture appear to be *Hartmannella* sp. An excellent summary of the history and problems in taxonomy of the free-living amoebae can be found in Culbertson (1971).

In the soil, the amoebae are found either in the cystic or vegetative form, where they feed on bacteria, yeasts, fungi, and other organic material. The presence of cysts in the air has been demonstrated by Kingston and Warhurst (1969), who have isolated in tissue culture or on agar plates 36 strains of amoebae from samples of the inside or outside air of a hospital obtained either through a slit sampler or on settle plates. Fifteen of the strains belong to the genus *Hartmannella* but a few *Naegleria* sp. have also been found. They observed a low rate of settling (0.026 amoeba per ft^2, or 0.28 per m^2/minute), which suggested that settling during laboratory manipulation would be infrequent. The values found for the particle diameters were high (33–37 μm) as compared to the diameter of the cyst (10–31 μm). One suggestion was that the cysts were associated with particles of soil and airborne particles of this size would be impacted in the nose.

Casemore (1969) recovered *Hartmannella* amoeba in large numbers from the air and soil. He also found that cysts maintained at low temperature (4°C to −70°C) vegetated readily when placed in a suitable medium. Thus, the possibility of airborne contamination is very likely.

Isolation of free-living amoebae from the human nasopharynx (Chi *et al.*, 1959; Wang and Feldman, 1967; Pereira *et al.*, 1966) suggests a second source of contamination; the possibility of contamination from the nasopharynx of the individual handling the cultures.

Evidence for a third possible source is suggested by the pathogenicity of these strains for mice, monkeys, and rabbits (Culbertson, 1961) and for humans (Cerva *et al.*, 1968; Culbertson *et al.*, 1968). Culbertson *et al.* (1958) report that sections of the kidney of infected monkeys show the presence of amoebae in the glomerular capillaries. This suggests an alternate explanation to airborne contamination for the finding of amoebae in monkey kidney tissue culture. It seems possible that amoebae or cysts in the monkey kidney may have been the source in some instances.

IV. Methods of Detection, Prevention, and Cure

A. Methods of Detection

1. Microscopic

Tissue cell cultures in Leighton tubes and plastic flasks or on agar plates should be closely examined by both phase and light microscopy and at both low and high magnification for cysts and/or sluggishly motile trophozoite forms of the amoeba. If the amoebae are predominantly present in the cystic form, the discrete, shrivelled cysts can be mistaken for dead tissue cells. The size and shape of the trophozoite varies considerably but even at 100 times magnification, blunt or filiform pseudopodia can be seen to protrude, change in shape, and retract. Maintaining a 37°C temperature during observation or suspending the culture in sterile 5% hog gastric mucin enhances the motility (Culbertson, 1961). Although not always visible in living cultures, the typical nucleus with its large central karyosome and the contractile vacuole aid in identifying the trophozoite form of the amoebae.

2. *Staining Methods*

For histological study of the trophozoite form, fixing coverslips with formalin or Bouin fixative and staining with Hematoxylin and Eosine is satisfactory. To bring out the eosinophilic karyosome in the nucleus, it is better to use methyl alcohol for fixation followed by either Wright's. Giemsa or May Grunwald stains. Culbertson (1961) and Culbertson *et al.* (1966, 1968) list other suitable stains. Staining of the cysts is more difficult. Jahnes *et al.* (1957) found that the Bauer reaction (Lillie, 1954) stained the entire cyst red.

3. *Culture*

The amoebae grow well when seeded on a 1.5% agar plate containing 0–0.4% NaCl that has been mixed with or spread with a lawn of living or dead *Escherichia coli* organisms (Culbertson, 1968, 1971) over the agar surface. Incubation of plates at several temperatures (25°C, 33°C, and 37°C) is advisable. Growth can also be obtained in nutrient broth media; some cultures grow without the addition of bacteria and others require the bacteria. Similarly, some *Hartmannella* species can grow in tissue culture medium in the absence of tissue cells (*H. glebae* and *H. rhysodes*), whereas others require the presence of tissue cells (*H. agricola*; Feldman, 1971). For inoculation into tissue cell cultures, 0.1 ml of a suspension of cysts or amoebae is inoculated onto 4- to 7-day-old cell sheets and examined frequently for cytopathology.

Although most of the strains isolated as contaminants of tissue culture have been identified as species of the genus *Hartmannella*, the possibility of a contamination with other genera of amoeba or other parasites must be considered. Culbertson (1968) described a pathogenic *Neagleria* sp. isolated from human cerebrospinal fluid. He has generously permitted us to reproduce some figures from this publication, showing photographs of both living and stained representative amoebae and cysts of two species, *Naegleria* (HB-1) and *Hartmannella* (A-1 and HN-3). Figures 1 and 2 demonstrate the typical lobate pseudopodia of the living trophozoite forms of the *Naegleria* sp. contrasted with the filiform pseudopodia of the *Hartmannella* sp. Figures 3 and 4 show the respective cysts. Figures 5 and 7 are photographs of the stained *Naegleria* amoebae showing the typical morphology in MK_2 tissue culture and the characteristic mitotic figure with large polar masses and in-

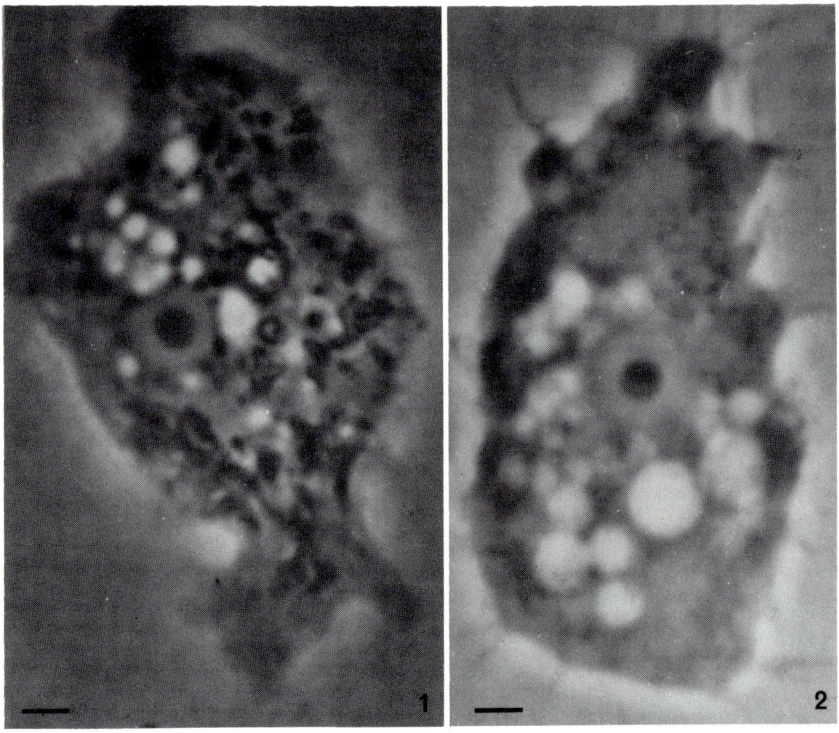

Fig. 1. HB-1 strain transplanted from MK_2 tissue culture to water agar plate with *E. coli*, covered with cover slip, and incubated 24 hours at 37°C. Note large size, distinct nucleus, and lobate pseudopods. Phase. × 1800. The line indicates the scale of 4 μm. (From Culbertson, 1968.)

Fig. 2. *Hartmannella* (A-1) under same conditions as Fig. 1. Note filiform pseudopods. Phase. × 1800. The line indicates the scale of 4 μm. (From Cubertson, 1968.)

terzonal body. Figure 6 is a similar photograph of *Hartmannella* amoebae in MK_2 tissue culture showing a mitotic figure without polar masses.

B. Prevention and Cure

Culbertson *et al.* (1965) reported that mice infected with *Hartmannella* amoeba were protected by the addition of sulfadiazine to the diet, whereas relatively large doses of amphotericin B in combination with large doses of sulfadiazine in the diet had activity against experimental infection of mice with a *Naegleria* sp. (Culbertson, 1968). Casemore (1969) found that 10 μg/ml of metronidazole added to the culture medium prevented the contamination of cell cultures for a period of about 6 months in a laboratory

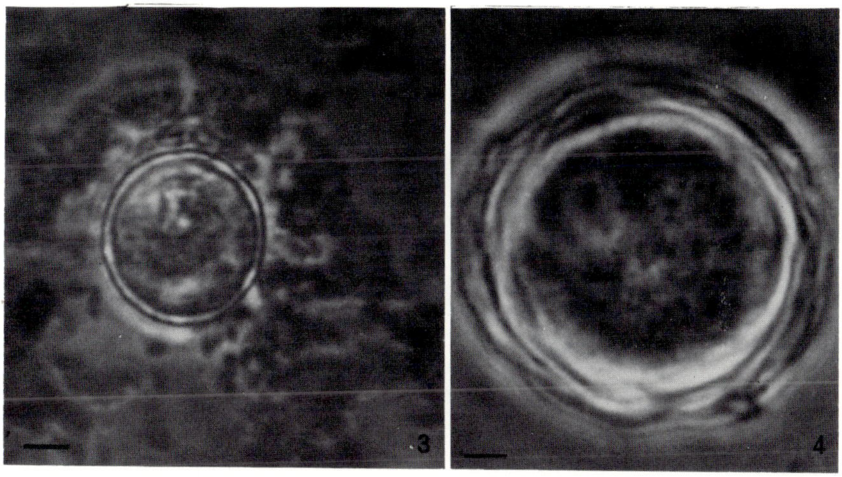

Fig. 3. Naegleria Hb-1 cyst. Phase. × 1500. The line indicates the scale of 4 μm. (From Culbertson, 1968.)

Fig. 4. *Hartmanella* HN-3 cyst with same environment as Fig. 3. Phase. × 1800. The line indicates the scale of 4 μm. (From Culbertson, 1968.)

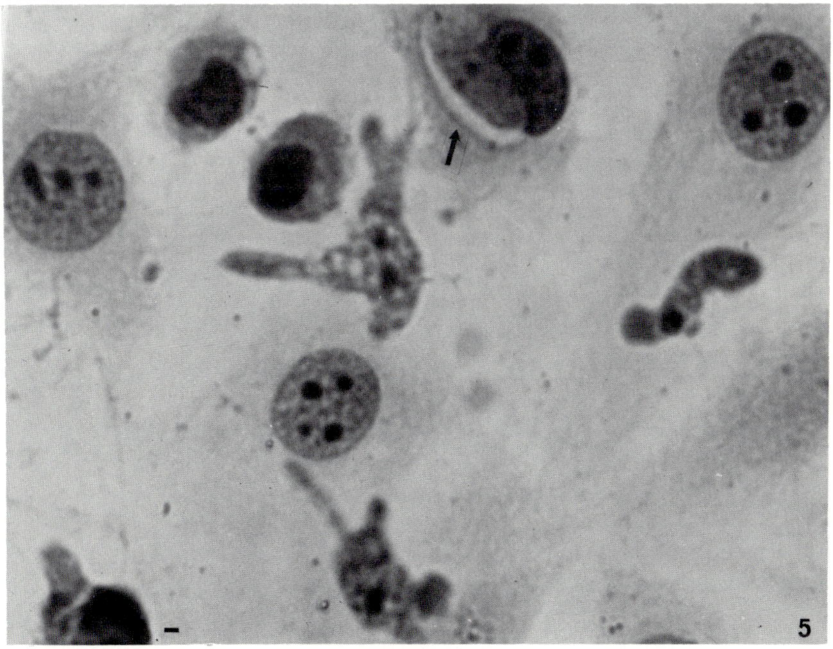

Fig. 5. HB-1; late effects on MK_2 tissue culture. Note lobate amoebas and tissue cells shrinking and becoming loose from glass. Amoeba at arrow is closely applied to surface of cell. H and E; × 525. The line indicates the scale of 4 μm. (From Culbertson, 1968.)

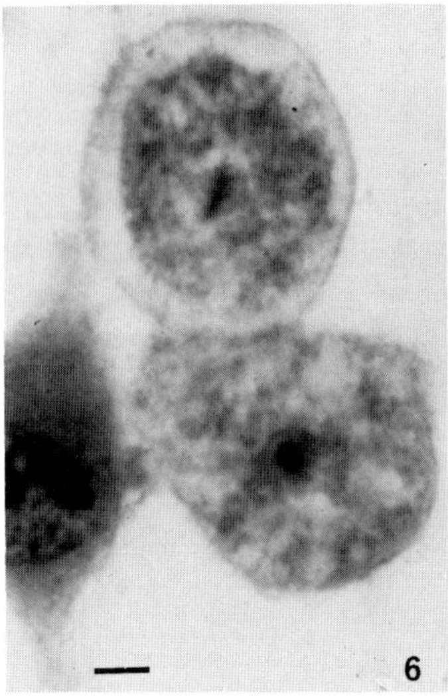

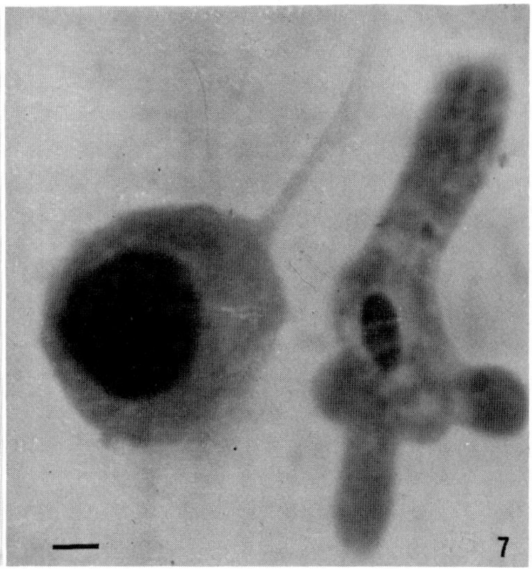

Fig. 6. *Hartmannella* (A-1) in Mk_2 tissue culture. Lower amoeba has intermitotic nucleus. Upper one shows metaphase mitotic figure without polar masses. H and E. × 1350. The line indicates the scale of 4 μm. (From Culbertson, 1968.)

Fig. 7. HB-1 in mitosis; MK_2 tissue culture Note large polar masses and interzonal body. H and E. × 1350. The line indicates scale of 4 μm. (From Culbertson, 1968.)

favoring contamination with amoebae from the environment. Thorough cleaning of the laboratory did not in itself prevent further contamination.

However, kanamycin or neomycin at a concentration of 100 μg/ml in the medium did not inhibit a *Hartmannella* sp. (Pereira *et al.*, 1966), nor was therapy of 16 human cases of amoebic meningoencephalitis successful with sulfonamides, penicillin, tetracycline, oxytetracycline, chlortetracycline, oleandomycin, erythromycin, or chloromycetin (Cerva *et al.*, 1968).

V. Summary

This review of the literature on contamination in tissue culture by parasites stresses, above all, the importance of alerting investigators to the possibility of contamination with an amoeba, particularly because of the close resemblance of the cysts and trophozoites to damaged cells. Microscopic examination at higher magnification should enable a presumptive identification of the trophozoite by motility, characteristic nucleus, and contractile vacuole. Culturing on agar or liquid medium containing edible bacteria (Singh, 1960) should confirm the diagnosis. The source appears to be either aerial contamination or the tissue of origin, particularly because these parasites encyst under adverse conditions and readily excyst in suitable culture medium. Droplet infection from the human nasopharynx cannot be ruled out as a possible source. For prevention, aseptic technique and care to prevent an outside source of cysts from entering the laboratory seem paramount. Preliminary studies indicate that the use of such chemosuppressants as metronidazole may help to prevent infection in areas where airborne contamination is difficult to control. Finally, it emphasizes the necessity of keeping healthy stocks of cells preserved in the frozen state to resort to in case of an overwhelming contamination with amoebae.

References

Armstrong, J. A., and Pereira, M. S. (1967). *Brit. Med. J.* **1**, 212

Casemore, D. P. (1969). *J. Clin. Pathol.* **22**, 254.

Castellani, A. (1930). *J. Trop. Med. Hyg.* **33**, 188.

Cerva, L., Novak, K., and Culbertson, C. G. (1968). *Amer. J. Epidemiol.* **88**, 436.

Chang, R. S. (1961). *Proc. Soc. Exp. Biol. Med.* **107**, 135.

Chang, S. L. (1971). *Curr. Top. Pathobiol.* **1**, 241.

Chi, L., Vogel, J. E., and Shelokov, A. (1959). *Science* **130**, 1763.

Culbertson, C. G. (1961). *Ameri J. Clin. Pathol.* **35**, 195.

Culbertson, C. G., Smith, J. W., and Minner, J. R. (1958). *Science* **127**, 1506.

Culbertson, C. G., Holmes, D. H., and Overton, W. M. (1965). *Amer. J. Clin. Pathol.* **43**, 361.

Culbertson, C. G., Ensminger, P. W., and Overton, W. M. (1966). *Amer. J. Clin. Pathol.* **46**, 305.

Culbertson, C. G., Ensminger, P. W., and Overton, W. M. (1968). *J. Protozool.* **15**, 353.

Culbertson, C. G. (1971). *Annu. Rev. Microbiol.* **25**, 231.

Douglas, M. (1930). *J. Trop. Med. Hyg.* **33**, 258.

Dunnebacke, T. H., and Williams, R. C. (1967). *Proc. Nat. Acad. Sci. U.S.A.* **57**, 1363.

Feldman, H. A. (1971). Personal communication.

Hull, R. N., Minner, J. R., and Mascoli, C. C. (1958). *Amer. J. Hyg.* **68**, 31.

Jahnes, W. G., Fullmer, H. M., and Li, C. P. (1957). *Proc. Soc. Exp. Biol. Med.* **96**, 484.

Jézéquel, A. M., Shreeve, M., and Steiner, J. (1967). *Lab. Invest.* **16**, 287.

Kingston, D., and Warhurst, D. C. (1969). *J. Med. Microbiol.* **2**, 27.

Kudo, R. R. (1966). "Protozoology" 5th ed. Thomas, Springfield, Illinois.

Lillie, R. D. (1954). "Histopathologic Technic and Practical Histochemistry," 2nd ed. McGraw-Hill (Blakiston), New York.

Moore, A. E., and Hlinka, J. (1968). *J. Nat. Cancer Inst.* **40**, 569.

Pereira, M. S., Marsden, H. B., Corbitt, G., and Tobin, J. O'H. (1966). *Brit. Med. J.* **1**, 130.

Singh, B. N. (1952). *Phil. Trans. Roy. Soc. London, Ser. B.* **236**, 405.

Singh, B. N. (1960). *Proc. Indian Sci. Congr., 47th Session* Part 2, p. 145.

Wang, S. S., and Feldman, H. A. (1967). *N. Engl. J. Med.* **277**, 1174.

CHAPTER **9**

Identification of Tissue Culture Contaminants by Electron Microscopy

ETIENNE DE HARVEN

I. Introduction 206
II. Examples of Tissue Culture Contamination Observed under the Electron Microscope 206
A. Particles Resembling Murine Leukemia Viruses (MuLV) 206
B. Intracisternal Particles (IC) 211
C. Hamster Cell R Particles 211
D. Reoviruses 214
E. Mycoplasmas 214
III. Normal Particulate Cell Organelles of Viral Size 217
A. Perichromatin Granules 217
B. Centriole Satellites 221
C. Coated Vesicles 221
IV. Abnormal Cell Structures of Possible Viral Origin 221
A. Nuclear Bodies 221
B. Intracytoplasmic Crystalline Aggregates 225
V. Methods for Electron Microscopic Identification of Viral Contaminants 226
A. Cell Monolayers 226
B. Cell Suspensions 227
C. Culture Fluids 227
D. Negative Staining 228
E. Other Methods 228
References 229

I. Introduction

In this chapter, the diagnostic value of electron microscopy (EM) in detecting virus contamination in tissue culture will be illustrated. Bacterial contamination, which is easily recognized under the light microscope, will not be considered. Among viral contaminants, particular attention will be given to some of those which do not produce cell destruction, or CPE, and otherwise may remain undetected. It is indeed in the study of cryptic infections that electron microscopy has proven of special diagnostic value. This is reinforced by the fact that several viral agents have been recognized solely on the basis of their fine structure because a system sensitive enough to demonstrate their pathogenicity has yet to be devised. Comparable situations have been encountered in *in vivo* studies. For example, germ-free mice are no longer considered as virus-free on the basis of electron microscopic observations of typical virus particles (de Harven, 1964), which, however, have never been concentrated in sufficient quantities for any attempt to characterize their biochemical and biological properties.

In Section III, cultured cells will be examined from an ultrastructural point of view because viruses identification cannot be reliably achieved without an extensive knowledge of the fine structure of the infected cells. Normal cells contain particulate organelles of viral size (Haguenau and Hollmann, 1963) but of poorly understood function. Some of them are observed within the nucleus, others within the cytoplasm. It is important to precisely identify these structures and to be aware of their possible significance in cell pathology. In Section IV, two examples will be given of abnormal structures of possible viral significance.

Section V will deal with the various techniques available to optimize the chances of identifying viral contaminants under the electron microscope.

II. Examples of Tissue Culture Contaminations Observed under the Electron Microscope

A. Particles Resembling Murine Leukemia Viruses (MuLV)

Electron microscopic observations have indicated that the multiplication of murine leukemia virus (MuLV) is not restricted to the leukemic cells. It is therefore not surprising to identify particles of the same morphology in cultured mouse fibroblasts, such as the established line of L cells (Dales and Howatson, 1961; Kindig and Kirsten, 1967). Since these observations were made, many cell lines of murine origin have been studied under the EM, frequently with the hope of selecting acceptable virus-negative cell lines that can serve as controls in MuLV studies. Unfortunately, it appears extremely difficult to find murine cell lines unquestionably free of viruses of MuLV morphology. An example of such findings in a cell line derived from mouse lung is given in Fig. 1. In this case, large numbers of viruses have been released into the culture supernatant, facilitating purification as illustrated in Fig. 2, which represents a thin section of a high-speed pellet of culture fluid.

Cells contaminated with MuLV-like agents very rarely show cytopathic alteration. They release viruses in culture fluids by a process of budding from plasmalemma (de Harven, 1968b). The viral infection is probably transmitted vertically by a virogene associated with the cell genome (Aaronson *et al.*, 1969; Huebner and Todaro, 1969; Huebner *et al.*, 1970) that completely escapes electron microscope recognition. Complete viral expression (i.e., budding and release) therefore need not be a prerequisite for the transmission to consecutive cell generations of the viral infection. This may therefore lead to false negative electron microscope reports, because budding and extracellular particles are the only EM-recognizable viral structures. This consideration has limiting implications in most efforts to identify similar agents in tissue cultures derived from human neoplasia. However, it appears that EM identification of oncornaviruses (Nowinski *et al.*, 1970) can be facilitated by certain treatments of the cultured cells. Friend leukemia cultured cells growing in presence of 2% DMSO (Friend *et al.*, 1971) show a significant increase in the number of budding viruses (Sato *et al.*, 1971). Whether or not this effect is restricted to Friend leukemia cultured cells await further studies. Of possibly more general significance are recent reports (Lowy *et al.*, 1971; Aaronson *et ai.*, 1971) on the effects of 5-bromodeoxyuridine and 5-iododeoxyuridine, under the influence of which a fuller expression of the viral genome is induced, resulting in increased virus release. One can anticipate that the use of these drugs or of other inducers will also enhance the possibility of identifying viral agents of MuLV type under the EM. Because it has long been recognized that identification of viruses under the electron microscope depends on the observations of large numbers of similar particles (de Harven, 1965), one can expect that the use of inducers will facilitate correct ultrastructural recognition.

Even at the high level of resolution of today's electron microscopes, viruses from the same group look very much alike. Other methods are needed for identifying individual viral types. This has been well illustrated in the ESP-1 cell line (Priori *et al.*, 1971), originating from a case of human lymphoma, in which particles of MuLV morphology have been observed (Fig. 3). The question of the origin of this virus could not have been answered by electron microscopy but rather by immunological methods (Gilden *et al.*, 1971). Interspecies cell contamination was already suspected in the case of McCoy cells (Kajima *et al.*, 1967), whereas interspecies virus infections were

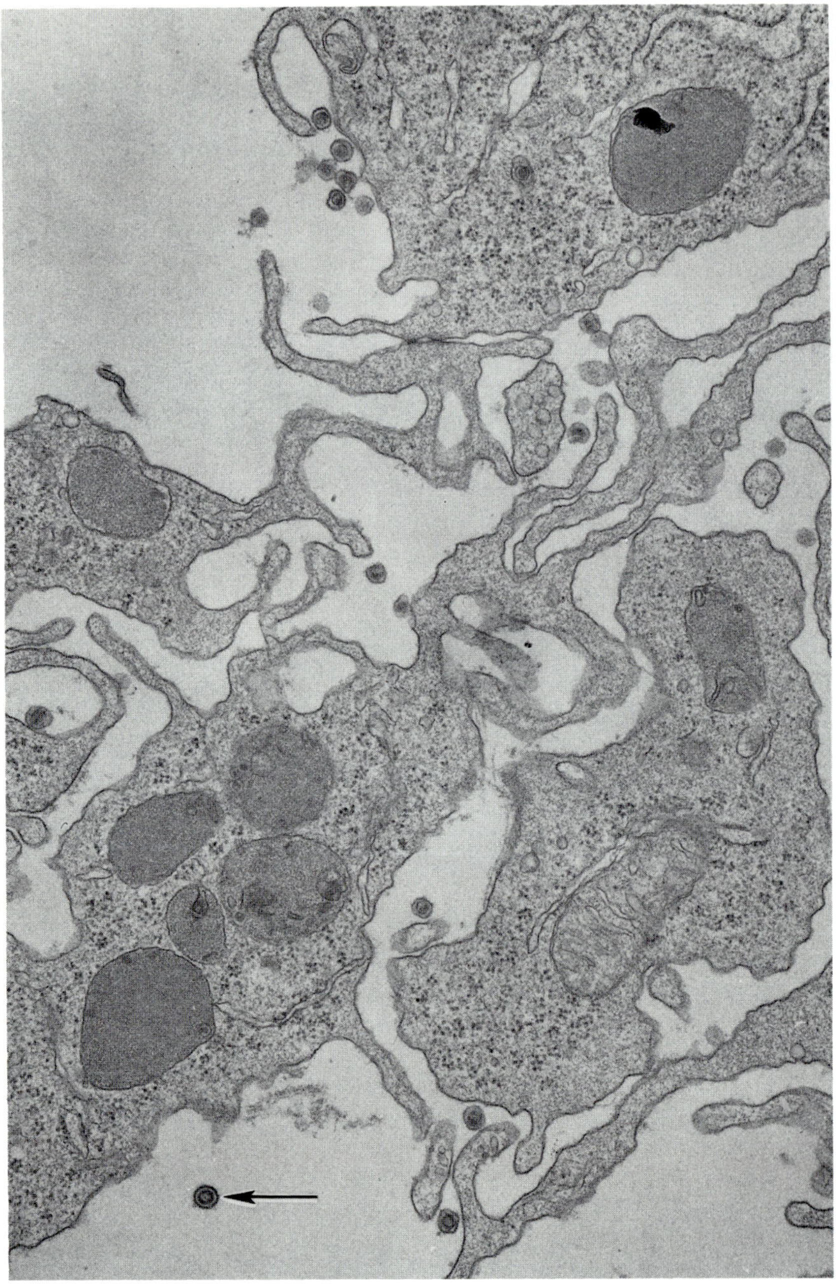

Fig. 1. Cultured cells from an established line of murine (lung) origin contaminated with a viral agent morphologically similar to murine leukemia viruses. Several extracellular type C particles are recognizable. One enveloped A-type particle is also seen (arrow). ×25,500.

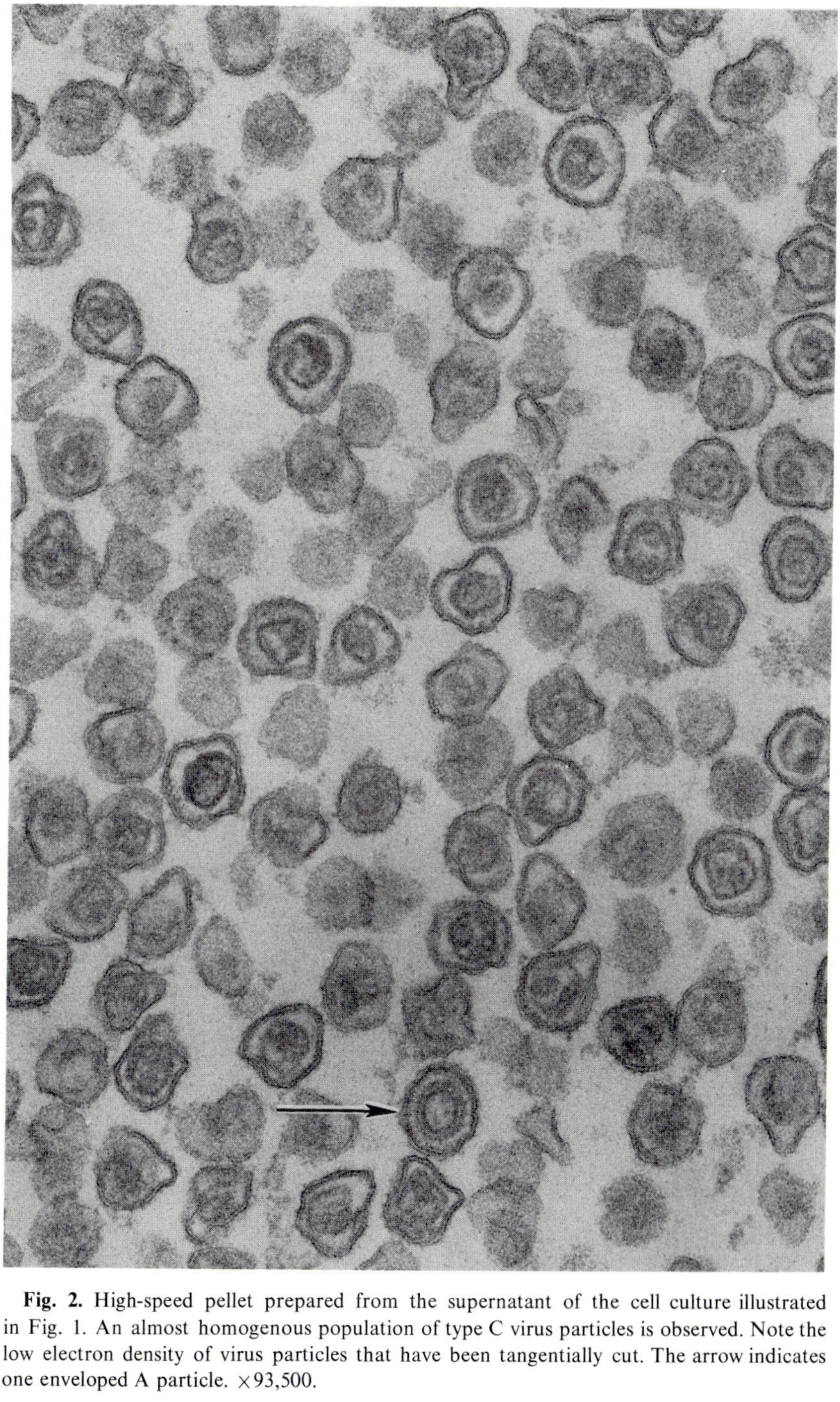

Fig. 2. High-speed pellet prepared from the supernatant of the cell culture illustrated in Fig. 1. An almost homogenous population of type C virus particles is observed. Note the low electron density of virus particles that have been tangentially cut. The arrow indicates one enveloped A particle. ×93,500.

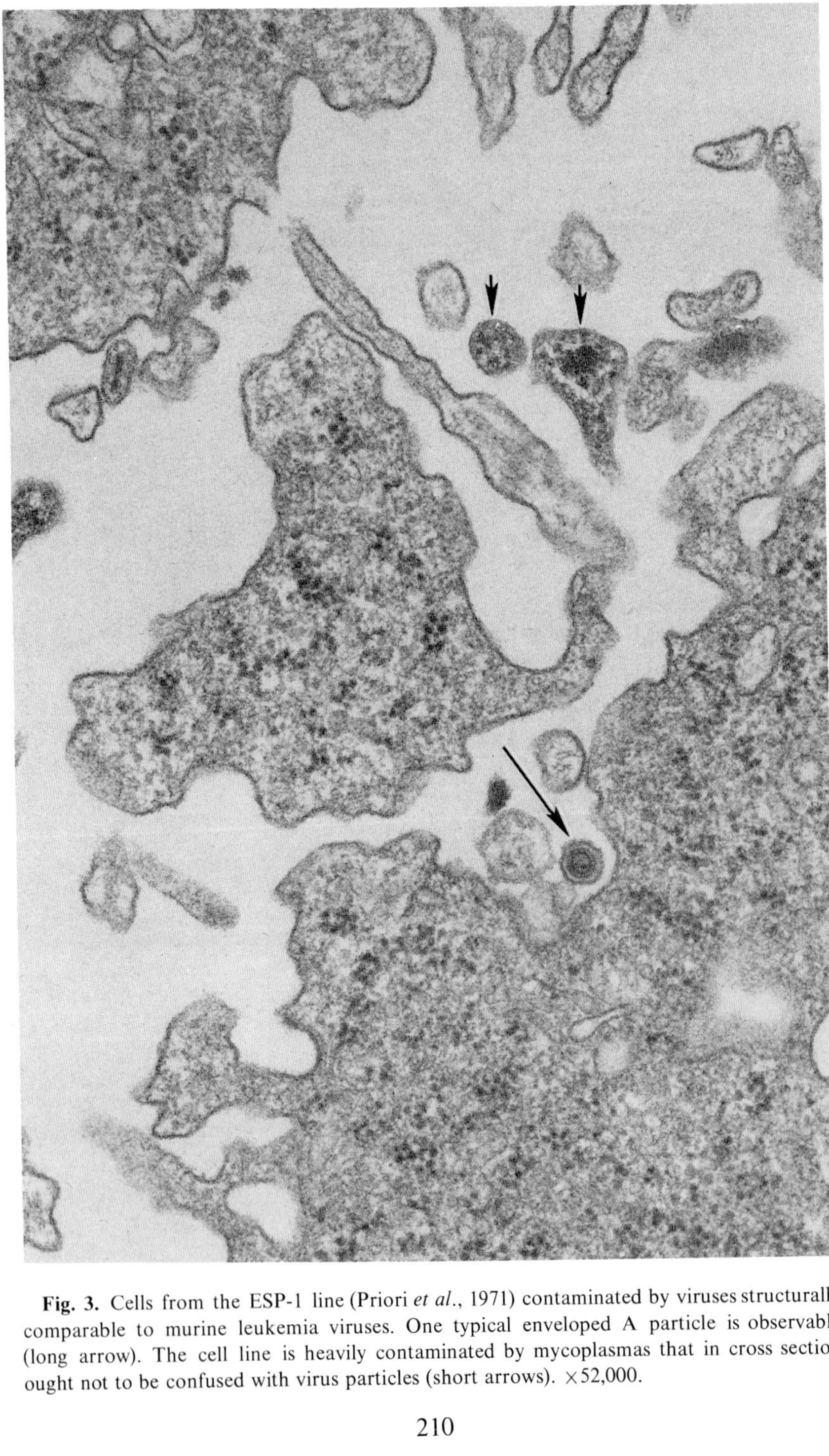

Fig. 3. Cells from the ESP-1 line (Priori *et al.*, 1971) contaminated by viruses structurally comparable to murine leukemia viruses. One typical enveloped A particle is observable (long arrow). The cell line is heavily contaminated by mycoplasmas that in cross section ought not to be confused with virus particles (short arrows). ×52,000.

demonstrated in Rauscher leukemia virus (Wright and Korol, 1969) and feline leukemia viruses (Jarret *et al.*, 1969).

B. Intracisternal Particles (IC)

At a conference on the nomenclature of MuLV (Anonymous, 1966), it was recomended that particles usually referred to as *intracisternal* should be considered separately from MuLV, although they are frequently observed within MuLV-infected cells.

The viral nature of IC particles has not yet been established. Morphologically, they differ from MuLV by a slightly smaller diameter (70–100 nm) and by their intracytoplasmic localization within the cisternae of the rough endoplasmic reticulum (Fig. 4). They have been observed *in vivo*, especially in transplanted murine neoplasms (Friedlander and Moore, 1956), and in a large number of murine cell lines (Claude and Augusti-Tocco, 1970; Volkman *et al.*, 1971). Despite their successful purification from a microsomal fraction of a plasma-cell tumor line (Kuff *et al.*, 1968), their biological activity is so far unknown. Their antigenic makeup seems to indicate their independence from the MuLV group (Wivel *et al.*, 1971). The same investigators, however, consider the possibility that IC particles result from partial expression of the MuLV genome (Wivel and Smith, 1971). Common antigens have been reported (Watson *et al.*, 1970) but only on preparations of questionable purity. Obviously, the group appartenance of IC particles is not clear at the present time. The importance of the question stems from (*1*) the widespread occurrence of IC particles in murine cells, as in EL-4 cells frequently used as G^- controls (Aoki *et al.*, 1970); and (*2*) their presence in such other species as rats (Chopra *et al.*, 1970) and even possibly in cells derived from human sarcomas (Morton *et al.*, 1969).

C. Hamster Cell R Particles

First identified under the electron microscope by Bernhard and Tournier (1964), the viral nature of R particles still awaits biological demonstration. They can be easily distinguished from MuLV and IC particles on the basis of their fine structure (Fig. 5). Their localization within endoplasmic reticulum (ER) cisternae is similar to that of IC particles. However, their characteristic feature, the spoked profile of their core, gives a radiating appearance from which the terminology has been proposed. The particles are always enclosed in a smooth envelope that seems to derive from a budding process along ER membranes (Thomas *et al.*, 1968). R particles were found in all BHK cell lines (Shipman *et al.*, 1969) and in many other hamster cell lines (De Petris and Harvey, 1969). They do not bud from the cell surface and are therefore not released in the culture supernatant, at least in forms recognizable as R

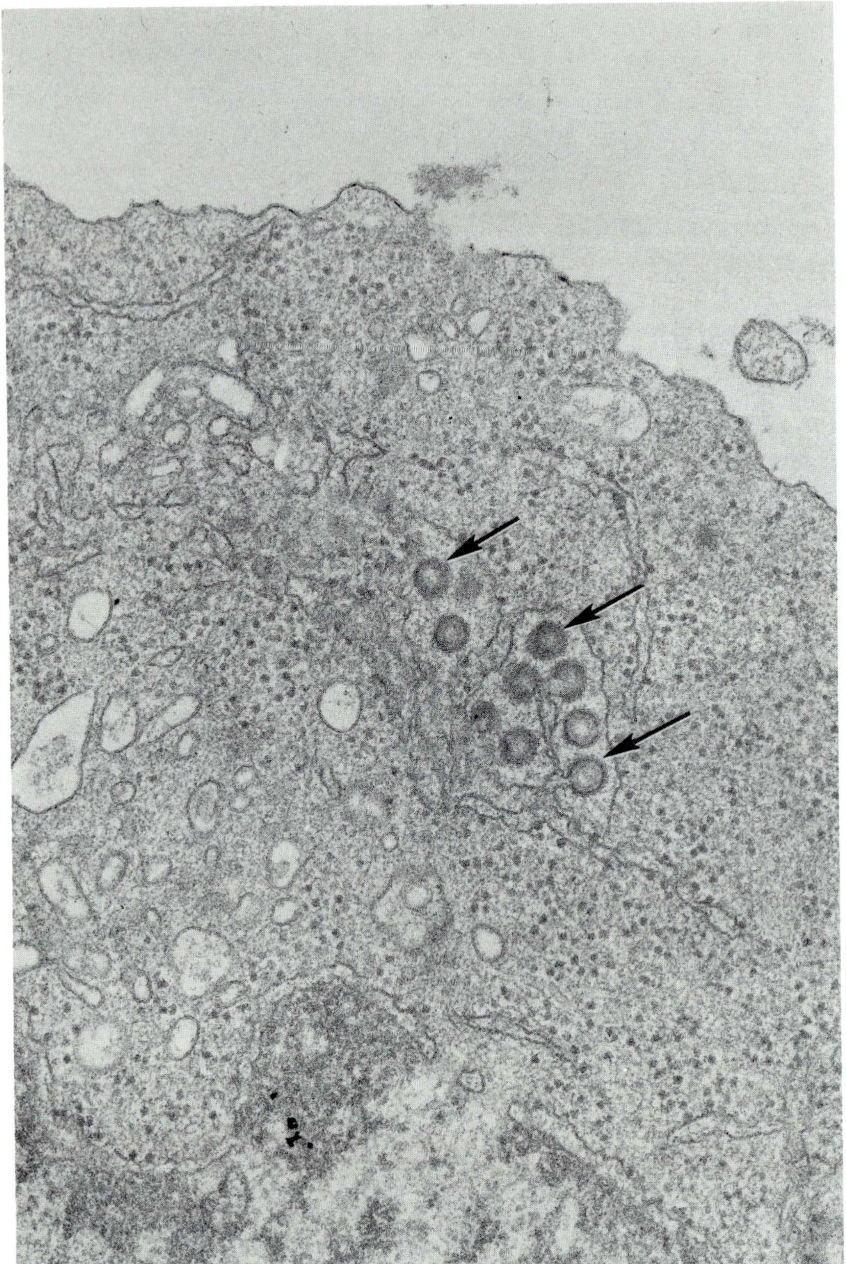

Fig. 4. Cell of cultured Friend murine leukemia cell line showing a cluster of intracisternal particles (arrows) within the cisternae of the rough endoplasmic reticulum. The particles are surrounded by the envelope with unit membrane structure. These envelopes are in direct contact with the core of the particle. ×55,000. (By courtesy of Dr. T. Sato.)

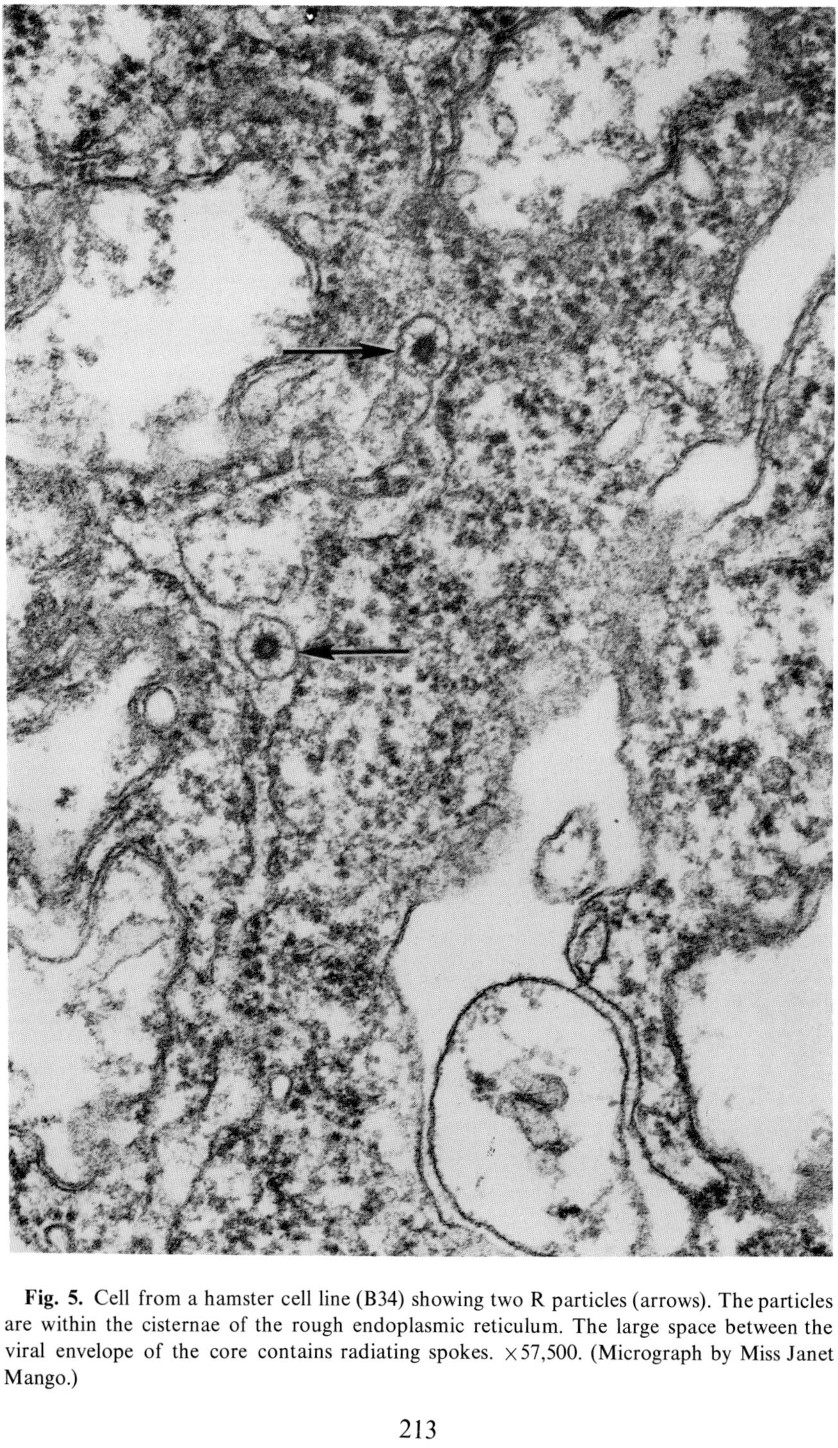

Fig. 5. Cell from a hamster cell line (B34) showing two R particles (arrows). The particles are within the cisternae of the rough endoplasmic reticulum. The large space between the viral envelope of the core contains radiating spokes. ×57,500. (Micrograph by Miss Janet Mango.)

particles. Their purification, which could be attempted from microsome fractions of the cells, may be difficult in view of the small number of particles. The biological activity of R particles is not known at the present time. Like MuLV, they do not seem to affect the viability of the infected cells. In certain hamster tumors, the simultaneous presence of R and MuLV-like particles has been reported (Stenback *et al.*, 1966). Although R particles are most prevalent in hamster cells, particles of similar morphology have been described in a bovine cell line (Mussgay *et al.*, 1969), and it has been recently hypothesized that hamster cells might be highly permissive and systematically contaminated by fetal calf serums (Cesarini, 1971).

D. Reoviruses

Acute accidental infection of a cell line is sometimes rapidly diagnosed by electron microscopy, facilitating the serological typing of the contaminating agent. For example, a murine leukemia cell had shown intriguing cytoplasmic inclusions. Electron microscopy oriented the diagnosis toward reoviruses (Fig. 6). Reovirus type III was later serologically identified.

E. Mycoplasmas

As reviewed in other chapters of this book, mycoplasma (PPLO) frequently induce cytopathic changes (Fogh *et al.*, 1965). Complete loss of the culture or an inapparent infection characterized by "pacific coexistence" with the infected cells can both be observed. It is in these last cases that electron microscopy can helpfully reveal cryptic levels of contamination (Fig. 7).

Several points should be made regarding the localizations and fine structure of mycoplasmas. They appear to adsorb to cell surfaces; there is little evidence suggesting their capacity to penetrate inside the infected cells. When occasionally seen within cytoplasmic vacuoles (Edwards and Fogh, 1960), the images may correspond to cross sections of cell surface invaginations. They are rarely seen within phagocytic vacuoles, except after treatment with antimycoplasma antibody (Jones and Hirsch, 1971). Cytopathic effects are believed often to result from essential amino acid depletion in the culture medium (arginine) (Stanbridge, 1971) rather than from the release of cytotoxic substances that may hypothetically permeate the cells at the level of the sometimes very close contact between the mycoplasma and the cell plasmalemma. Chromatin clumping that gives a leopard-spot appearance to the contaminated cells has been described under the light microscope (Stanbridge *et al.*, 1969). This alteration has not yet, to my knowledge, been studied under the EM, and is probably different from the nucleolar segregation described earlier in mycoplasma-infected cells (Jézéquel *et al.*, 1967).

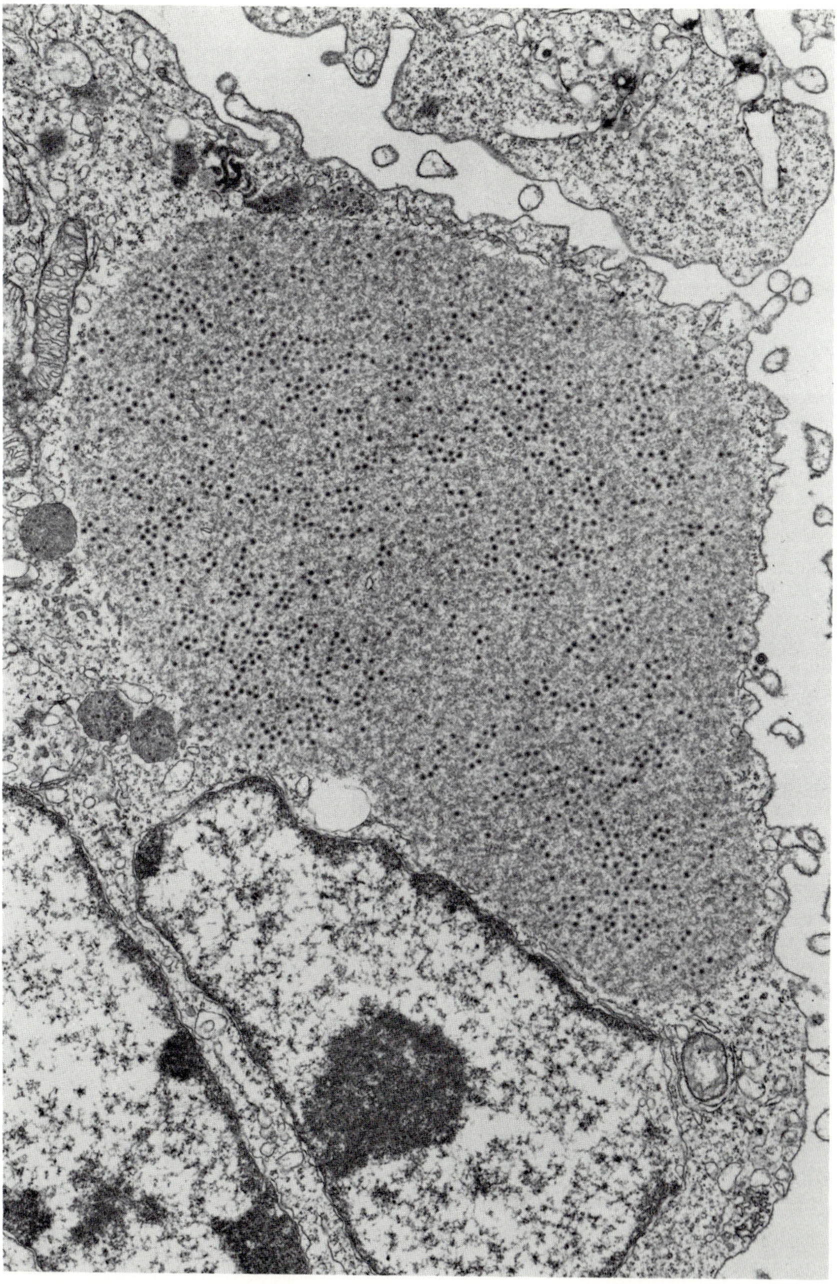

Fig. 6. Cell of cultured Friend murine leukemia cell line grown in suspension culture and accidentally contaminated with reovirus type 3. A large intracytoplasmic inclusion body is seen, displacing all normal cytoplasmic structures and containing a large number of electron-dense particles having the morphology of reovirus. ×12,500.

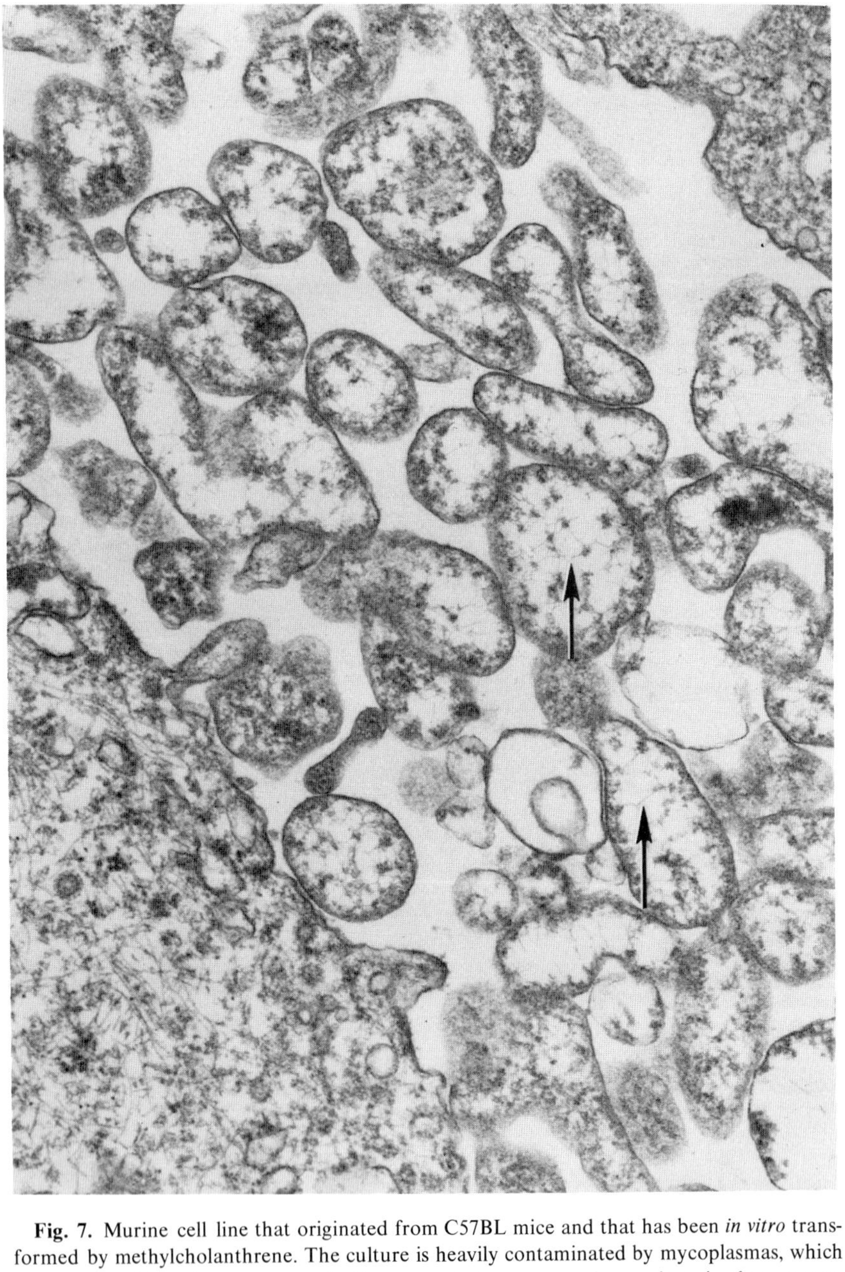

Fig. 7. Murine cell line that originated from C57BL mice and that has been *in vitro* transformed by methylcholanthrene. The culture is heavily contaminated by mycoplasmas, which are recognizable in large numbers in the extracellular spaces and sometimes in close contact with cell surfaces. They are limited by a single membrane and contain ribosomes. Their differential diagnosis from cytoplasmic blebs is facilitated by the observation of DNA–protein filaments (arrows). ×36,000.

The resemblance between some mycoplasmas and L forms has long been recognized (Dienes, 1963; Brown and Officer, 1968). Mycoplasmas cannot be confused with viruses on the basis of their size. They could, however, be occasionally confused with cross sections of cytoplasmic blebs. The latter are also surrounded by a membrane and contain ribosomes, although in optimum conditions of fixation, their general electron density is somewhat different. Moreover, filaments approximately 3 nm thick and believed to represent DNA protein complexes are easily seen in mycoplasmas but not in cytoplasmic blebs.

Mycoplasmas have been classified on the basis of their serological specificity or metabolic requirements, which have not so far been correlated with differences in ultrastructural type. Differences in size and shape are well demonstrated, for example, by the variety of mycoplasmas contaminating ESP-1 cells (Priori *et al.*, 1971). One type is small and somewhat resembles the *mycoplasmas pneumoniae* described previously (Biberfeld and Biberfeld, 1970). It has a characteristic club or pear shape (Fig. 8) and contains a dense skein of filaments, probably containing DNA, continuous with a rodlike structure within the axis of the elongated portion. This sharpened portion is frequently oriented toward the cell and in very intimate contact with its plasmalemma. This particular type of mycoplasma has been cultured on agar and in broth (Fogh, 1972), and a pellet prepared from the broth culture showed a pure collection of pear-shaped mycoplasmas (Fig. 9) similar in structure to those associated with ESP-1 cells. Other types of mycoplasmas in ESP-1 cells also cultured on artificial media seem to resemble those illustrated in Fig. 7.

In negative staining, mycoplasmas show slender armlike extensions that when viewed in thin sections produce images of circular profiles (Fig. 8), not to be confused with viruses or with mycoplasmal elementary particles. The existence of elementary bodies 80 to 100 nm in diameter reported by Anderson and Manaker (1966) has been difficult to confirm, because ultrafiltrates of cultures frequently contain mycoplasmas capable of passing through filters with average pore size diameters considerably smaller than the diameters of the mycoplasmas themselves (de Harven, 1965).

III. Normal Particulate Cell Organelles of Viral Size

A. Perichromatin Granules

At the margin between hetero- and euchromatin, electron-dense particles 40–45 nm in diameter and surrounded by an electron-lucent halo (Watson, 1962) are frequently observed, *in vivo* as well as in cultured cells (Fig. 10). Their number increases after treatment with actinomycin D (Smetana and

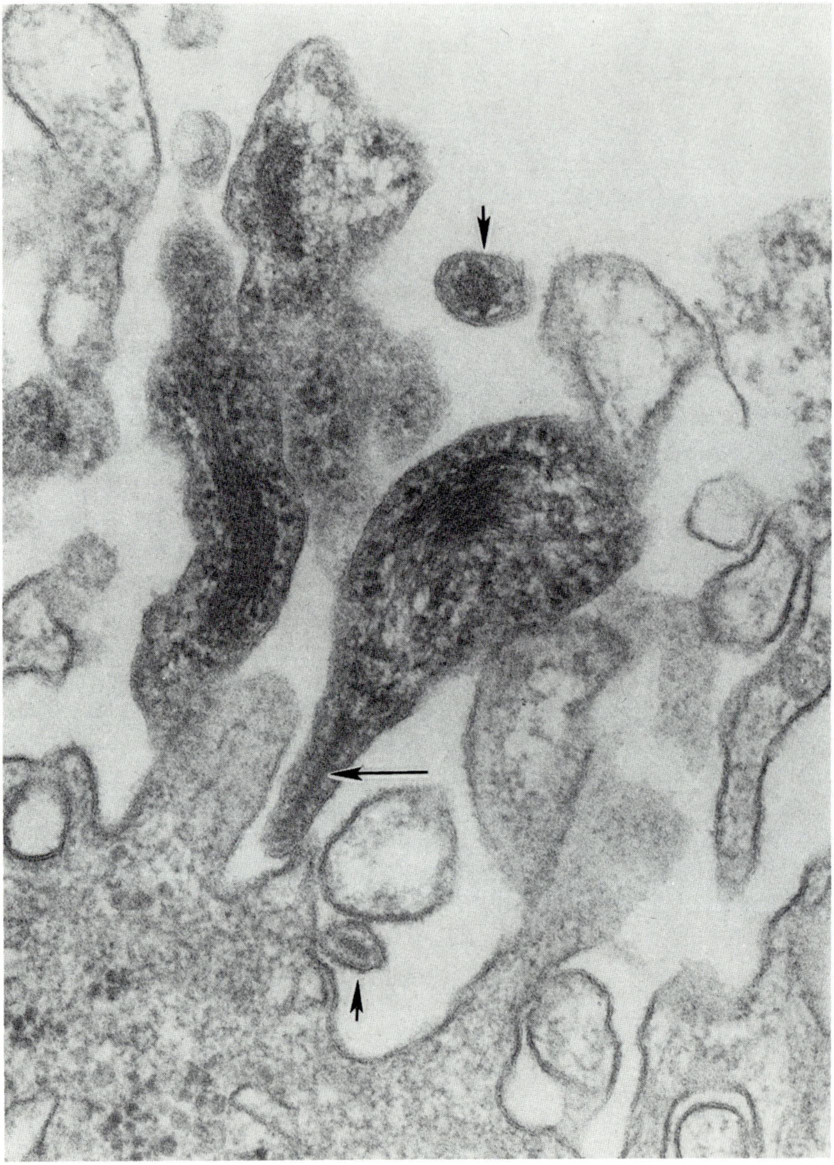

Fig. 8. Same ESP-1 cell line as in Fig. 3. One type of mycoplasma contaminating this cell line has a characteristic pear-shaped profile. The large extremity contains some ribosomes and an electron-dense skein of fibers that probably represent DNA–protein complexes. The pointed extremity of the mycoplasma contains a condensed central structure (long arrow). Cross sections of the pointed end of such microorganisms (short arrows) ought not to be confused with viral type C particles. ×92,500.

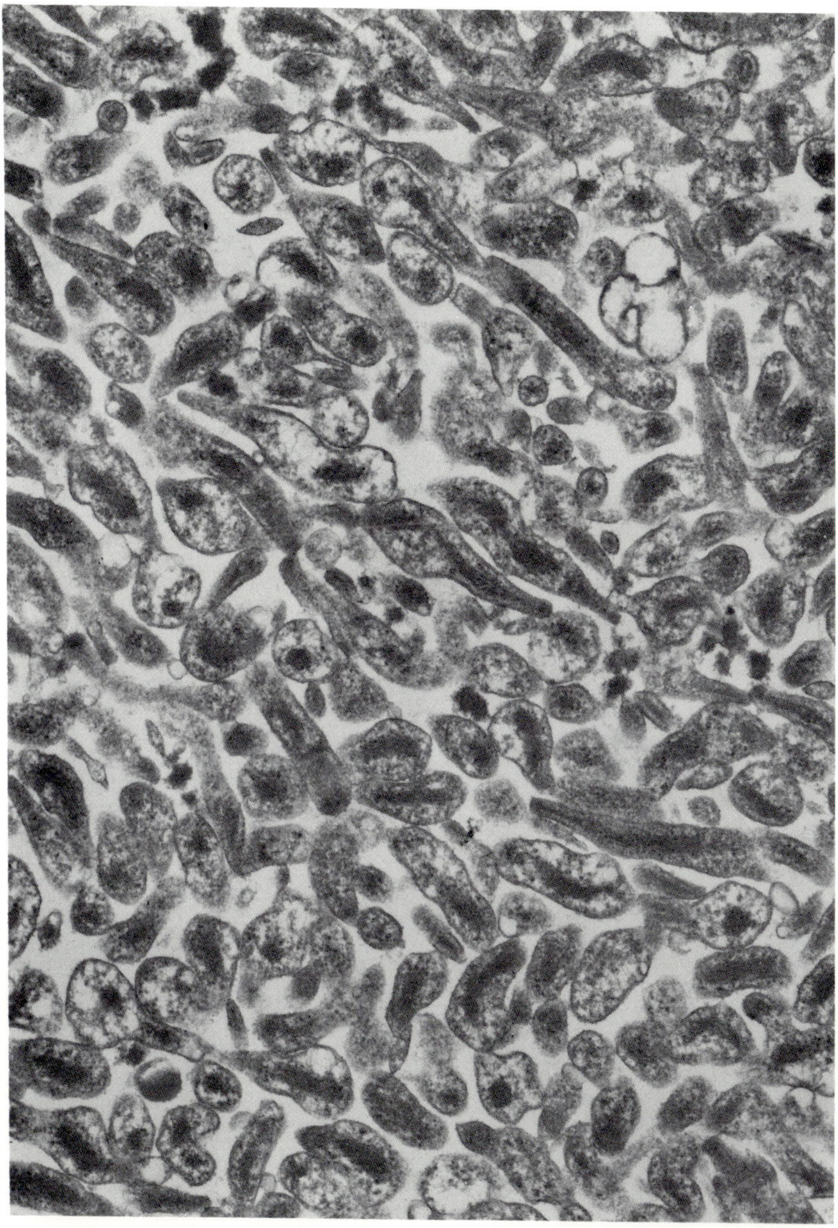

Fig. 9. The pear-shaped mycoplasma contaminating the ESP-1 cell line was isolated in Dr. Fogh's laboratory and grown in broth. A pellet (62,000 rpm for 15 minutes) shows a homogenous population of pear-shaped microorganisms comparable to those seen in Fig. 8. ×28,300.

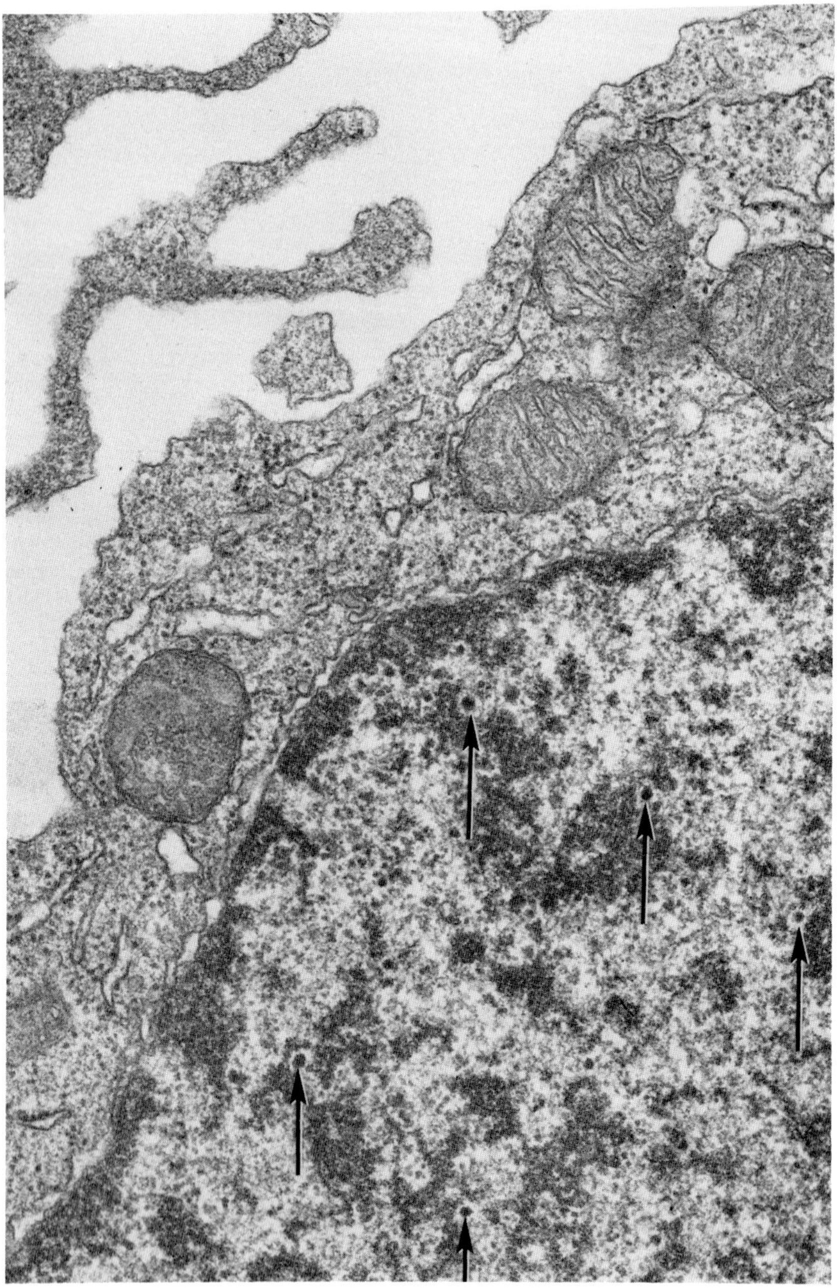

Fig. 10. Cells grown in suspension culture and that originate from a biopsied node of a Hodgkin's disease patient. The cell nuclei show a large number of perichromatin granules (arrows). These structures are considered as normal cell components. They are typically located at the margin between hetero- and euchromatin and are surrounded by a halo of low electron density. $\times 36{,}000$.

Potmesil, 1968) and in cells grown at supranormal temperatures (Heine *et al.*, 1971). They contain RNA and were hypothetically considered as possible carriers of mRNA (Monneron and Bernhard, 1969). Although similar granules appear in large numbers in the nuclei of virus-infected cells, this might be caused by nonspecific effects of the infection. Perichromatin granules are normal constituents; their presence does not suggest latent viral infection.

B. Centriole Satellites

Cytoplasmic granules of high electron density and approximately 70 nm in diameter, the centriole satellites (Bessis *et al.*, 1958) are normal cell components (Fig. 11), present throughout the cell cycles but vanishing during mitosis (de Harven, 1968a). Their size range is the only factor that may lead to a confusion with a viral contaminant. The function of centriole sattelites in mitosis is related to microtubule attachments.

C. Coated Vesicles

Coated vesicles (Roth and Porter, 1962) (Fig. 12) are normal cell components, frequently observed near the cell surface or near Golgi membranes, and are probably implicated in phenomena of intracytoplasmic transport and of membrane infolding. They have been successfully fractionated by Kaneseki and Kadota (1969), who have analyzed their structure in considerable detail. The diameter of vesicle averages 60–70 nm, plus 15 to 20 nm for the bristles. Similar particles have been mistakenly described as viruslike particles (Chapman *et al.*, 1967). However, the structure of coated vesicles is not very different from that of a foamy viruses (Clarke *et al.*, 1969), which have similar diameters and bristle coats. Foamy viruses grow in many cell species (Parks *et al.*, 1971) and have been tentatively identified in cultures that originated from a case of nasopharyngeal carcinoma (Achong *et al.*, 1971). Confusion between foamy viruses and coated vesicles is unlikely when the morphology of intact cells is studied but may occur during the study of cell fractions or of concentrates of particulate structures. The same probably holds true for the bovine syncitial virus that has been isolated from cattle with lymphosarcoma (Malmquist *et al.*, 1969).

IV. Abnormal Cell Structures of Possible Viral Significance

A. Nuclear Bodies

Nuclear bodies (Weber *et al.*, 1964) (Fig. 13), a group of pleomorphic, intranuclear, spherical inclusions ranging in diameter from 0.5 to 1.8 μm,

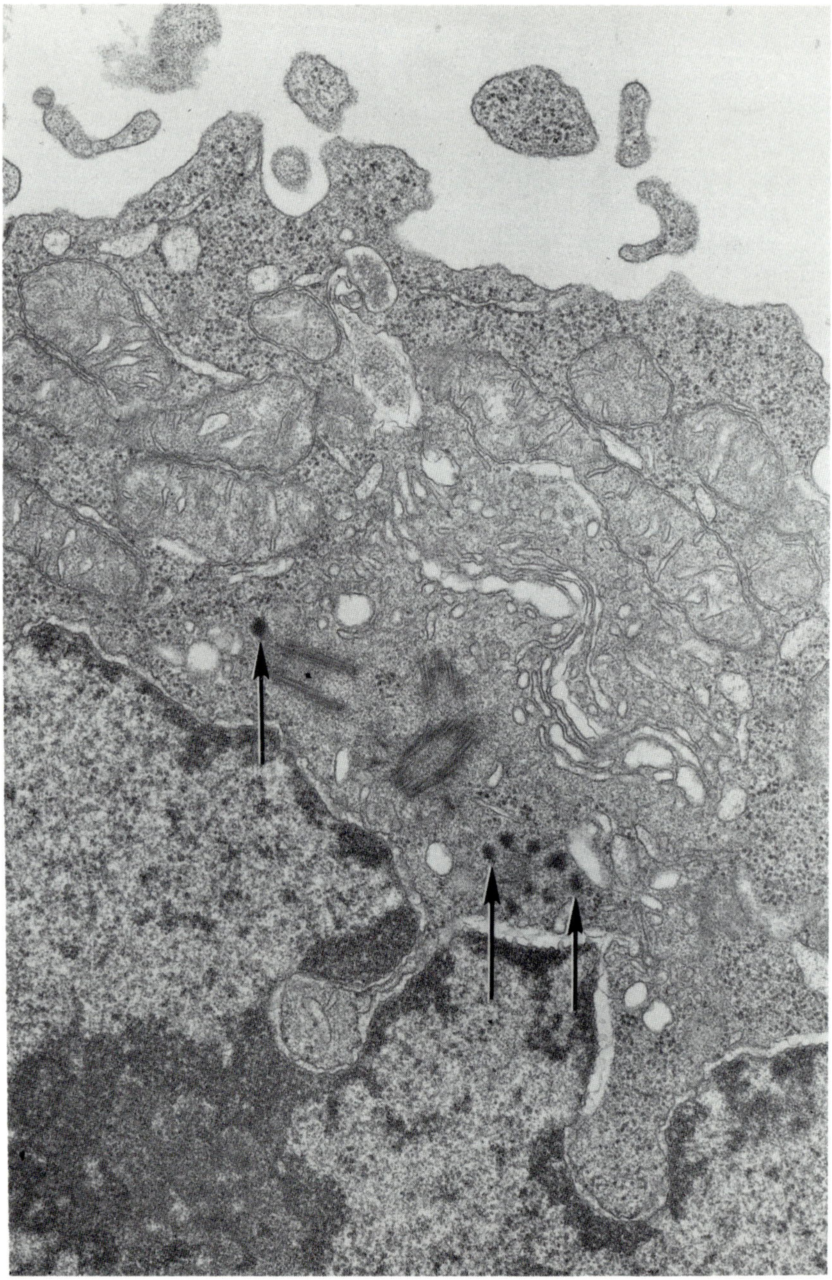

Fig. 11. A cell of the Friend murine leukemia cell line grown in suspension culture, showing a typical centrosphere area with three centrioles and a few typical centriole satellites (arrows). $\times 25{,}800$. (Electron micrograph by Miss Janet Mango.)

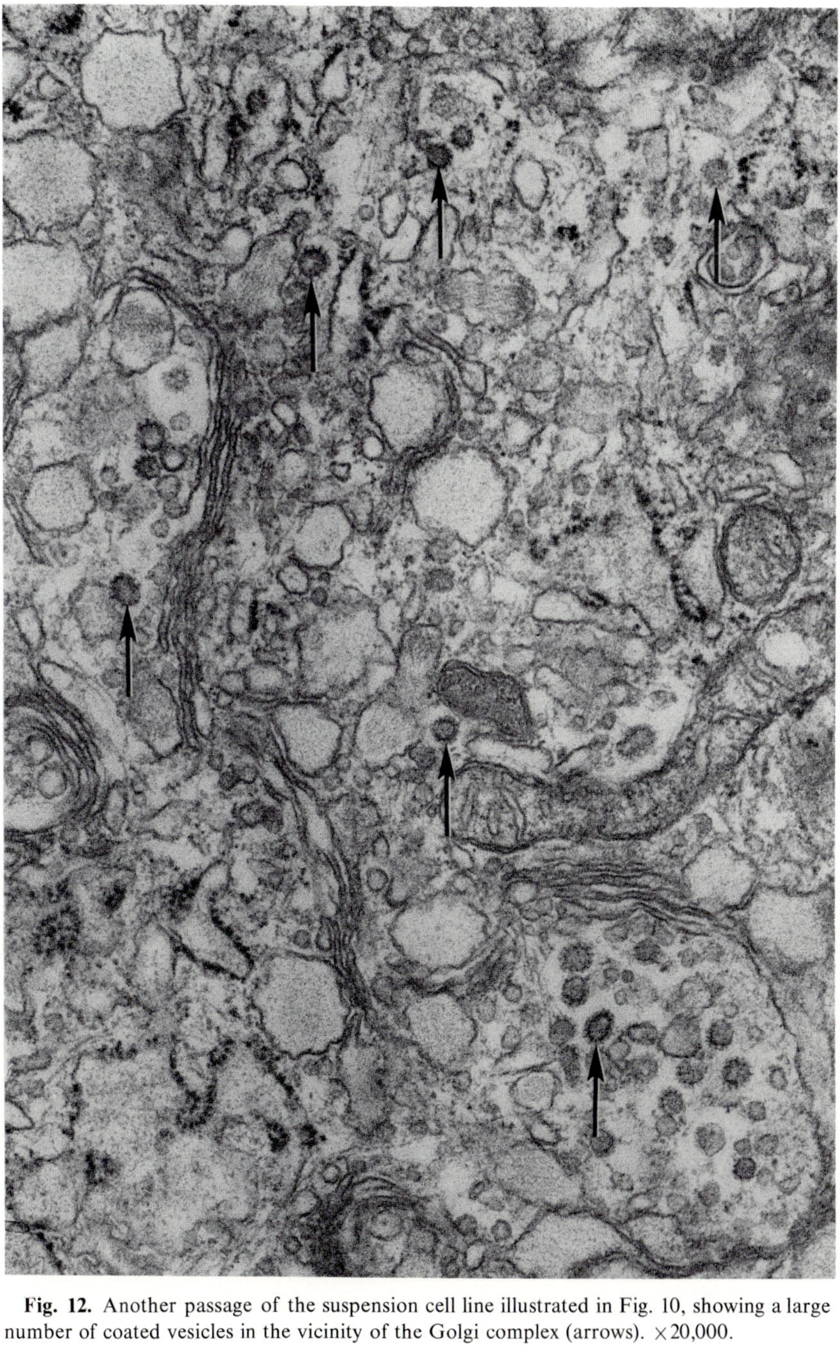

Fig. 12. Another passage of the suspension cell line illustrated in Fig. 10, showing a large number of coated vesicles in the vicinity of the Golgi complex (arrows). ×20,000.

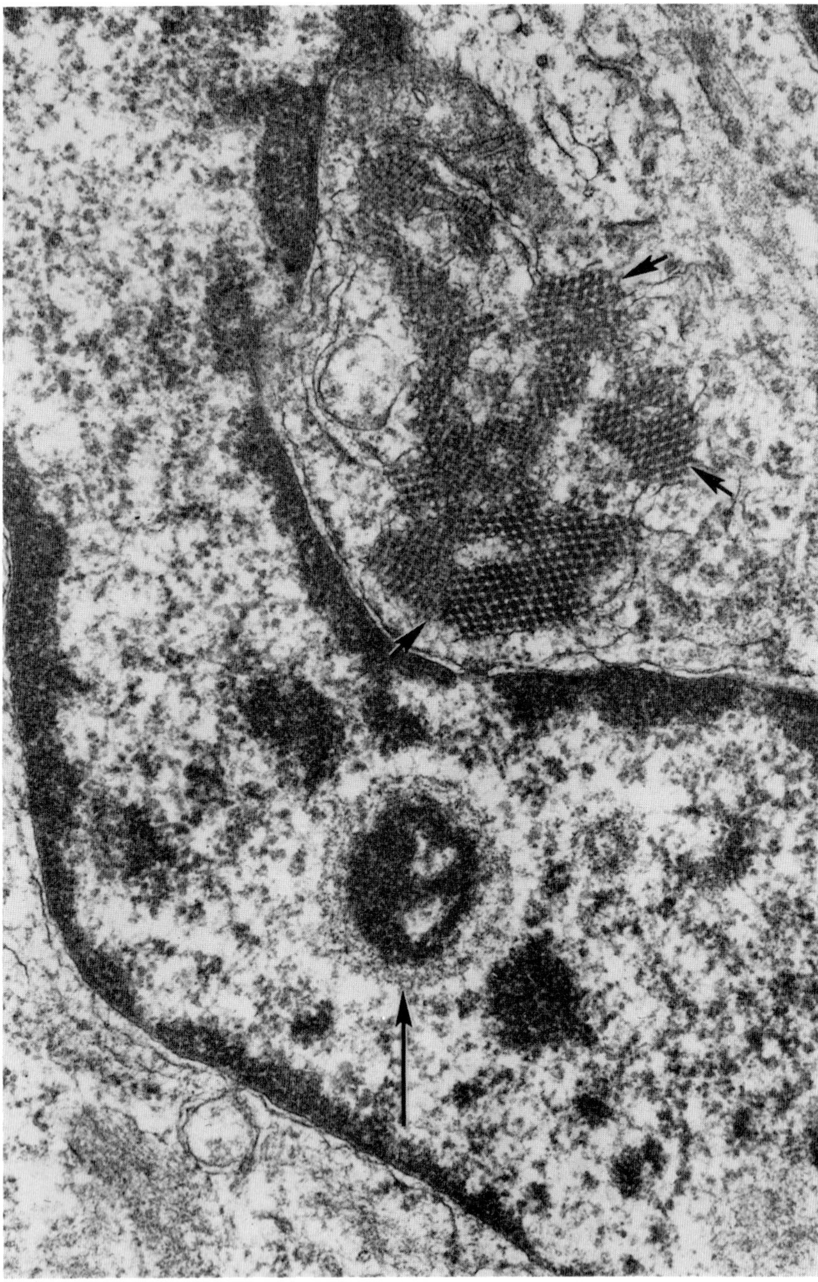

Fig. 13. A cell from the spleen of a dog with transplanted leukemia, showing a large nuclear body (long arrow) surrounded by a layer of fibrillar material. The same cell shows a large intracytoplasmic inclusion (short arrow), corresponding to the crystalline aggregates frequently observed in canine cells, both *in vivo* and *in vitro*, and having some resemblance to the crystalline arrays of particles seen in polio virus-infected monkey cells. $\times 39,000$.

raise more problems of interpretation. Several different structural types have been characterized (Bouteille *et al.*, 1967; Popoff and Stewart, 1968) and the claim has been made that they represent a normal cell constituent (Brooks and Siegel, 1967; Dupuy-Coin *et al.*, 1969; Henry and Petts, 1969; Ishi and Mori, 1969), although they have been frequently recognized in tumor cells (Krishan *et al.*, 1967). However, they have been observed in cultured cells infected with vacinia virus by Patrizi and Middelkamp (1969), who suggested that nuclear bodies are frequently seen in cells in which DNA viruses are being replicated. It is of interest to recall that de Thé, who recognized nuclear bodies for the first time (de Thé *et al.*, 1960), observed them in rabbit tumor cells suspected of carrying the genome of a papova virus, the Shope papilloma virus. However, typical nuclear bodies have been described by Périer (1967) in a human demyelinating disease known to be associated with an RNA virus (measles). The presence of nuclear bodies should be regarded as nonspecific nuclear alteration, possibly but not necessarily caused by virus infection. The fact that nonviral agents can induce nuclear bodies has been indicated recently in a study of hamster peritoneal macrophages (Dumont and Robert, 1971).

B. Intracytoplasmic Crystalline Aggregates

A regular arrangement of ± 22 nm granules is sometimes observed within the cisternae of the endoplasmic reticulum of cultured cells as well as *in vivo* in a variety of pathological conditions. The affinity of these granules for lead and uranyl salts as well as their tendency to take a cubic lattice configuration are suggestive of their viral nature. They were first observed in monkey tumors induced by Rous sarcoma virus (RSV) (Munroe *et al.*, 1964). Interestingly enough, tumors induced similarly in dogs by different strains of RSV showed the very same crystalline aggregates in leptomeningiomas (Bucciarelli *et al.*, 1967) and in gliomas (Haguenau *et al.*, 1971). We have observed very similar structures in transplantable dog lymphomas (Fig. 13), confirming an observation made by Kakuk *et al.* (1968). Because dogs seem to frequently harbor such particles, distemper virus had to be suspected. Although there are some similarities between replicating distemper virus and the crystalline aggregates, distemper virus inclusions have been described as tubular aggregates, without precise relationship with the endoplasmic reticulum (Koestner and Long, 1970), or as nucleocapsid filaments (Lawn, 1970). More intriguing is the resemblance between crystalline aggregates and images interpreted as representing developmental stages of poliovirus in a recent study by Blinzinger *et al.* (1969) where a similar cubic lattice and a similar relationship with the endoplasmic reticulum, although perhaps a somewhat larger particle size (27 nm), has been described.

Identical ER-bound crystalline aggregates have been observed in several lines of human lymphoblastoid cells (Moore *et al.*, 1968; Chandra, 1968) and in human sarcomas (Györkey *et al.*, 1971). The interpretation of these findings should perhaps be reviewed, taking into consideration the different classes of viruses that have been shown in association with crystalline aggregates.

V. Methods of Electron Microscopic Identification of Viral Contaminants

Inadequate methods, especially in cell sampling, can easily lead to false negative EM reports. The following recommendations apply to the study of cell monolayers, cell suspensions, and culture fluids.

A. Cell Monolayers

1. If a cell monolayer appears under the phase-contrast microscope as a homogeneous population of cells (no foci, no CPE, no giant cells, no visible inclusion bodies), the best sampling will be a pellet of $1–2 \times 10^6$ cells obtained by low-speed sedimentation (800 rpm/5 minutes) immediately after having detached the cells from the glass surface, either by trypsin (0.05% to 0.25% for ± 10 minutes) or mechanically with a rubber "policeman." The cell pellet is then immediately covered with ± 3 ml of 2% glutaraldehyde (pH 7.4) for a minimum for 30 minutes.

The choice between trypsin or the policeman will depend on the material under study. If suspected contaminants are prone to cell surface adsorption (like mycoplasmas) trypsin could wash them off the cell surfaces and should perhaps be avoided (although Fig. 7 represents a pellet of trypsinized cells). Detaching the cells with the policeman results in unequal preservation. A compromise consists of covering the monolayer with aldehyde fixatives for 15–30 minutes and then detaching them with the policeman. This results in excellent preservation of cell structure. Unfortunately, in pellets of prefixed cells the cell density is low, because of the difficulty of packing them close to each other, and the number of cells may be insufficient for a significant observation.

2. If cytopathic alteration is recognized under phase contrast, preparing a cell pellet as outlined under (*1*) will be an unforgivable mistake. In these cases the monolayer should be fixed and embedded *in situ* (Brinkley *et al.*, 1967) so that foci or giant cells or inclusions can be identified under phase contrast in the plate of epon-embedded monolayer for selective trimming. Thin sections will then correspond to the altered cells that will have been missed if a randomized cell pellet has been prepared as outlined under (*1*).

3. The age of the monolayer sometimes influences virus maturation. In the case of MuLV, the budding phenomenon is seen in light but not in heavy confluent monolayers (Lambertenghi *et al.*, 1972). Finally, if contaminants alter the cell shape, this will be recognized in *in situ* fixation, not in the pellet of detached cells, which all become round as soon as they leave the glass surface.

B. Cell Suspensions

Few sampling problems can be expected in cell suspensions. Providing the viability of the cultures is found acceptable by the Trypan Blue exclusion test, 1–2 $\times$ 10^6 cells are pelleted and fixed as in Section V,A (*1*). Fixation in suspension may give better preservation of cell structures and therefore increase the chances of identifying contaminants. However, the dehydration and embedding of fixed cells may necessitate preembedding in 2% agar (Hirsch and Fedorko, 1968). Suspension cultures with high viability counts are not always the best samples for the purpose of identifying viruses. In the case of EB viruses, it has been generally recognized that the viruses are seen in larger number in moderately starved cells (de Harven, 1967).

C. Culture Fluids

1. If virions are constantly released in the culture fluid by the contaminated cells, a high-speed pellet (30,000 *g* for 1 hour) of the fluid represents adequate sampling for virus identification (Fig. 2). However, total harvest appears difficult in many cases. Because of aggregation and/or some stickiness on cell surfaces, many viruses sediment together with their host cells at low speed. This is constantly observed in the case of MuLV and inevitably results in incomplete harvest. In addition, even if all viruses are in free suspension, their complete sedimentation probably requires higher acceleration (Offers and Bentvelzen, 1971) than those currently used.

A significant technical improvement has been the introduction by Smith and Gehle (1969) of the conical capsule (Beem* Catalog number 1001 SPC) for *in situ* fixation and embedding of pelleted viruses. With proper adapters, 2 ml of fluid can be spun directly into conical capsules at accelerations up to 30,000 *g*.

2. Pelleting viruses at high speed is usually preceeded by clarifying the sample either by ultrafiltration or by equilibration in density gradients. To identify contaminants under the electron microscope, the study of ultrafiltrates is frequently adequate because the structure of the virus is well preserved. In contrast, in hypertonic gradients, the structure of viruses can be altered beyond the point of recognition, especially in the case of fragile en-

*Beem Better Equipment for Electron Microscopy, Inc., P.O. Box 132, Jerome Ave Station, Bronx, N.Y. 10468.

veloped particles. In addition, ultrafiltration gives a better chance to recover all particles, providing a filter of large enough porosity is used (0.65 μm, for example). Viral pellets can lack homogeneity if particles of different densities are present. If, in addition, the pellet is too large (i.e., much larger in diameter than the side of a thin section), this can lead to false negative EM reports. It is therefore essential that the pellet be as small as possible to increase the sampling significance of thin sections randomly prepared in such pellets.

D. Negative Staining

The techniques of negative staining have been recently reviewed by Haschemeyer and Myers (1972). They apply to the identification of viral contaminants within certain limits. For example, naked virions with cubic symmetry will be precisely identified in negatively stained preparations, whereas enveloped viruses can be damaged or even lost under the same preparatory conditions (Luftig and Kilham, 1971). Special techniques of fixation, however, are not required if large numbers of enveloped viruses are present and if surface spikes (like those in myxoviruses, MTV, and foamy viruses) make identification of the envelopes unmistakable and preclude confusion with small vesicular structures of cell origin so frequently present in viral concentrates.

E. Other Methods

1. Freeze Etching

The application of the freeze-etching method to virology is recent (Easterbrook and Rozee, 1970; Brown *et al.*, 1971). So far, the method does not permit the characterization of viruses at a resolution comparable to that regularly obtained by thin sectioning or negative staining. It seems unlikely that the method will lead to significant improvement in the EM identification of virus contaminants in tissue cultures.

2. Scanning Electron Microscopy

The study of cell surfaces at a resolution approaching 10 nm is possible with the scanning electron microscope, especially if the critical point method of Anderson (1966) is used to dry the sample with minimum distortion. At that resolution, budding viruses should theoretically be visible. In face view of the entire cell surface, the "bird's eye" of the scanning microscope will probably see hundreds of viruses when thin sections of the same cells usually show a very small number of them. This might be of considerable significance in monitoring tissue cultures for possible contaminants associated with cell surfaces, such as budding particles or mycoplasms.

Acknowledgments

I am grateful to Dr. Charlotte Friend and Dr. Stephen S. Sternberg for their critical review of the manuscript. The samples for the illustration of this chapter were kindly provided by Mr. Anthony Albino, Dr. George Barski, Dr. Ellen Borenfreund, Dr. Magdalena Eisinger, Dr. Jørgen Fogh, Dr. Charlotte Friend, Dr. William D. Hardy Jr., and Dr. Graziella Moldovanu. The skillful assistance of Miss Janet Mango and Miss Fay Pointner for the electron microscopy, Mr. Juan Marchese for the photography, and Mrs. Dorothy Saltzer for the typescript is gratefully acknowledged. This investigation was supported in part by the Health Research Council of the City of New York, Contract I-325, and by NCI Grant CA-08748.

References

Aaronson, S. A., Hartley, J. W., and Todaro, G. J. (1969). *Proc. Nat. Acad. Sci. U.S.* **64**, 87.
Aaronson, S. A., Todaro, G. J., and Scolnick, E. M. (1971). *Science* **174**, 157.
Achong, B. G., Mansell, P. W. A., Epstein, M. A., and Clifford, P. (1971). *J. Nat. Cancer Inst.* **46**, 299.
Anderson, D. R., and Manaker, R. A., (1966). *J. Nat. Cancer Inst.* **36**, 139.
Anderson. T. F. (1966). *Phys. Tech. Biol. Res.* **3**, 319–387.
Anonymous. (1966). *J. Nat. Cancer Inst.* **37**, 395.
Aoki, T., Stuck, B., Old, L. J., Hammerling, U., and de Harven, E. (1970). *Cancer Res.* **30**, 244.
Bernhard, W., and Tournier, P. (1964). *Ann. Inst. Pasteur, Paris* **107**, 447.
Bessis, M., Breton-Gorius, J., and Thiéry, J-P. (1958). *Rev. Hematol.* **13**, 363.
Biberfeld, G., and Biberfeld, P. (1970). *J. Bacteriol.* **102**, 855.
Blinzinger, K., Simon, J., Magrath, D., and Boulger, L. (1969). *Science* **163**, 1336.
Bouteille, M., Kalifat, S. R., and Delarue, J. (1967). *J. Ultrastruct. Res.* **19**, 474.
Brinkley, B. R., Murphy, P., and Richardson, L. C. (1967). *J. Cell Biol.* **35**, 279.
Brooks, R. E., and Siegel, B. V. (1967). *Blood* **29**, 269.
Brown, A., and Officer, J. E. (1968). *In* "Methods in Virology" (K. Maramorosch and H. Koprowski, eds.), Vol. 4, pp. 531–564. Academic Press, New York.
Brown, D. T., Waite, M. R., and Pfefferkorn, E. R. (1971). *29th Annu. EMSA Meet.* p. 382.
Bucciarelli, E., Rabotti, G. F., and Dalton, A. J. (1967). *J. Nat. Cancer Inst.* **38**, 359.
Cesarini, J.-P (1971). *C. R. Acad. Sci.* **272**, 901.
Chandra, S. (1968). *Lab. Invest.* **18**, 422.
Chapman, A. L., Bopp, W. J., Brightwell, A. S., Cohen, H., Nielsen, A. H., Gravelle, C. R., and Werder, A. A. (1967). *Cancer Res.* **27**, 18.
Chopra, H. C., Woodside, N. J., and Bogden, A. E. (1970). *Cancer Res.* **30**, 1544.
Clarke, J. K., Attridge, J. T., and Gay F. W. (1969). *J. Gen. Virol.* **4**, 183.
Claude, P., and Augusti- Tocco, G. (1970). *J. Cell Biol.* **47**, 36.
Dales, S., and Howatson, A. F. (1961). *Cancer Res.* **21**, 193.
de Harven, E. (1964). *J. Exp. Med.* **120**, 857.
de Harven, E. (1965). *In* "Methodological Approaches to the Study of Leukemias" (V. Defendi, ed.), pp. 147–156. Wistar Inst. Press, Philadelphia, Pennsylvania.
de Harven, E. (1967). *Cancer Res.* **21**, 2447.
de Harven, E. (1968a). *In* "The Nucleus" (A. J. Dalton and F. Haguenau, eds.), pp. 197–227. Academic Press, New York.
de Harven, E. (1968b). *In* "Experimental Leukemia" (M. Rich, ed.), pp. 97–129. Appleton, New York.
De Petris, E., and Harvey, J. J. (1969). *J. Gen. Virol.* **5**, 561.

de Thé, G., Rivière, M., and Bernhard, W. (1960). *Bull. Cancer* **47**, 569.
Dienes, L. (1963). *Recent Prog. Microbiol., Symp. Int. Congr. Microbiol., 8th, 1962*. pp. 511–517.
Dumont, A., and Robert, A. (1971). *J. Ultrastruct. Res.* **36**, 483.
Dupuy-Coin, A. M., Lazar, P., Kalifat, S. R., and Bouteille, M. (1969). *J. Ultrastruct. Res.* **27**, 244.
Easterbrook, K. B., and Rozee, K. R. (1970). *Proc. Electron Microsc. Soc.* **28**, 184.
Edwards, G. A., and Fogh, J. (1960). *J. Bacteriol.* **79**, 267.
Fogh, J. (1972). Personal communication.
Fogh, J., Hahn, E., and Fogh, H. (1965). *Exp. Cell Res.* **39**, 554.
Friedlander, M., and Moore, D. H. (1956). *Proc. Soc. Exp. Biol. Med.* **92**, 828.
Friend, C., Scher, W., Holland, J. G., and Sato, T. (1971). *Proc. Nat. Acad. Sci. v.s.* **68**, 378.
Gilden, R. V., Parks, W. P., Huebner, R. J., and Todaro, G. J. (1971). *Nature* (*London*) **233**, 102.
Györkey, F., Sinkovics, J. G., and Györkey, P. (1971). *Cancer* **27**, 1449.
Haguenau, F., and Hollmann, K. H. (1963). *Bull. Cancer* **50**, 29.
Haguenau, F., Rabotti, G. F., Lyon, G., and Moraillon, A. (1971). *J. Nat. Cancer Inst.* **46**, 539.
Haschemeyer, R. H., and Myers, R. J. (1972). *In* "Principles and Techniques of Electron Microscopy: Biological Applications" (M. A. Hayat, ed.), Vol. 2, pp. 101–147. Van Nostrand-Reinhold, Princeton, New Jersey.
Heine, U., Sverak, L., Kondratick, J., and Bonar, R. A. (1971). *J. Ultrastruct. Res.* **34**, 375.
Henry, K., and Petts, V. (1969). *J. Ultrastruct. Res.* **27**, 330.
Hirsch, J. G., and Fedorko, M. E. (1968). *J. Cell Biol.* **38**, 615.
Huebner, R. J., and Todaro, G. J. (1969). *Microbiology* **64**, 1087.
Huebner, R. J., Kelloff, G. J., Sarma, P. S., Lane, W. T., Turner, H. C., Gilden, R. V., Oroszlan, S., Meier, H., Myers, D. D., and Peters, R. I. (1970). *Proc. Nat. Acad. Sci. U.S.* **67**, 366.
Ishi, Y., and Mori, M. (1969). *J. Electron microsc.* **18**, 128.
Jarret, O., Laird, H. M., and Hay, D. (1969). *Nature* (*London*) **224**, 1208.
Jézéquel, A. M., Shreeve, M. M., and Steiner, J. W. (1967). *Lab. Invest.* **16**, 287.
Jones, T. C., and Hirsch, J. G. (1971). *J. Exp. Med.* **133**, 231.
Kajima, M., Sharon, N., and Pollar, M. (1967). *J. Ultrastruct. Res.* **18**, 661.
Kakuk, T. J., Hinz, R. W., Langham, R. F., and Conner, G. H. (1968). *Cancer Res.* **28**, 716.
Kaneseki, T., and Kadota, K. (1969). *J. Cell Biol.* **42**, 202.
Kindig, D. A., and Kirsten, W. H. (1967). *Science* **155**, 1543.
Koestner, A., and Long, J. F. (1970). *Lab. Invest.* **23**, 196.
Krishan, A., Uzman, B. G., and Hedley-Whyte, E. T. (1967). *J. Ultrastruct. Res.* **19**, 563.
Kuff, E. L., Wivel, N. A., and Lueders, K. K. (1968). *Cancer Res.* **28**, 2137.
Lambertenghi, G., de Harven, E., Sato, T., and Tennant, J. R. (1972). *Cancer Res.* **32**, 1108.
Lawn, A. M. (1970). *J. Gen. Virol.* **8**, 157.
Lowy, D. R., Rowe, W. P., Teich, N., and Hartley, J. W. (1971). *Science* **174**, 155.
Luftig, R. B., and Kilham, S. (1971). *Virology* **46**, 277.
Malmquist, W. A., Van Der Maaten, M. J., and Boothe, A. D. (1969). *Cancer Res.* **29**, 188.
Monneron, A., and Bernhard, W. (1969). *J. Ultrastruct. Res.* **27**, 266.
Moore, G. E., Kitamura, H., and Toshima, S. (1968). *Cancer,* **22**, 245.
Morton, D. L., Malmgren, R. A., Hall W. T., and Schidlovsky, G. (1969). *Surgery* **66**, 152.
Munroe, J. S., Shipkey, F., Erlandson, R. A., and Windle, M. S. (1964). *Nat. Cancer Inst., Monogr.* **17**, 365.
Mussgay, M., Reczko, E., and Ahl, R. (1969). *J. Gen. Virol.* **4**, 445.
Nowinski, R. C., Old, L. J., Sarkar, N. H., and Moore, D. H. (1970). *Virology* **42**, 1152.

Offers, S., and Bentvelzen, P. (1971). *Euro. J. Cancer* **7**, 357.
Parks, W. P., Todaro, G. J., Scolnick, E. M., and Aaronson, S. A. (1971). *Nature* (*London*) **229**, 258.
Patrizi, G., and Middelkamp, J. N. (1969). *J. Ultrastruct. Res.* **28**, 275.
Périer, O. (1967). *Acta Neuropathol.* **8**, 362.
Popoff, N., and Stewart, S. (1968). *J. Ultrastruct. Res.* **23**, 347.
Priori, E. S., Dmochowski, L., Myers, B., and Wilbur, J. R. (1971). *Nature* (*London*), *New Biol.* **232**, 61.
Roth, T. F., and Porter, K. R. (1962). *Electron Microsc. Proc. Int. Cong., 5th, 1962* Vol. 2, Art. LL-4.
Sato, T., Friend, C., and de Harven, E. (1971). *Cancer Res.* **31**, 1402.
Shipman, C., Jr., Vander Velde, G. C., and Booe, I. M. (1969). *Virology* **38**, 707.
Smetana, K., and Potmesil, M. (1968). *Z. Zellforsch. Mikrusk. Anat.* **92**, 62.
Smith, K. O., and Gehle, W. D. (1969). *Proc. Soc. Exp. Biol. Med.* **130**, 1117.
Stanbridge, E. (1971). *Bacteriol. Rev.* **35**, 206.
Stanbridge, E., Onen, M., Perkins, F. T., and Hayflick, L. (1969). *Exp. Cell Res.* **57**, 397.
Stenback, W. A., Van Hoosier, G. L., Jr., and Trentin, J. J. (1966). *Proc. Soc. Exp. Biol. Med.* **122**, 1219.
Thomas, A., Henry, M., Hollande, E., and Vilain, C. (1968). *C. R. Acad. Sci.*, **266**, 2129.
Volkman, L. E., Smucker, E. A., and Krueger, R. G. (1971). *J. Nat. Cancer Inst.* **46**, 953.
Watson, J., Raplh, P., Sarkar, S., and Cohn, M. (1970). *Proc. Nat. Acad. Sci. U.S.* **66**, 344.
Watson, M. L. (1962). *J. Cell Biol.* **13**, 162.
Weber, A., Whipp, S., Usenik, E., and Frommes, S. P., (1964). *J. Ultrastruct. Res.* **11**, 564.
Wivel, N. A., and Smith, G. H. (1971). *Int. J. Cancer* **7**, 167.
Wivel, N. A., Lueders, K. K., and Kuff, E. L. (1971). *Fed. Proc., Fed. Amer. Soc. Exp. Biol.* **30**, No. 3, Part II of Two (62nd Annu. Meet.), 1099 (abstr.).
Wright, B. S., and Korol, W. (1969). *Cancer Res.* **29**, 1886.

CHAPTER **10**

Endogenous Virus Contaminants in Fetal Bovine Serum and Their Role in Tissue Culture Contamination

ALEXIS J. KNIAZEFF

I. Introduction . . . 233
II. Target Viruses . . . 234
A. Bovine Herpesvirus 1 (BHV-1) . . . 234
B. Bovine Virus Diarrhea (BVD) . . . 235
C. Parainfluenza 3 Virus (PI-3) . . . 237
III. Procedure for Isolating and Identifying Viral Contaminants . . . 237
A. Cell Culture . . . 238
B. Serums for Virus Isolation . . . 238
C. Concentration of Serum Sediments . . . 239
D. Virus Detection . . . 239
E. Identification of Virus Isolates . . . 239
IV. Results . . . 240
V. Discussion and Conclusions . . . 241
References . . . 241

I. Introduction

The possibility of endogenous virus contaminants in fetal bovine serum and their potential hazard for virological and cytological studies were publicized as early as 1967 by various investigators. Some attention was eventually directed to the problem through a publication that dealt with viral antibodies in fetal bovine serum (Kniazeff *et al.*, 1967). In this investigation, it was demonstrated that appreciable levels of bovine virus diarrhea antibody and traces of bovine herpesvirus 1 antibody were present in a significant number of fetal bovine serum samples collected sterilely by heart puncture. Because it was generally accepted that the transplacental transfer of maternal antibodies to the fetus normally does not occur in species of Bovidae, the authors concluded that the antibodies were a sequel to a specific viral infection of the fetus. Again, because the incidence of specific viral antibodies in our sampling was relatively high, the incidence of fetal infection with the two viruses should have been commensurate. In this circumstance, it appeared very likely that some of the fetal bovine serum pools collected for tissue culture propagation might contain infective virus. This possibility appeared to us sufficiently alarming to warrant further research.

II. Target Viruses

In the study mentioned in Section I, bovine herpesvirus 1 (BHV-1), bovine virus diarrhea virus (BVD), and parainfluenza 3 virus (PI-3) were selected as reagents for identifying the presence of viral antibodies in fetal bovine serum. These particular viruses were selected because of their very high incidence of infection in the cattle population and their property of eliciting generalized infections.

In outlining the protocols for the isolation of viruses from fetal bovine serum, the three viruses also qualified as the most likely serum contaminants because in addition to the above-mentioned properties, they also have been incriminated in bovine abortions. For these reasons, BHV-1, BVD, and PI-3 were selected as primary targets for isolation.

A. Bovine Herpesvirus 1 (BHV-1)

Bovine herpesvirus 1 has the typical morphological characteristics of the herpesviruses. It is the etiologic agent of infectious bovine rhinotracheitis (IBR), which is a common infection of cattle in feedlots and dairies. It also causes abortion, pustular vulvovaginitis, meningoencephalitis of calves, and conjunctivitis. Infection of cattle with the virus is almost worldwide. The incidence of infection is high, with viremia a common feature. The virus, however, is not found free in the bloodstream but is concentrated in the leukocytes. It is not clear whether the virus replicates in the leukocytes or is carried passively. It is also not known what effect, if any, the virus may have on the cells of the reticuloendothelial system or whether it can participate in the induction of cell alteration *in vitro*. Not much is known regarding the pathogenesis of the virus in common laboratory animals or in humans. Some English investigators did produce herpes-like lesions in rabbits inoculated intracutaneously with a strain of BHV-1 and conjunctivitis following topical virus exposure. In cases of abortion, the virus can also readily be isolated from most tissues of the fetus, including blood (Kendrick and Straub, 1967).

On the basis of existing studies, it is apparent that the virus is cytopathogenic to a large variety of cultured mammalian cells. In the majority of cell cultures that have been studied, the virus produces rather rapid and destructive CPE. Besides the bovine cells, the virus is known to replicate in cell cultures of elk, mule deer, sheep, L cells, chick embryo cells (York, 1957), pig (Schwarz *et al.*, 1957), human amnion cells (Cheatham and Crandall, 1957), HeLa cells (Cabasso *et al.*, 1957), and in primary monkey kidney cells (Rouhandeh, 1963). Although it is obvious that the virus will propagate in a variety of cultured cells, there is no adequate information on its behavior in such subculturable human cells as WI-38 or in the cultures used by various biological houses for the production of vaccines. From the studies of Cabasso (Cabasso *et al.*, 1957), it is evident that the agent may spontaneously establish inapparent chronic infections in cultured cells. In my laboratory, we have also observed a pronounced reduction, or very gradual progression, of CPE in BHV-1-infected bovine and wallaby cells maintained on a medium containing BHV-1 antibody. As we previously demonstrated that fetal bovine serum may contain appreciable levels of such antibody (Kniazeff *et al.*, 1967), it is conceivable that a chronic silent infection could be established unwittingly in cultured cells given a BHV-1-contaminated serum and specific antibody.

In bovine cells, infection with BHV-1 leads to rounding and shrinking of cells within 24 to 48 hours. The complete destruction of a monolayer within 4 days is not uncommon. Propagation of the virus under both *in vitro* and *in vivo* conditions leads to the formation of large eosinophilic intranuclear inclusions of the type also observed with other herpesviruses (Cheatham and Crandall, 1957).

B. Bovine Virus Diarrhea (BVD)

Bovine virus diarrhea (also termed *mucosal disease* by some investigators) is a very common infectious and contagious disease of cattle. The severity of the disease is variable, with the majority of animals apparently going through only a mild, frequently clinically unrecognizable infection. The virus may then become established chronically, persisting for several months. In a smaller percentage of cases, the infection may be very acute, with a high rate of mortality.

The disease has been recognized in many countries. Epidemiological studies in the United States suggest that it may occur in every state in the union. In those states where serological surveys for the presence of bovine virus diarrhea antibodies have been conducted, it has been found that anywhere from 44% to 73% of the cattle tested are infected by bovine virus diarrhea virus (Baker *et al.*, 1954; Kniazeff, 1961; Newberne *et al.*, 1961).

In natural outbreaks, the disease has been recognized only in cattle; however, virus diarrhea antibodies have frequently been found in sheep (Kniazeff and Pritchard, 1960). The virus has been propagated in sheep and goats (Huck, 1957) and also by virus adaptation in rabbits (Baker *et al.*, 1954).

Some strains of bovine virus diarrhea virus are completely noncytopathic in cell culture; others vary from mildly to highly pathogenic. The cytopathic and the noncytopathic strains can be propagated in cultured bovine cells. Among cells used successfully are bovine embryonic skin and muscle, embryonic bovine kidney (Lee and Gillespie, 1957), and bovine testicular culture (Kniazeff *et al.*, 1965).

The virus has been propagated in the cells of only a very limited number of heterologous species. These include human intestinal cells (Henley), human leukemia cells (Osgood J-113) (Kniazeff, 1963), and lamb testicular cells (Coggins, 1962).

The extent of cytopathic changes in cells infected with BVD depends on the strain of the virus. With strongly cytopathic strains, cytological changes may become apparent within 2 to 3 days but with less cytopathic viruses alterations in cultured cells may not be obvious until 7 days of cultivation, or even later. The characteristic changes include elongation and shrinking of individual cells, frequently very pronounced vacuolation, and pyknosis of the nucleus. The progression of the changes can frequently be accelerated by raising the pH of the culture medium to approximately 8.2. Strongly cytopathic viruses readily form plaques under an agar overlay. In susceptible cell cultures, most of the virus strains achieve virus titers of 1×10^5 to 1×10^7 within approximately 7 days. In a study employing EBTr culture, it was found that most efficient replication of BVD occurred in stationary rather than in rapidly multiplying cells (Groyon and Kniazeff, 1967). Replication of noncytopathic BVD does not affect the plating efficiency of embryonal bovine kidney cells or embryonal bovine testicular cells (Gratzek, 1961).

The quantitation of the noncytopathic viruses and serological comparison can be carried out by a virus interference system, employing cytopathic bovine virus diarrhea agent, or by vesicular stomatitis or vaccinia virus (Gillespie *et al.*, 1963; Kniazeff and McClain, 1963). The identification of the noncytopathic strains of the virus can also be accomplished by means of fluorescein-labeled specific bovine virus diarrhea antibody.

In the course of infection, the animal develops a viremia and the agent can readily be isolated from most tissues of the body, including the blood (Thompson and Savan, 1963; Gutekunst and Malmquist, 1963a). In studying BVD of the newborn calf, Gutekunst and Malmquist noted that in a certain percent of cases virus was present at birth without the presence of antibody. On the basis of these observations, they postulated that early infection of the fetus with BVD virus may lead to immune tolerance with consequent continual replication of the virus after birth.

The virus has been incriminated in abortions of cattle, and seroepizootiological investigations have shown that the incidence of abortion may be very high. Infection of the fetus apparently results in viremia and if fetal blood is harvested at this time, high concentrations of the virus may be present in the serum.

Bovine virus diarrhea has a particle size of 40 to 55 mμ and a sedimentation coefficient of approximately 80 to 90 S. It contains RNA. It is ether and chloroform sensitive, and it appears to multiply by budding (Coggins, 1964). Although a considerable amount of characterization work has been done on this group of viruses, the agents still have not been assigned to a definite viral group within the scheme of viral characterization. So far, they are described as being myxolike.

C. Parainfluenza 3 Virus (PI-3)

The bovine strain of parainfluenza virus designated SF-4 is a participant in a common respiratory syndrome of cattle known as *shipping fever*. The agent has frequently been recovered from both diseased and clinically well animals. Seroepizootiological studies show the incidence of infection to be high. In an early survey (Hoerline *et al*., 1959), 68% of the calves were found to have significant antibody levels to the virus. Strains of bovine parainfluenza 3 have been serologically differentiated from human strains. Studies by Reisinger (1962) have shown no evidence that cattle strains can infect humans or that human strains can infect cattle.

The virus has been incriminated as a cause of bovine abortion, having been isolated from the thoracic fluid and lung tissue of an aborted fetus (Sattar *et al.*, 1965). It has also been found to produce generalized infections in bovine fetuses when administered *in utero* (Sattar *et al.*, 1967). The virus can be propagated readily in a variety of bovine cell cultures, such as primary bovine kidney cells, subpassageable bovine embryonal tracheal culture (EBTr) (Kniazeff *et al.*, 1965), and an established bovine kidney cell line (MDBK) (Madin and Darby, 1957). The virus can be quantitated both by hemagglutination with guinea pig erythrocytes and by plaque assay.

III. Procedures for Isolating and Identifying Viral Contaminants

The studies on isolating and identifying endogenous virus contaminants of fetal bovine serum proceeded in two stages. In the initial stage (Molander *et al.*, 1972a), emphasis was placed principally on the development and evaluation of methods and procedures. The second stage consisted of isolating viruses, determining the overall incidence of viral contamination, and determining the incidence of viral contamination in serums obtained from fetal, newborn, and young calf and adult animals.

A. Cell Cultures

Bovine diploid embryonal tracheal culture (EBTr) (Kniazeff *et al.*, 1965) and a human diploid (uncharacterized) lung culture (HDL) obtained from the National Cancer Institute were the basic substrates used for isolating and identifying viruses during the first phase of the studies. Additional cultures, such as established bovine kidney (MDBK) (Madin and Darby, 1957); uncharacterized, diploid fetal bovine bone marrow; diploid fetal mink lung (Mv1Lu) (Kniazeff *et al.*, 1964); established rabbit cornea (SIRC) (Volkert, 1957); and diploid human lung (WI-38) (Hayflick and Moorhead, 1961), were also included. These were used either as additional screens for virus isolation during the initial phase of experimentation or for propagating control virus stocks. (In the second phase, only the EBTr culture was used, the others having been found superfluous.)

The EBTr and HDL were propagated in Eagle's MEM (Eagle, 1959), supplemented with nonessential amino acids, glutamine, antibiotics, and 10% fetal bovine serum. For maintenance, EBTr was on Eagle's MEM supplemented with 3% agamma horse serum, whereas the HDL cells were on MEM supplemented with 5% agamma calf serum. All of the other cell lines were propagated in L-15 Leibovitz medium (Leibovitz, 1963). Serums used in cell propagation were first screened for the presence of viral contaminants.

B. Serums for Virus Isolation

Two varieties of serum were screened in the initial trials for virus isolation. One was fetal bovine serum processed by Hyland Laboratories under contract to the National Cancer Institute (NCI). This was sterilely collected by cardiac puncture and was not filtered. There were ten lots of this serum. The other variety of serum, comprising 16 lots, was procured by NCI from various commercial suppliers of fetal bovine serum. The methods used to process these serums were not known to the author. Once received, all serums were handled identically.

In the second phase of the virus isolation studies, the following varieties of serum were processed: fetal bovine, newborn calf, agamma newborn calf, calf serum from 2- to 3-month-old animals, and adult bovine serum. All of the serums were collected and processed by Hyland Laboratories.

The fetal bovine serum was prepared as described earlier. The other types of serum were collected aseptically, allowed to clot, centrifuged, pooled, and then filtered through 0.22 μm filters. The agamma newborn calf serum was prepared by a modified Cohn fractionation procedure (Cohn *et al.*, 1946).

C. Concentration of Serum Sediments

Samples of serum for sedimentation were made up from two or three lots each. The total volume of serum was 10 ml per sample. These were then centrifuged at 198,000 *g* for 1 hour at 4°C. The liquid phase was then decanted and the pellet was resuspended in 1.2 ml of Eagle's MEM and briefly sonicated. The sediments were dispensed into three sets of five tubes each of HDL- and EBTr-washed monolayer cultures. The volume of inoculum per tube was 0.2 ml. After 2 hour adsorption at 35°C, maintenance medium was added and the cultures were further incubated at 35°C.

D. Virus Detection

The detection of viruses in cell cultures treated with serum sediments was based on three approaches: (*1*) appearance of cytopathic changes; (*2*) hemadsorption with guinea pig erythrocytes; and (*3*) interference with the propagation of an indicator cytopathic virus. In this scheme, one set of tubes of each of the inoculated cell cultures was observed for cytopathic changes for a period of 21 days, and for hemadsorption at 6 and 13 days postinoculation. The results of these observations were compared with appropriate sets of controls. For virus interference tests, two sets of sediment-treated EBTr cultures were inoculated with 10 and with 100 $TCID_{50}$ of the NADL strain of BVD virus (Gutekunst and Malmquist, 1963b), 4 days postinoculation. In parallel, sets of HDL cells were inoculated with 10 and 100 $TCID_{50}$ of Simbis virus, 11 days postinoculation. Appropriate EBTr and HDL cell controls that were not exposed to the serum sediments but that were inoculated with identical concentrations of challenge viruses were run in parallel to permit comparison of virus replication.

When serum-sediment-treated cultures did not show evidence of virus activity by the above three methods, the cultures were frozen and thawed, gently centrifuged, and the supernatant fluid dispensed into fresh cultures and then passaged twice, as described for passage one. In instances where the presence of a viral agent was recognized, the isolation was confirmed through reisolation of the virus from unopened samples of the serum involved.

E. Identification of Virus Isolates

The identification of isolated viruses was carried out principally by serum neutralization tests and the fluorescent antibody technique, using specific antiserums for various viruses.

The search for inclusion bodies was made in cells exposed to the virus and fixed with Bouin's solution, after an appropriate period of cell incubation, and stained with Hematoxylin.

IV. Results

Two viruses were isolated in the course of the initial studies. One was isolated from the specially prepared, unfiltered fetal calf serum, and the other from one of the lots of commercial serum. Both agents were cytopathic in EBTr cultures. Isolate 115 induced strongly cytopathic changes in the cultures in the first passage. Isolate D-2 had to be line passaged once before CPE became clearly evident. In both instances, confirmation of isolation was made by reisolating the virus from unopened samples of the serum. Using the above-described procedures for virus identification, isolate 115 was confirmed to be a BHV-1 by serological comparison, the presence of typical intranuclear inclusion bodies associated with herpes-like viruses, and electron microscopic examination.

The identification of isolate D-2 was considerably more difficult, principally because of its fluctuating growth characteristics and inconsistent production of CPE, which made the serum neutralization test quite unreliable. The CPE, however, when it did occur, resembled that usually seen with BVD. Further studies with fluorescein-labeled antibody did indicate that the virus was a member of the BVD group.

The isolation of the two agents irrevocably confirms the proposition that bovine viruses do act as endogenous contaminants of fetal bovine serum and unquestionably may contaminate cell cultures, not only of bovine origin, but also of other species, including the human.

It should be noted that isolate 115 had a titer of 150 $TCID_{50}$ in the serum from which it was isolated.

In the second phase of the studies (Molander *et al.*, 1972b), 14 viruses were isolated from 149 lots of serum comprised of large pools from the following five types of serums: fetal bovine, newborn calf, agamma newborn calf, calf serum from 2- to 3-month-old animals, and adult bovine serum. Thus, approximately 10% of the lots tested were contaminated. In addition, twenty-five 1 liter bottles of fetal bovine serum, each taken from a different lot, were tested; two viruses were isolated from these, making a total of 16 isolates.

In the preparation of serum samples for virus isolation, an effort was made to draw the serum proportionately to the total volume of the lots. Fetal bovine serum, which had the greatest volume, also had the highest incidence of virus contamination. The only serum found to be free of contaminants was that from the 2- to 3-month-old animals.

Twelve of the isolates were from fetal bovine serum, two were from newborn calf, and one each was from agamma newborn calf and from adult bovine serum.

Of the 16 isolates, two were detected by interference; the rest showed cytopathic changes.

Five of the viruses were confirmed as BHV-1 by serum neutralization tests. Five had the cytopathogenic characteristics of BVD viruses but were not further characterized. Three were hemadsorbing and formed intranuclear and intracytoplasmic inclusion bodies and, therefore, were assumed to be members of the parainfluenza group. Three were completely unknown. (Of the unknowns, two were noncytopathic and were discovered through interference; the third had some characteristics of BHV-1 virus but could not be neutralized by specific BHV-1 antiserum.)

V. Discussion and Conclusions

The results of these studies make it conclusive that the bovine viruses may often contaminate bovine serum. It is also quite clear that these viruses can readily contaminate a variety of cell cultures, perhaps including human cells, and that they present a serious threat in the production of at least the veterinary vaccines.

It must also be evident from these studies that our knowledge of the total extent of viral contamination through the medium of serum is by no means complete. Although the studies to date have been carefully executed, they cannot be considered definitive or exhaustive. Additional work needs to be done to identify all the types of viruses that may participate in cell culture contamination. Finally, it is also clear that routine serum testing cannot be relied upon to insure freedom from viral contamination. Consequently, other means must be found to guarantee sterility of serum. Some method of serum sterilization that will insure complete inactivation of all biological contaminants, without reducing the cell growth promoting capacity, appears to be the best answer.

References

Baker, J. A., York, C. J., Gillespie, J. H., and Mitchell, G. B. (1954). *Vet. Res.* **15**, 525.

Cabasso, V. J., Brown, R. G., and Cox, H. R. (1957). *Proc. Soc. Exp. Biol. Med.* **95**, 471.

Cheatham, W. J., and Crandall, R. A. (1957). *Proc. Soc. Exp. Biol. Med.* **96**, 536.

Coggins, L. (1962). Ph.D. Thesis, Cornell University, Ithaca, New York.

Coggins, L. (1964). Unpublished data.

Cohn, E. J., Strong, L. E., Hughes, W. L., Mulford, D. J., Ashworth, J. N., Melin, M., and Taylor, H. L. (1946). *J. Amer. Chem. Soc.* **68**, 459.

Eagle, H. (1959). *Science* **130**, 432.
Gillespie, J. H., Madin, S. H., and Darby, N. V. (1963). *Proc. Soc. Exp. Biol. Med.* **110**, 248.
Gratzek, J. B. (1961). Ph.D. Thesis, University of Wisconsin, Madison.
Groyon, R., and Kniazeff, A. J. (1967). Unpublished data.
Gutekunst, D. E., and Malmquist, W. A. (1963a). *Can. J. Comp. Med.* **27**, 121.
Gutekunst, D. E., and Malmquist, W. A. (1963b). *Can. J. Comp. Med.* **28**, 19.
Hayflick, L., and Moorhead, P. S. (1961). *Exp. Cell Res.* **25**, 585.
Hoerline, A. B., Mansfield, M. E., Abinanti, F. R., and Huebner, R. J. (1959). *J. Amer. Vet. Med. Ass.* **135**, 153.
Huck, R. A. (1957). *J. Comp. Pathol. Ther.* **67**, 2671.
Kendrick, J. W., and Straub, D. C. (1967). *Amer. J. Vet. Res.* **28**, 1269.
Kniazeff, A. J. (1961). *Fla. Vet. Bull.* **4**, 16.
Kniazeff, A. J. (1963). Unpublished data.
Kniazeff, A. J., and McClain, M. (1963). Annual Report to NC34 Technical Committee (cited by Pritchard, 1963).
Kniazeff, A. J., and Pritchard, W. R. (1960). *Proc., USLSA, 1960* p. 344.
Kniazeff, A. J., Nelson-Rees, W. A., and Darby, N. (1964). "American Type Culture Collection," Certified Cell Line CCL-64, Mv-1-Lu.
Kniazeff, A. J., Nelson-Rees, W. A., Groyon, R. M., and Darby, N. (1965). *Excerpta Med., Sect. 1* **19**, No. 10.
Kniazeff, A. J., Nelson-Rees, W. A., and Darby, N. (1965). "American Type Culture Collection," Certified Cell Line, CCL-44, EBTr.
Kniazeff, A. J., Reimer, V., and Gaeta, L. (1967). *Nature* (*London*) **214**, 805.
Lee, K. M., and Gillespie, J. H. (1957). *Amer. J. Vet. Res.* **18**, 952.
Leibovitz, A. (1963). *Amer. J. Hyg.* **78**, 173.
Madin, S. H., and Darby, N. (1957). "American Type Culture Collecion," Certified Cell Line, CCL-22, MDBK.
Molander, C. W., Kniazeff, A. J., Boone, C. W., Paley, A., and Imagawa, D. T. (1972a). In press.
Molander, C. W., Kniazeff, A. J., and Imagawa, D. T. (1972b). In press.
Newberne, J. W., Robinson, V. B., and Atler, M. J. (1961). *Vet. Med.* **56**, 395.
Pritchard, W. R. (1963). *Advan. Vet. Sci.* **8**, 2.
Reisinger, R. C. (1962). *Ann. N. Y. Acad. Sci.* **101**, 576.
Rouhandeh, H. (1963). *Nature* (*London*) **200**, 386.
Sattar, S. A., Bohl, E. H., and Saturk, M. (1965). *J. Amer. Vet. Med. Ass.* **147**, 1207.
Sattar, S. A., Bohl, E. H., Trapp, A. L., and Handy, A. H. (1967). *Amer. J. Vet. Res.* **28**, 45.
Schwarz, A. J. F., York, C. J., Zirbell, L. W., and Estela, L. A. (1957). *Proc. Soc. Exp. Biol. Med.* **96**, 453.
Thompson, R. G., and Savan, M. (1963). *Can. J. Comp. Med.* **27**, 207.
Volkert, M. (1957). "American Type Culture Collection," Certified Cell Line, CCL-60, SIRC.
York, C. J. (1957). Unpublished data.

CHAPTER **11**

The Presence of Viruses in Uninoculated Tissue Cultures: Sources and Methods of Detection

F. KINGSLEY SANDERS

I.	Introduction	243
II.	Passenger, Indigenous, and Cryptic Viruses	244
III.	The Relationship of Viruses to an Ability of Cells to Grow in Culture	248
IV.	Significance of the Presence of Viruses Inherent to Cultured Cells	250
V.	Detection of Viruses Indigenous to Cultured Cells	252
	References	254

I. Introduction

It is a commonplace of *in vitro* studies of cells that, besides mycoplasmas (Barile, 1968), tissue cultures often contain extraneous viruses. The latter can come from medium ingredients that have a biological origin, such as serum (Kniazeff, this volume) or trypsin (Hallauer *et al.*, 1971), can result from accidental contamination of the cultures by the technicians who handle them (Collier, 1957), or can be introduced in association with the cells used to initiate the cultures (Hsiung, 1968), the latter being particularly likely to occur when primary tissue cultures are used (Kalter and Heberling, 1968). The first two sorts of viral contamination can be considered trivial from a biological point of view, because they constitute a peculiar hazard of the circumstances in which tissue culture is carried out and can be expected to be eliminated when a foolproof methodology is adopted. The third source from which viruses can be unknowingly introduced, namely those viruses inherent to the cultured tissues themselves, is the one with which this chapter is concerned. It cannot be considered trivial, because the large extent of its probable occurrence raises the question of whether the presence of viruses can be considered an inevitable, or perhaps essential, accompaniment of the propagation of cells *in vitro*, especially when continuous cell lines are to be established.

II. Passenger, Indigenous, and Cryptic Viruses

Very many animal species have hidden virus infections, that is, viruses that appear to exist in close association with particular tissues or organs of the species involved, without causing overt symptoms. It follows that primary cultures prepared from almost any tissue may be found infected with viruses *in vitro* simply because the organ from which they were prepared was infected *in vivo* (Hsiung, 1968). In this way, primary cultures from chickens (Burke *et al*., 1965), chimpanzees (Rogers *et al*., 1967), dogs (Smith *et al*., 1970), ducks (Luginbuhl, 1968), guinea pigs (Hsiung and Kaplow, 1969), hamsters (Stenback *et al*., 1966), horses (Hsiung *et al*., 1969), man (Hsiung, 1968), rabbits (Nesburn, 1969), rats (Melendez *et al*., 1967), monkeys (Hsiung, 1968), and swine (Huygelen and Peetermans, 1967) have all been found to yield viruses spontaneously. More than one agent has been found in the same primary culture (Hsiung, 1968). Primates have been more intensively studied in this respect that any other group of species, perhaps because of the widespread use of viral vaccines. Tumor cell cultures are also a rich source of adventitious viruses and, in view of what is often an obvious lack of relationship to the etiology of the tumor from which they have been derived, have been termed *passenger viruses*. By reason of their capacity for continuous synthesis and replication, tumor cells have been assumed to provide a favorable milieu for the multiplication of viruses and therefore to be especially likely to harbor such passengers. At the present time, it may be preferable to restrict the use of the term *passenger* to viruses carried into culture with the explanted tissue in the form of fully infectious extracellular virions that subsequently become capable of invading and replicating within cells of the culture. In the most obvious cases, this type of association will eventually be manifested as a frank cytopathic effect, as seems to occur with passenger adenoviruses in human adenoids (Rowe *et al*., 1956) and the foamy viruses that are often found in monkey kidney cultures (Hsiung, 1968). The cytopathic effects observed *in vitro* may sometimes result from the acquisition of susceptibility to infection and lysis by cells that were originally resistant to the virus *in vivo*. A striking illustration of an effect on the virus–cell relationship brought about by altering the cells' environment has been given by Kaplan (1955). When monkey kidneys are infected *in vivo* with polio virus, no cytopathic effect can be observed and no virus can be recovered following subsequent perfusion of the kidney or homogenization of its tissue. However, if parts of an infected kidney found negative for virus by such tests are reduced to cell suspensions by trypsin treatment and used to set up conventional monolayer cultures, cell destruction eventually takes place, accompanied by the appearance of fully infective virus. When the kidney cells are removed from their *in vivo* situation to an alien environment, they somehow acquire susceptibility to polio virus. The nature of the mechanisms involved—whether an actual change in the cells themselves or merely their removal from a protected environment, such as would be provided by antibody—remains unknown. The so-called *conditional lethal mutants* are further examples of how extraneous conditions can affect the response of cells to infection. Many strains of polio virus (Cooper, 1964) are now known that multiply normally at 37°C but do not complete their multiplication cycle when incubated at 39.5°C. Such a virus will not be apparent at the higher temperature but will emerge, accompanied by a cytopathic effect, when the temperature is lowered. It is not suggested that temperature-dependent viruses of this type are widespread in nature—most of those known are selected laboratory mutants—but they are mentioned here to illustrate the point that appropriate culture conditions may allow hidden viruses to emerge that would not otherwise be detected.

A passenger association of a similar kind, although a less obvious one, occurs when only a small number of the cells in the culture are or become susceptible to the virus, although the latter is carried into the culture in its fully infectious form. Virus replication will occur, although sufficient cells will not be destroyed at any one time for the culture to show a cytopathic effect. However, the virus never enters a truly latent form. An association of this type between the paramyxovirus SV5 and human fibroblasts has been described by Wong and Kilbourne (1961). Three different kinds of carrier cultures of this type have also been established deliberately.

1. In the first, almost all the cells in the culture are genetically resistant to the virus, which then replicates only in a minority of susceptible cells. Examples of this type are Coxsackie virus A9 in HeLa cells (Takemoto and Habel, 1959) and polio virus type 3 in HeLa cells (Vogt and Dulbecco, 1958).

2. In this type of association, all the cells remain susceptible but the spread of the virus is restricted by antibody in the medium, e.g., herpes simplex virus in HeLa cells (Wheeler, 1960).

3. A third type is one in which the cells also retain genetic susceptibility but most of them become refractory through the action of interferon produced within the culture, as with Newcastle disease virus infection of murine L cells (Henle, 1963).

It is doubtful, though, whether the term *passenger* should be extended to include every virus that appears fortuitously in primary cultures of tissues. A tabulation of recent reports of such isolations given by Smith (1970) shows that they include agents, such as members of the herpes group, that are often present *in vivo* in a latent noninfectious form, even in the presence of circulating humoral antibody (Andrewes, 1967). Latent herpes simplex virus infection can be activated by a number of nonspecific stimuli, after which completed virions are released by infected skin cells that show a cytopathic effect (Fenner, 1968). It is unlikely that the herpes-type viruses that appear predominantly in primary tissue cultures all represent populations stemming from complete virions introduced to the cultures along with the cells. Many may represent latent herpesvirus genomes activated as a result of the stress of culture initiation.

In a number of cases, tissue culture propagation for many transfers is necessary before a virus emerges. Primary cultures of cells from children with Burkitt's African lymphoma rarely release Epstein-Barr virus (EBV); its presence cannot be demonstrated with the electron miscroscope, either in the tumor itself or in the early stages of its culture history. Nor do the cells at this stage contain viral antigens detectable by indirect immunofluorescence using EBV-positive human sera. However, following a number of transfers *in vitro*, cells from many such cultures can be shown to contain microscopically identifiable viral structures and a proportion of them now show positive immunofluorescence (Henle and Henle, 1966). The virus contained within such cells can also be caused to increase in amount by subjecting the cultured cells to stresses of various kinds, such as arginine-depleted media (Henle and Henle, 1968) or a suboptimal temperature of cultivation (Horoszewicz *et al*., 1968).

Acquisition of an identifiable herpes-like virus following passage is not a unique property of cells from Burkitt's African lymphoma and has been shown to occur with many floating lymphoid cell lines of varied origin (Jensen *et al*., 1967). Indeed, it has been stated that all such continuously transferable lines carry a herpes-like virus (Gerber *et al*., 1970). In this instance, the recent establishment from cases of Hodgkin's disease (Eisinger *et al*., 1971) of lymphoblastoid cell lines, capable of growth in suspension and carrying a herpes-like virus resembling, but not identical with, Epstein-Barr virus, is of interest, although there is as yet no evidence of a connection between this virus and the etiology of Hodgkin's disease. In certain other human tumor culture systems, prolonged cultivation may be necessary before antigenic and other changes, possibly indicative of the emergence of viruslike agents, can take place (Sabin, 1965). It is therefore proposed that the term *indigenous* rather than *passenger* be used to describe this group of viruses, indigenous viruses being defined as those that are frankly pathogenic in the appropriate circumstances [Epstein-Barr virus has been identified as the cause of infectious mononucleosis (Henle *et al*, 1968; Hirshaut *et al.*, 1969); herpesvirus B is latent in the rhesus monkey but causes fatal disease in man (Andrewes, 1964)] but that are normally introduced into tissue cultures within latently infected cells, later to be induced to proliferate by stimuli at present unknown.

Cells in tissue culture may have viruses associated with them in a third way, which we may term *cryptic*. Cryptic viruses remain integrated within their host cells throughout their life in culture. They do not cause a cytopathic effect under normal circumstances and do not often become induced or activated under the normal conditions of culture. Naturally, this is a broad class that includes associations between viruses and cells of widely varying degrees of intimacy. On the one hand, there are cryptic viruses that closely resemble indigenous viruses in that prolonged cultivation causes their emergence in identifiable form, although without cytopathic effect. Examples are the type C viral particles that frequently appear in cultures of mouse tissues following prolonged cultivation (Aaronson *et al.*, 1969). On the other hand, are cells infected by viruses, the only evidence of whose presence is an alteration of cellular properties, such as the acquisition of malignancy [Nil 2E cells transformed by the hamster sarcoma virus HaSV (Albino and Sanders, 1972)]. An intermediate category is typified by the association between SV40 virus and nonpermissive mouse cells (Green, 1970) or between polyoma virus and hamster kidney cells (BHK_{21}) (Dulbecco, 1968) in which the viral presence is signaled not only by altered cellular properties but also by the manufacture of virally specified antigens and the possibility of rescuing complete extracellular virus by fusion with permissive cells (Koprowski *et al.*, 1967). Temperature-sensitive variants of polyoma virus are also known that affect the expression of this virus when in the cryptic state (Dulbecco, 1969). Cells incubated at the nonpermissive temperature appear normal; they do not have the transformed morphology and mode of growth that stamp them as carrying the viral genome. When shifted to the permissive temperature, they transform but, unlike the conditional lethal mutants of polio virus mentioned earlier, do not proceed to the production of complete virus. Treatment of cultured cells with certain nucleoside analogs (bromo- or iododeoxyuridine) can cause the activation of apparently cryptic viral RNA genomes, resulting in the appearance of products varying from special antigens thought to be virally specified (Weiss *et al.*, 1971) up to complete infective type C oncornavirus particles (Lowy *et al.*, 1971; Aaronson *et al.*, 1971). Similar chemical treatment of cells carrying cryptic DNA virus genomes (SV40-transformed 3T3 mouse fibroblasts; Watkins, 1970) causes many more cells to become capable of making virus upon fusion with permissive cells than when they are not so treated.

Most of the above examples of a cryptic association between viruses and host cells have been taken from cases in which the association has been deliberately established by treating cells with viruses in tissue culture and then showing later that a cryptic type of association has been set up. There is, however, no *a priori* reason for assuming that associations of this kind do not occur *in vivo* and that cells detached from whole organisms and subsequently cultivated *in vitro* will not prove to be cryptically infected. That this may be so is indicated by the observation that cells derived from tumors caused by the inoculation of polyoma or SV40 viruses into infant hamsters, and subsequently cultured do not manufacture virus but can be shown, nevertheless, to contain viral genomes because complete virus may be rescued by fusion with permissive cells (Green, 1970). The transfer of cells from the *in vitro* to an *in vivo* situation can show a similar effect. Syrian (Vigier, 1970) or Chinese hamster (Hlozanek *et al.*, 1966) or rat cells (Svoboda, 1964) transformed morphologically *in vitro* by means of Rous sarcoma virus show a cryptic association between virus and cell, for their injection into chickens results in the growth of a tumor consisting of chicken cells that release the same sarcoma virus.

III. The Relationship of Viruses to an Ability of Cells to Grow in Culture

For over 10 years it has been widely believed that cell cultures initiated from normal tissues have a limited lifespan *in vitro* (Hayflick, 1965). Two main hypotheses have been advanced to account for this condition.

1. It can result from inadequacy of the medium used and/or the unnatural stresses to which cells are subject during manipulation in culture (Puck *et al.*, 1958; Todaro and Green, 1964), as well as from the particular schedule adopted for subculture (Hay, 1970).

2. Cell cultures exhibit aging *in vitro*, the time of onset of which reflects the physiological age *in vivo* of the cells of the tissue from which the cultures were initiated. Each tissue cell is conceived as having an inborn potential for a finite number of cell divisions. Once the limit is reached, the cell dies, supposedly owing to accumulated copying errors in its DNA that lead to associated mistakes in protein synthesis (Orgel, 1963; Hayflick, 1966).

At the present time, it is still uncertain whether either of these hypotheses is correct. There is a clear difference between species with regard to the ability of their cells to propagate indefinitely in culture. Human diploid fibroblasts appear generally to have a limited lifetime *in vitro* (Hayflick, 1965); those from rats appear capable of long-term cultivation without loss of growth potential (Petursson *et al.*, 1964). Cells from mice, hamsters, and many other species frequently give rise to cell lines capable of continuous cultivation, the properties of which, such as saturation density and oncogenic potential, depend on the seeding density and transfer interval adopted for their propagation (Macpherson and Stoker, 1962; Todaro and Green, 1963). Many have a tendency to become heteroploid (e.g., mouse) at the time a capacity for continuous transfer is acquired.

Tissues from other species, such as the Chinese hamster (Yerganian and Leonard, 1961; Hsu *et al.*, 1964), also readily give cell lines that can be continuously propagated *in vitro*, although here the cells tend to become pseudodiploid rather than heteroploid when continuously maintained.

The possibility of establishing continuous lines with cells from these species does not seem to depend on their physiological age at the time of explantation. With the mouse, as with both the Chinese and the Syrian hamster, cell lines have been readily established from embryos (Todaro and Green, 1963; Diamond, 1967), new borns (MacPherson and Stoker, 1962), and adults (Earle, 1943; Hsu *et al.*, 1964) with equal facility.

The mode of origin of the abnormal chromosome complements of continuous cell lines, whether by selection or by some other means, is at present uncertain. However, it has been claimed that it can itself be a consequence of employing defective media and that by using the proper regimen, continuous lines of euploid cells can be established, even from humans (Puck *et al.*, 1958).

It has not been ruled out, however, that the ability of tissues to generate continuous cell lines may be linked with the occurrence of indigenous or cryptic viruses. As already mentioned, many human continuous lymphoid cell lines carry a virus of the herpesvirus groups (Gerber *et al.*, 1972). A few that contain no virus detectable by immunofluorescence or electron microscopy [e.g., Raji cells derived from a Burkitt lymphoma (Pulvertaft, 1965), $NHDL_3$ cell cultures from normal human leukocytes (Gerber and Monroe, 1968), NC_{37} from leukocytes of a patient with pneumonia (see Gerber, 1972)] have been shown to contain cryptic virus by cloning (Maurer *et al.*, 1969), nucleic acid hybridization (Zur Hausen and Scholte-Holthausen, 1970), complement fixation (Vonka *et al.*, 1970), or activation by bromodeoxyuridine (BrdU) (Gerber, 1972).

The readiness with which murine tissues give rise to continuous cell lines can be correlated with the presence of cryptic genomes of type C RNA viruses. Activation of the latter can take place spontaneously in cells propagated *in vitro* (Aaronson *et al.*, 1969) and in mice as the result either of aging or of certain carcinogenic stimuli (Kaplan, 1967). Treatment with halogenated pyrimidines can be used to reveal the presence of such viruses in clonal lines of cells from both high and low leukemia incidence strains of mice (Lowy *et al.*, 1971; Aaronson *et al.*, 1971). Indeed, cells from embryos of the high leukemic AKR strain of mice normally grow in culture as virus-negative cell lines. *In vivo* they develop a capacity to produce virus, for cells from newborns or older mice can be shown to release virus when cultured.

Although the above information indicates strongly that mouse cells may contain cryptic type C viral genomes that can be activated by various means, there is as yet no direct evidence that activation of the latter confers upon the cells a capacity to propagate indefinitely *in vitro*. An indirect suggestion that this may be so, however, is given by the work of Aaronson *et al.* (1969), who carried cultures of murine fibroblasts according to the 3T3 schedule of culture transfer. Most of their cultures began to produce type C virus particles after about 35 passages in culture, which represents roughly 50–70 cell generations, the time at which cell cultures are said to enter crisis and then either die out or give rise to continuous cell lines. The question thus remains open as to whether indigenous or cryptic viruses invariably accompany an ability to grow continuously in cell culture, and its answer must await further experiment. In the meantime, it must be kept in mind that the presence of viral particles of abnormal morphology, such as the corona viruses common to hamster cells (Bernhard and Tournier, 1964), may be a normal feature of cells in culture and not the result of contamination by some extraneous agent.

It is also necessary to be cautious regarding the relationship between the acquisition by cells of hetero- or pseudodiploidy and the onset of the ability to propagate continuously in culture. Viral infection of cells has been shown in many instances to be accompanied by karyotypic changes (Hampar and Ellison, 1963; Nichols *et al.*, 1962; Shein and Enders, 1962; Koprowski *et al.*, 1962; Black and Rowe, 1963), so that the chromosomal alterations observed at the initiation of continuous cell lines may be secondary to the real cause of the phenomenon.

IV. Significance of the Presence of Viruses Inherent to Cultured Cells

Apart from the possible role of viruses in the establishment of normal tissue cultures, it is necessary to consider the extent to which the presence of passenger, indigenous, or cryptic viruses can invalidate experimental observations. To the virologist, in particular the viral geneticist, the absence of passenger viruses would seem essential.

The widespread adoption of tissue culture methods in the 1950's brought a curious paradox into modern virology in that we now often have to look for a disease to fit a given virus, rather than the other way round. A great many viruses are now known that can be associated with no known disease, and with some of them it is quite uncertain whether they originated from the tissues used to set up the cultures used or from the materials that were later used to inoculate them.

Not only are passenger viruses an obvious source of error when tissue cultures are employed in attempts to isolate viruses from clinical material but extraneous agents can also confuse the results of experiments undertaken to study the behavior of human viruses in tissue culture, by suppressing (Hoggan *et al.*, 1966) or even stimulating (Rabson *et al.*, 1964) the multiplication of the agent under study. It is, however, in genetic experiments that their presence is most likely to lead to false conclusions. Hybridization, at the level of the genetic material, between viruses growing in the same cell is well known and subsequent replication of the hybrid agent can give rise to virions of altered properties (Rapp *et al.*, 1965). Virus stocks containing virions with drastically changed properties can also arise by transcapsidation (Hanafusa *et al.*, 1964), a condition where the genetic material of one virus becomes enclosed within the protein coat of another. Compared with passenger viruses, indigenous and cryptic agents are more difficult to detect. Indeed, in some cases they may remain undetected during many tissue culture passages. The recovery of viruses following prolonged cultivation of material from human tumors or as a result of exposure of the latter's cells to an alien environment, such as that provided by the tissues of another species (McAllister *et al.*, 1971), can give rise to false hopes that a tumor etiologic agent has been isolated, whereas in reality an indigenous latent virus has been stimulated to proliferate. Positive chemical stimulation of tumor cells to release hitherto cryptic agents is likewise of doubtful significance, unless it can be shown that normal cells from the same tissue in the same individual do not also respond by the production of virus and/or antigens when so stimulated. Like passenger viruses, stimulated indigenous or activated cryptic viruses have potential to affect the replication of known viruses added to the culture and can also be concerned in hybridization and transcapsidation (Rapp *et al.*, 1965). The experimental virologist thus has the need to know that his cells are free from all three classes of agents before he can make an unequivocal interpretation of his experimental findings. Methods for the recognition of indigenous and cryptic virus infections have been suggested by Fogh *et al.* (1971) and will be mentioned in Section V.

The possibility that cryptic viral genomes are already present in the cells used to support virus growth raises difficulties in the interpretation of cell transformation and viral replication brought about by type C oncornaviruses, in particular when an attempt is made to discover the part played by RNA-dependent DNA polymerases (Baltimore, 1970; Temin and Mizutani 1970) with respect to these processes. Following demonstration in cell-free systems, it is widely accepted that these enzymes, by means of an RNA–DNA hybrid intermediate, can transcribe part or all of the information in the viral RNA into a DNA copy that then becomes integrated into the host cell DNA, causing cell transformation followed by maintenance of the transformed state when this material is replicated in unison with the cellular DNA. In productive infection, it may also serve as a template for the production of further copies of the viral RNA. In support of the theory originally put forward by Huebner and Todaro (1969), there is now evidence that normal mouse cells, even from strains with a low incidence of leukemia (Lowy *et al.*, 1971; Aaronson *et al.*, 1971), already contain the information for the production of MuLV virions for whose replication they are commonly used as substrate. Similarly, cells from uninfected embryos of the kind used for studying the behavior of avian sarcoma viruses can be induced to form virally specified antigens by treatment with IUdr (Weiss *et al.*, 1971). Infection of these cells with the appropriate oncornaviruses constitutes a kind of superinfection of cells already cryptically carrying a related viral genome, and the mode of replication of these viruses becomes hard to visualize at a biochemical level. In the case of the type C RNA viruses, it is even possible to suggest that perhaps cancer causes viruses rather than the other way around! The presumably common occurrence of indigenous and cryptic infection among cells from continuous cell lines as well as early passage cell strains suggests that it is probably safer for the experimenter to assume their presence rather than otherwise.

V. Detection of Viruses Indigenous to Cultured Cells

Viruses defined as passengers by the criteria adopted prove easy to detect. Once the presence of a particular virus is presumed, inoculation of homogenates of suspected cultures into cells known to respond cytopathically should reveal it. In some instances, such cells may be hard to find. Murine leukemia viruses multiply exuberantly in cultures of mouse fibroblasts without cytopathic effect and could thus remain undetected. Cocultivation of the cells of suspected cultures with rat XC fibroblasts (transformed by and cryptically carrying the genome of Rous sarcoma virus) reveals the presence of this virus by means of a cytopathic effect on the latter cells (Klement *et al.*, 1969). Direct examination of both cultured cells and gradient-centrifuged tissue culture supernatants with the electron microscope, using both thin sectioning and negative staining techniques, should also serve to reveal passenger viruses. However, the relative insensitivity of such methods may fail to reveal minimal contamination. Passenger viruses that do not have an obvious cytopathic effect, such as certain myxoviruses, might be detected by their ability to induce interferon production, causing the appearance of resistance to infection by an interferon-sensitive cytopathic agent, such as vesicular stomatitis or encephalomyocarditis virus. Myxo- or paramyxoviruses may also be demonstrated directly through hemadsorption (Kilbourne, 1969; Marcus and Carver, 1969).

The revelation of indigenous or cryptic viruses may require more sophisticated methods. Prolonged cultivation has already been mentioned as an aid to the evocation of indigenous viruses. Other techniques might include the cocultivation and/or fusion (using β-propiolactone-inactivated Sendai virus) of cells with indicator cells thought to be permissive for the virus suspected. In this instance, properties of the virus other than its ability to replicate and cause cell destruction could be used to devise a detection system; for example, such cells as skin fibroblasts from human cases of Fanconi's anemia (Todaro *et al.*, 1966), Down's syndrome (Todaro and Martin, 1967) or Bloom's syndrome (German *et al.*, 1965) may be sensitive indicators for transforming viruses. Cocultivation and cell fusion, combined with chemical activation by nucleoside analogs and, perhaps, water-soluble mutagens and carcinogens (Sanders, 1971; Igel *et al.*, 1969), may prove useful for the detection of cryptic viruses, as will the use of helper viruses to call forth hidden viral genomes from cultured cells. Feline leukemia virus (FeLV) would be an obvious choice, from its ability to multiply without cytopathic effect in many types of mammalian cells. Its most important property is that it can form a hybrid with murine sarcoma virus that can then infect cat cells and be sarcomagenic for this species. FeLV may thus be used to call forth cryptic sarcoma virus genomes that could then be detected by their effect in cats. More direct tests for indigenous or cryptic viruses or their intracellular manifestations will be to search, by immunofluorescence, for the presence of antigens associated with a specific group of viruses, using selected antiserums. Immunofluorescence may also be used to detect the results of activating cryptic viral genomes by any of the methods suggested in the last paragraph. Less certain, but giving suggestive results, is the technique of molecular hybridization. It can be used in two ways.

1. The first is to see whether messenger RNA (mRNA) from cells suspected of harboring a cryptic virus hybridizes with nucleic acids from viruses already known. Using this method, an RNA specific for adeno- and papovaviruses has been found in tumors induced by these viruses.

2. In the case of type C oncornaviruses, the viral RNA-dependent DNA polymerase has been used to make a "DNA" *in vitro*, using the viral RNA itself as a template. This DNA is then used as a probe to search for cellular RNA's resembling the viral RNA in uninfected cells (Hehlmann *et al.*, 1972).

Possibly the best way to determine whether a passenger, indigenous, or activated cryptic virus in a particular tissue culture system is complicating the results of experiments is to determine whether the viral yields comprise a mixed population of virions. This can be done by analyzing the infective dose–response curve to see whether it obeys single- or multiple-hit kinetics. In the case of such systems as MuLV–MSV or RSV–RAV, where the presence of a helper virus is obligatory, double-hit kinetics is the rule. However, it should be pointed out that even in these systems, the dose–response curve has the appearance of being single-hit once the helper virus is present in great excess, so that its presence will not be suspected if this is the only means of testing used.

References

Aaronson, S. A., Hartley, J. W., and Todaro, G. J. (1969). *Proc. Nat. Acad. Sci. U.S.* **64**, 87.
Aaronson, S. A., Todaro, G. J., and Scolnick, E. M. (1971). *Science* **174**, 157.
Albino, A., and Sanders, F. K. (1972). Unpublished work.
Andrewes, C. H. (1964). "Viruses of Vertebrates." Baillère, London.
Andrewes, C. H. (1967). "The Natural History of Viruses," p. 163. Weidenfeld & Nicolson, London.
Baltimore, D. (1970). *Nature* (*London*) **226**, 1209.
Barile, M. F. (1968). *Nat. Cancer Inst., Monogr.* **29**, 201.
Bernhard, W., and Tournier, P. C. (1964). *Ann. Inst. Pasteur, Paris* **107**, 447.
Black, P. H., and Rowe, W. P. (1963). *Virology* **19**, 107.
Burke, C. N., Luginbuhl, R. E., and Helmboldt, C. F. (1965). *Avian Dis.* **9**, 31.
Collier, L. H. (1957). *Nature* (*London*) **180**, 757.
Cooper, P. D. (1964). *Virology* **22**, 186.
Diamond, L. (1967). *Int. J. Cancer* **2**, 143.
Dulbecco, R. (1968). *Science* **166**, 962.
Earle, W. R. (1943). *J. Natl. Cancer Inst.* **3**, 555.
Eisinger, M., Fox, S. M., de Harven, E., Biedler, J. L., and Sanders, F. K. (1971). *Nature* (*London*) **233**, 104.
Fenner, F. (1968). "The Biology of Animal Viruses," Vol. 2, p. 632. Academic Press, New York.
Fogh, J., Holmgren, N., and Ludovici, P. P. (1971). *In Vitro* **7**, 26.
Gerber, P. (1972). *Proc. Nat. Acad. Sci. U.S.* **69**, 83.
Gerber, P., and Monroe, J. H. (1968). *J. Nat. Cancer Inst.* **40**, 855.
Gerber, P., Whang-Peng, J., and Monroe, J. H. (1970). *Bibl. Haematol.* **36**, 739.
German, J., Archibald, R., and Bloom, D. (1965). *Science* **148**, 506.
Green, M. (1970). *Annu. Rev. Biochem.* **39**, 701.
Hallauer, C., Kronauer, G., and Siegl, G. (1971). *Arch. Gesamte Virusforsch.* **35**, 80.
Hampar, B., and Ellison, S. A. (1963). *Proc. Nat. Acad. Sci. U.S.* **49**, 474.
Hanafusa, H., Hanafusa, T., and Rubin, H. (1964). *Proc. Nat. Acad. Sci. U.S.* **51**, 41.
Hay, R. (1970). *In* "Aging in Cell and Tissue Culture" (E. Holečková and V. J. Cristofalo, eds.), p. 7. Plenum, New York.
Hayflick, L. (1965). *Exp. Cell Res.* **37**, 614.
Hayflick, L. (1966). *In* "Topics in the Biology of Aging," p. 83. Wiley (Interscience), New York.
Hehlmann, R., Kufe, D., and Speigelman, S. (1972). *Proc. Nat. Acad. Sci. U.S.* **69**, 435.
Henle, G., and Henle, W. (1966). *J. Bacteriol.* **91**, 1248.
Henle, G., Henle, W., and Diehl, V. (1968). *Proc. Nat. Acad. Sci. U.S.* **59**, 94.
Henle, W. (1963). *J. Immunol.* **91**, 145.
Henle, W., and Henle, G. (1968). *J. Virol.* **2**, 182.
Hirshaut, Y., Glade, P. R., Moses, H., Manaker, R., and Chessin, L. N. (1969). *Amer. J. Med.* **47**, 520.

Hlozanek, I., Donner, L., and Svoboda, J. (1966). *J. Cell Physiol.* **68**, 221.

Hoggan, M. D., Blacklow, N. R., and Rowe, W. P. (1966). *Proc. Nat. Acad. Sci. U.S.* **55**, 1467.

Horoszewicz, J. S., Dunkel, V. C., and Grace, J. T. (1968). *Fed. Proc., Fed. Amer. Soc. Exp. Biol.* **27**, 262.

Hsiung, G. D. (1968). *Bacteriol. Rev.* **32**, 185.

Hsiung, G. D., and Kaplow, L. S. (1969). *J. Virol.* **3**, 355.

Hsiung, G. D., Fischman, H. R., Fong, C. K. Y., and Green, R. H. (1969). *Proc. Soc. Exp. Biol. Med.* **130**, 80.

Hsu, T. C., Humphrey, R. M., and Somers, C. E. (1964). *J. Nat. Cancer Inst.* **32**, 839.

Huebner, R. J., and Todaro, G. J. (1969). *Proc. Nat. Acad. Sci. U.S.* **64**, 1087.

Huygelen, C., and Peetermans, J. (1967). *Arch. Gesamte Virusforsch.* **20**, 260.

Igel, H. J., Huebner, R. J., Turner, H. C., Kotin, P., and Falk, H. L. (1969). *Science* **166**, 1624.

Jensen, E. M., Korol, W., Dittmar, S. L., and Medrek, T. J. (1967). *J. Nat. Cancer Inst.* **39**, 745.

Kalter, S., and Heberling, R. (1968). *Nat. Cancer Inst., Monogr.* **29**, 149.

Kaplan, A. S. (1955). *Virology* **1**, 377.

Kaplan, H. S. (1967). *Cancer Res.* **27**, 1325.

Kilbourne, E. D. (1969). *In* "Fundamental Techniques in Virology" (K. Habel and N. P. Salzman, eds.), p. 146. Academic Press, New York.

Klement, V., Rowe, W. P., Hartley, J. W., and Pugh, W. E. (1969). *Proc. Nat. Acad. Sci. U.S.* **63**, 753.

Koprowski, H., Pontén, J. A., Jensen, F., Ravdin, R. G., Moorhead, P., and Saksela, E. (1962). *J. Cell. Comp. Physiol.* **59**, 281.

Koprowski, H., Jensen, F., and Steplewski, Z. (1967). *Proc. Nat. Acad. Sci. U.S.* **58**, 127.

Lowy, D. R., Rowe, W. P., Teich, N., and Hartley, J. W. (1971). *Science* **174**, 155.

Luginbuhl, R. E. (1968). *Nat. Cancer Inst., Monogr.* **29**, 109.

McAllister, R. M., Nelson-Rees, W. A., Johnson, E. Y., Rongey, R. W., and Gardner, M. B. (1971). *J. Nat. Cancer Inst.* **47**, 603.

Macpherson, I., and Stocker, M. (1962). *Virology* **16**, 147.

Marcus, P. I., and Carver, D. H. (1969). *In* "Fundamental Techniques in Virology" (K. Habel and N. P. Salzman, eds.), p. 161. Academic Press, New York.

Maurer, B. A., Mamura, T., and Minowada, J. (1969). *Bacteriol. Proc.* p. 154.

Melendez, L. V., Hunt, R. D., King, N. W., Garcia, F. G., Like, A. A., and Miki, G (1967). *Lab. Anim. Care* **17**, 302.

Nesburn, A. B. (1969), *J. Virol.* **3**, 59.

Nichols, W. W., Levan, A., Hall, B. and Ostergren, G. (1962). *Hereditas* **48**, 367.

Orgel, L. E. (1963). *Proc. Nat. Acad. Sci. U.S.* **49**, 517.

Petursson, G., Coughlin, J. I., and Meylan, C. (1964). *Exp. Cell. Res.* **33**, 60.

Puck, T. T., Cieciura, S. J., and Robinson, A. (1958). *J. Exp. Med.* **108**, 945.

Pulvertaft, R. J. V. (1965). *J. Clin. Pathol.* **18**, 261.

Rabson, A. S., O'Conor, G. T., Berezesky, I. K., and Paul, F. J. (1964). *Proc. Soc. Exp. Biol. Med.* **116**, 187.

Rapp, F., Butel, J. S., and Melnick, J. L. (1965). *Proc. Nat. Acad. Sci. U.S.* **54**, 717.

Rogers, N. G., Basnight, M., Gibbs, C. J., Jr., and Gajdusek, D. C. (1967). *Nature* (*London*) **216**, 446.

Rowe, W. P., Hartley, J. W., Waterman, S., Turner, H. C., and Huebner, R. J. (1956). *Proc. Soc. Exp. Biol. Med.* **92**, 418.

Sabin, A. B. (1965). *Isr. J. Med. Sci.* **1**, 93.

Sanders, F. K. (1971). *Med. Clin. N. Amer.* **55**, 653.

Shein, H. M., and Enders, J. F. (1962). *Proc. Nat. Acad. Sci. U.S.* **48**, 1164.
Smith, K. O. (1970). *Progr. Med. Virol.* **12**, 302.
Smith, K. O., Dunlap, R. C., Thiel, J. F., Newman, J. T., and Palmer, A. E. (1970). *Proc. Soc. Exp. Biol. Med.* **133**, 560.
Stenback, W. A. Van Hoosier, G. L., Jr., and Trentin, J. J. (1966). *Proc. Soc. Exp. Biol. Med.* **122**, 1219.
Svoboda, J. (1964). *Nat. Cancer Inst., Monogr.* **17**, 277.
Takemoto, K. K., and Habel, K. (1959). *Virology* **7**, 28.
Temin, H., and Mizutani, S. (1970). *Nature* (*London*) **226**, 1211.
Todaro, G. J., and Green, H. (1963). *J. Cell. Biol.* **17**, 299.
Todaro, G. J., and Green, H. (1964). *Proc. Soc. Exp. Biol. Med.* **116**, 688.
Todaro, G. J., and Martin, G. (1967). *Proc. Soc. Exp. Biol. Med.* **124**, 1232.
Todaro, G. J., Green, H., and Swift, M. (1966). *Science* **153**, 1252.
Vigier, P. (1970). *Int. Symp. Tumor Viruses, 2nd 1969* p. 205.
Vogt, M., and Dulbecco, R. (1958). *Virology* **5**, 425.
Vonka, V., Beyesh-Melnick, M., and McCombs, R. M. (1970). *J. Nat. Cancer Inst.* **44**, 865.
Watkins, J. (1970). *J. Cell Sci.* **6**, 721.
Weiss, R. A., Friis, R. R., Katz, E., and Vogt, P. K. (1971). *Virology* **46**, 920.
Wheeler, C. E. (1960). *J. Immunol.* **84**, 394.
Wong, S. C., and Kilbourne, E. D. (1961). *J. Exp. Med.* **113**, 95.
Yerganian, G., and Leonard, M. J. (1961). *Science* **133**, 1600.
Zur Hausen, H., and Schulte-Holthausen, H. (1970). *Nature* (*London*) **227**, 245.

CHAPTER **12**

Summary

DONALD J. MERCHANT

I. Introduction 257
II. Types and Sources of Contamination 258
III. Sterility Tests of Media and Solutions 259
IV. Detection and Identification of Contaminants 261
V. Prevention of Contamination 264
VI. Treatment of Contaminated Cultures 266
VII. Use of Antibiotics 267
VIII. Effects of Contaminants on Cell Cultures 268
IX. Concluding Remarks 268
Reference 269

I. Introduction

Contamination, and its control, has played a key role in the development of tissue culture. In the early decades of the century bacterial and fungal contamination was a sufficiently formidable problem to discourage many investigators in the use of *in vitro* methods (see Chapter 2). Others, some of them eminent in the field, built up an aura of mystique around the use of tissue culture that was an added hindrance to full development and use of the methods. Several factors contributed to these problems.

It generally was believed that most tissues taken from apparently normal or healthy animals by aseptic procedures were sterile. More recently, it was recognized that such tissues frequently may harbor bacteria, mycoplasmas, viruses, or other microorganisms. It was likely that many attempts at primary culture by early investigators were doomed to failure from the beginning.

Explant techniques, which prevailed in the early days of tissue culture, tended to be relatively tedious and required working in a semiexposed environment for extended periods. Therefore, the chance for airborne contaminants to gain entrance was high. In addition, the tissue culture systems were exquisitely sensitive as growth media for microorganisms. The difference in sensitivity between common bacteriologic media and some tissue culture systems may be as great as an order of magnitude.

Until the advent of antibiotics in the 1940's no means of control of contamination was available to those using tissue culture methods. Moreover, media used for growth of cells was composed largely of natural products, such as serum and tissue extracts. As we now know, these may not be free of contamination even when prepared aseptically (Chapter 6, Chapter 10). At that time no adequate means of sterilizing natural products was available.

Since 1950, there has been a rapid expansion of our knowledge of the distribution of contaminants in tissues and body fluids. The importance of viruses as contaminants was recognized first with the discovery of SV40 and other agents in primary cultures used for virus isolation and propagation (Chapter 11). More recently, the presence of viruses in serum used for tissue culture has been documented (Chapter 10).

In the late 1950's two additional forms of contamination were recognized as not only existing but as being relatively prevalent. The emergence of mycoplasmas as a major contamination problem was related to the widespread and routine use of antibiotics. These agents also were found to be common contaminants in serum (see Chapters 2, 5, and 6). Development of a number of continuous cell lines and their widespread dissemination and use played an important role in the occurrence of contamination by extrinsic cells of higher animals (Chapter 1).

Fortunately the recognition of viral, mycoplasmal, and animal cell contamination problems led to intensive study with the development of methods for detection, prevention, and control. Although much remains to be done the progress in the past 20 years is very encouraging. It is the development of our awareness of the problem and efforts to overcome it to which the authors of this volume have addressed themselves. In almost 70 years of tissue culture practice we only now have reached the point where the subject of contamination can be treated authoritatively and in depth.

II. Types and Sources of Contamination

The listing by Armstrong (Table V, Chapter 3) of organisms commonly associated with different normal human tissues should provide a very useful reference to those workers preparing primary cultures from these sources.

Compilation of similar information on other animal species and extension to include mycoplasmal, viral, and protozoan agents would be of great benefit.

Although Coriell (Chapter 2) presents a thorough review of sources of contamination in a general sense, it is worthwhile to review briefly the information presented in this volume as it relates to various agents of contamination. Bacteria and fungi, which are ubiquitous in nature and relatively resistant to environmental factors, such as temperature, drying, radiation, etc., may gain entrance to cultures as (1) airborne contaminants on dust particles, (2) in aerosols from the tissue culturist or during manipulations, or (3) as contaminants in cells, medium, or on nonsterile equipment.

Mycoplasmas and viruses, which are associated in nature primarily with cells and body fluids and which are relatively much more sensitive to environmental factors, gain entrance to cell cultures from contaminated cells or tissues, body fluids such as serum, or from aerosols. The most important sources of mycoplasmal contamination are aerosols and contaminated serum used in culture media (Chapters 5 and 6), whereas viruses gain entrance by way of infected tissues, serum, or aerosols. To fully understand the problems of viral contamination it is necessary to recognize at least three levels of host–parasite relationships. These have been defined by Sanders (Chapter 11) as passenger, indigenous, or cryptic viruses.

Although relatively few instances of protozoan contamination are reported, it is known that the amebae species involved are present in soil in the encysted form and it appears likely that contamination may occur by airborne dust particles, aerosols, or as a result of the use of infected tissues (Chapter 8). Metazoan cells as contaminants may be spread in aerosols or by improper or careless culture techniques (Chapter 1).

III. Sterility Tests of Media and Solutions

Three important considerations determine the effectiveness of sterility test procedures: (1) the sensitivity and range of media used, (2) conditions and time of incubation, and (3) size and method of sampling.

As discussed by Coriell (Chapter 2), the media used for sterility tests must be sensitive and must be sufficiently broad in range to detect aerobic and anaerobic bacteria, fungi, and mycoplasmas on routine testing. Armstrong (Chapter 3) presents a range of media capable of growing all but the most unusual or fastidious bacteria and fungi, including bacterial L forms. Kenny and Barile (Chapters 5 and 6) give a thorough discussion of the isolation of mycoplasmal contaminants and give details of culture media found to be quite sensitive for this purpose. Routine use of the types of media suggested provides a high rate of protection in most laboratories against contamination from media and solutions. Barile (Chapter 6) suggests the use of a semisolid broth medium as a single system with the potential for growth of a range of microorganisms with varying nutritional and oxygen requirements. This medium appears to have considerable promise and ought to be evaluated carefully by several laboratories.

As indicated above, the conditions and time of incubation also are important. Cultures for bacteria and mycoplasmas must be incubated both aerobically and anaerobically to avoid missing a significant number of organisms. Cultures for mycoplasmas also grow better in the presence of 5% CO_2 (Chapter 6). Most fungi and some mycoplasmas grow slowly so that more prolonged periods of incubation are required. However, if broth cultures of mycoplasmas are being subcultured to agar for colony isolation, the time of incubation in broth is relatively critical as the organisms may undergo autolysis after prolonged incubation (Chapter 6).

Coriell (Chapter 2) presents a table in which he lists recommended inoculum sizes for various sterility test media. The indicated volumes are for tubed media and are adequate provided there is a moderately high level of contamination. However, microorganisms often are present in medium and solutions in very small numbers and the sampling error becomes an important factor. It should be reemphasized, also, that cell culture systems may be much more sensitive than our bacteriologic test media because of increased nutritional properties and the conditioning effects of living cells. For these reasons, it is recommended that the inoculum for sterility testing should be 1% of the material under test. Although this will miss some contaminants it is the largest practical sample to use routinely. This is equivalent to the large broth culture procedure suggested by Barile (Chapter 6). Thus, a 10 ml sample from 1 liter of medium (1%) is inoculated into a bottle containing 100 ml of sterility test medium. When media and solutions are sufficiently stable, the entire batch can be incubated prior to making sterility tests, as an enriching procedure.

Coriell (Chapter 2) recommends taking samples for sterility testing at the beginning, near the midpoint, and at the end of the use of a large batch of medium. If, as is the practice in many laboratories, large batches are prepared in a central media kitchen and these are dispensed in small quantities for subsequent use, the same principal prevails. As an added precaution, in this case, the medium or solution bottles ought to be numbered serially and each batch ought to be dated. Any investigator using the medium ought to record the batch and bottle number to simplify subsequent tracing of difficulties.

Armstrong (Chapter 3) reports the isolation of a bacterial L form from phosphate-buffered saline and both he and Coriell (Chapter 2) refer to the production of bacterial L forms as the result of inhibition of cell wall production of bacterial contaminants in the presence of penicillin. This is a real problem and, as Coriell suggests, is best avoided by not using antibiotics routinely. However, as emphasized by Armstrong, L-form contaminants are encountered and the inclusion of hypertonic media in sterility tests permits the recognition of these organisms.

IV. Detection and Identification of Contaminants

In the case of many bacteria and some fungi that grow readily in the culture medium, the presence of contamination is quickly manifest by such obvious changes as the development of turbidity, rapid changes in pH, and/or significant alteration or destruction of the cells. Such contaminations, although troublesome, are of the least concern. They are obvious quickly and the cultures are discarded. A similar picture may be seen with some viruses.

More fastidious or slow growing bacteria, many fungi, and some mycoplasmas and viruses also produce observable changes but much more slowly. Cultures may be fed and subcultured several times before the contamination is obvious. This not only causes a greater loss of time and effort but often leads to the spread of the contamination through a number of cultures.

The most troublesome contaminants are those that are well adapted to a parasitic existence and that produce only minimal or no observable effect for long periods of time. Often, such contaminants coexist for extended intervals, only to become manifest when some stress is applied, such as a failure to feed or transplant at adequate intervals. In other cases, obvious signs are never seen. Bacteria, fungi, mycoplasmas, viruses and protozoa all may fall in this category.

A related problem is the production of bacterial spheroplasts, intermediate bacterial forms resulting from partial destruction or incomplete synthesis of cell walls. These forms are very fragile and do not survive in normal isotonic solutions. It is not an infrequent occurence for bacteria that gain entrance to cultures to be phagocytized and converted, intracellularly, to spheroplasts. Such forms may be carried for extended periods and are capable of reconverting to the parental form and breaking out as fulminating contaminants. The presence of spheroplasts can be detected by disruption of cells and plating on hypertonic media containing a high concentration of cations (Kourany and Kendrick, 1966).

As emphasized by several of the authors in this book (Chapters 2, 3, 5, and 6) it is desirable, whenever possible, to confirm contamination by cultivation and identification of the agent. Methods for isolating and identifying bacteria (including L forms) and fungi (Chapters 2 and 3) mycoplasmas (Chapters 5 and 6), and amoebae (Chapter 8) have been presented. These methods detect the majority of contaminants.

Fogh (Chapter 4) describes an enrichment method for detecting bacterial, fungal, and mycoplasmal contamination of cell cultures based on methods largely developed in his laboratory over the past several years. The method consists of inoculating material from cultures suspected of being contaminated into cultures of FL (human amnion) cells. After suitable incubation these indicator cultures are prepared and examined either directly in the living state by Nomarski interference contrast microscopy or stained with Orcein and examined with phase optics.

The sensitivity of the method appears to be quite high and, according to Fogh, is superior to other methods for the detection of mycoplasmas. No indication is given of the comparative sensitivity of this method and other methods for detecting bacterial or fungal contaminants except to indicate that slow growing fungi may show up more rapidly than in conventional isolation procedures.

Taking into account the significant difference in sensitivity between tissue cultures and common bacteriological media, it seems logical that the procedures described by Fogh are of considerable value. It probably is wise to compare other cell cultures with the FL cell for sensitivity and perhaps to use more than one culture system, particularly for mycoplasmas, to increase the possibilities of isolating more fastidious strains.

Although the above method of Fogh may be somewhat more sensitive than conventional culture methods for detecting bacterial and/or fungal contaminants, it is questionable whether the added advantage with these agents justifies the use of the method as a routine laboratory procedure in most laboratories because it requires maintaining yet another cell line(s) as an indicator system. In the case of mycoplasmas, it is too early to be certain but the evidence suggests that the method may be preferable to current procedures although maintenance of an indicator line is still a drawback.

Not emphasized by Fogh or other contributors to this volume is the desirability of routine, direct microscopic surveillance of all cell cultures. This procedure, which should be obvious and routine, is all too often neglected in laboratories where cells are used primarily as a test system. It ought to include, but extend beyond, examination of flasks at $10\times$ to $100\times$ magnification. Examination of flasks or tubes at magnification up to $100\times$ can indicate subtle changes in morphology and growth patterns but may fail to reveal contaminants that produce no changes.

It is our experience that examination of cells in the living state by phase-contrast microscopy often yields clues of contamination before it is suspected otherwise. This is true particularly of intracellular contamination. In any situation in which unexplained changes in pH, growth rate, microscopic morphology, or other parameters occur, it is wise to remove a small sample of cells (a sterile bacteriologic loop is adequate), suspend them in a small drop of medium on a slide, cover with a coverglass, and seal to prevent evaporation. Such a preparation, examined first at 100 × and then with oil-immersion phase optics, often provides the earliest solid evidence of intracellular contamination.

An obvious extension of microscopic techniques for detecting contaminants is electron microscopy. This subject is reviewed by de Harven (Chapter 9), who points out that it is of particular value in the study of cryptic viral infections, the agents of which may first be recognized on the basis of their fine structure. It must be pointed out that the methods of electron microscopy are relatively tedious and that the sample that can be examined is necessarily very small. For these reasons it is not a method applicable to the problems of most laboratories. However, it is a very sensitive method that is capable of yielding invaluable evidence under the proper circumstances.

Because viruses cannot be grown in the absence of cells, the detection of viral contaminants is a different type of problem than that posed by other agents. Sanders (Chapter 11) presents a series of detection methods beginning with the passenger virus and extending to the cryptic viruses. It is obvious that all but the most sophisticated virus laboratories are limited to tests for passenger viruses. In fact, the methods required to detect the majority of passenger viruses (growth on indicator cell lines, immunofluorescence, hemadsorption, viral interference, production of interferon, and electron microscopy) generally are unavailable to laboratories not actively engaged in virologic studies.

It is highly recommended that any laboratory intending to use established cell lines obtain these from the American Type Culture Collection. They then will start with cells known to be free of the majority of passenger viruses and many indigenous viruses (as well as free of mycoplasmas and other agents). The cells ought to be tested at intervals if they are maintained in continuous passage. However, if a frozen seed stock of the virus-free cells is kept and if experiments are performed on cells within a few passages of this stock, the probability of introducing viruses from serum or in culture procedures can be minimized.

Laboratories maintaining cell lines in continuous passage or that derive cultures from primary sources ought either to develop the capability of routine screening for passenger viruses or to arrange for periodic tests by a qualified laboratory.

The above considerations largely hold for contamination by metazoan cells. As outlined by Stulberg (Chapter 1), contaminating cells may be detected by serologic or karyologic methods or by isoenzyme analyses. These methods are not difficult and can be performed in most laboratories provided the proper reagents (antiserum) are available. Again, the utilization of characterized cell lines from the American Type Culture Collection may help to avoid many problems.

If contamination by any extraneous agent occurs, it must be identified, if possible, and the source of the contamination must be determined. In Chapter 3, Armstrong states that a search for the source of contamination is fruitless in a majority of cases. In contrast, Coriell (Chapter 2) states that ". . . every contamination that occurs is the result of an error and if all the recommended procedures have been followed it should be possible to identify where the break occurred." In my opinion, neither of these positions is entirely correct although I agree with Coriell that it ought to be possible to trace most contaminations to their source. Certainly, the procedures that he recommends go a long way toward making this possible. One can go even further by maintaining a record of contamination such as that presented in Table I.

From this sample record it is possible to draw several conclusions concerning the contamination occuring in this hypothetical situation. RJT, who is carrrying HeLa cells in serial subculture, either is carrying along an intracellular contaminant that is breaking through sporadically or he is using a batch of medium or glutamine containing a very low grade contamination. Methods are readily available to distinguish between the two possibilities.

On the other hand, it is quite clear that DES has poor technique, as several different cell types growing in different media are contaminated and different organisms are involved. Contaminations experienced by FGP, ELR, and JLM appear to be caused by chance and probably result from airborne organisms. The last three listings on the chart almost certainly reflect a break in sterilization procedures.

Although the above examples may have been oversimplified in order to make the points, anyone who maintains an accurate record of this type will find it to be well worth the effort.

V. Prevention of Contamination

Coriell (Chapter 2) presents a thorough discussion of prevention of contamination. This is illustrated graphically as a series of barriers, including aseptic technique, sterilization, surveillance procedures, quarantine, and proper use of antibiotics. Throughout his presentation he has emphasized the prime importance of good aseptic technique. This is endorsed by the other authors of this volume as well. However, there is little or no detailed discussion of individual points of technique.

TABLE I
Record of Contaminations

Initials of individual	Date	Culture	Passage number	Medium	Additives[a]	Culture vessels	Units contaminated/ units in experiment	Type of contaminant: Morphology and motility	Type of contaminant: Gram stain reaction
RJT	4/6/71	HeLa	345	MEM_{90} FCS_{10}	Glutamine	250 ml Flasks	1/10	Motile rod	−
FGP	4/8/71	Monkey kidney	Primary	MEM_{80} FCS_{20}	Glutamine	Tubes	2/150	Coccus	+
								Nonmotile rod	+
ELR	4/10/71	L-M	2640	199P	Methocel	Spinner flasks	1/6	Coccus	+
JLM	4/14/71	WI-38	20	MEM_{90} FCS_{10}	Glutamine	30 ml Flasks	1/7	Nonmotile rod	+
RJT	4/14/71	HeLa	346	MEM_{90} FCS_{10}	Glutamine	250 ml Flasks	1/12	Motile rod	−
DES	4/15/71	HEp2	96	BME_{90} FCS_{10}	Glutamine	Leighton tubes	3/12	Coccus	+
DES	4/17/71	Chick embryo	Primary	199_{90} CS_{10}	—	30 ml Flasks	5/7	Nonmotile rod	+
DES	4/19/71	L-M	237	199P	—	4 oz Bottles	4/6	Motile rod	−
RJT	4/20/71	HeLa	347	MEM_{90} FCS_{10}	Glutamine	30 ml Flasks	2/12	Motile rod	−
FGP	4/22/71	Monkey kidney	Primary	MEM_{80} FCS_{20}	Glutamine	Tubes	50/50	Mixed organisms	
ELR	4/22/71	L-M	2641	199P	Methocel	Spinner flasks	6/6	Mixed organisms	
DES	4/22/71	HEp2	97	BME_{90} FCS_{10}	Glutamine	Leighton tubes	12/12	Mixed organisms	

One common technique that deserves special mention is pipetting. It now is recognized that mouth pipetting is a significant hazard (see Chapter 6). The common practice of triturating a cell suspension with a pipet is particularly liable to result in dispersal of an aerosol from the mouth and nose. As reported in Chapters 5, 6, and 11, this is a common source of contamination by mycoplasmas and viruses. It also is a source of bacterial contamination. Whenever practical, a propipet or similar device should be used to eliminate mouth pipetting.

Good technique includes the development of confidence in the methods of tissue culture being used so that each manipulation is carried out quickly and easily, thus minimizing the possibilities of contaminants gaining entrance.

Because airborne contamination probably is the most common overall source, the development of laminar-flow clean air systems has been a highly significant development in perfecting of adequate control measures. In Chapter 2, Coriell discusses these systems and their importance in providing a germ-free environment for tissue culture work. Although the development of laminar flow rooms and hoods constitutes one of the major technological advances influencing tissue culture, I find it difficult to agree with Coriell that no one ought to attempt to work with cell cultures without such facilities. However, I cannot agree with Armstrong (Chapter 3) that the decision as to whether or not to use antibiotics should be determined by ". . . the skills of the team and the equipment in the laboratory." Both Coriell and Armstrong emphasize the problems created by routine use of antibiotics, particularly when culturing cell lines. Many laboratories have grown cultures in large quantities over a period of years without the advantage of laminar flow and without resort to the use of antibiotics. Good aseptic technique combined with proper surveillance, as outlined by Coriell (Chapter 2), minimizes contamination. In the absence of antibiotics, an occasional contaminant usually manifests itself quickly and the culture is discarded. If possible, laminar flow systems ought to be utilized, but if they are not available a great deal of good tissue culture work still can be done and without dependence on antibiotics. The cost of laminar flow equipment is sufficiently low at present to be in the budget range of most research programs and represents a very wise investment.

VI. Treatment of Contaminated Cultures

This topic is seperable into two categories. In those instances in which it is necessary to start with contaminated tissue or in which contaminated materials (nasal washings, feces, etc.) must be added, it is necessary to use a combination of washing and high concentrations of appropriate antibiotics initially. This is covered very adequately by Armstrong (Chapter 3). It must be emphasized that the proper combination of antibiotics is important and that large doses are required for a sufficient time to insure destruction of the contaminants and to avoid emergence of resistant forms.

Washing of tissues initially can be of considerable importance as dilution of contaminants makes the antibiotic treatment more effective and washing may dilute some contaminants below the level required for continued survival. Washing is not without danger, however, as essential nutrients may be leaked from the cells (e.g., free amino acid pools and intermediates of the Krebs cycle), thus reducing the chances of tissue cell survival.

In Chapter 3, Armstrong discusses the use of antibiotics for treatment of contaminated cultures. Unless a cell line has properties that are unique and cannot be duplicated readily, it is advisable to discard contaminated cultures and start over. Even with unique cell lines, the practice of freezing and storing characterized batches of cells usually frees the investigator from the hazard of loss caused by contamination.

In those rare instances in which an attempt to salvage a contaminated cell line is indicated, the high levels and prolonged use of antibiotics necessary to clear up an infection may lead to population selection and permanent alteration of the culture.

VII. Use of Antibiotics

Problems resulting from the use of antibiotics are thoroughly discussed by Coriell (Chapter 2). He points out the major contribution that antibiotics have made in permitting growth of tissues from frankly contaminated sources but rightly emphasizes the false sense of security that has been generated and the difficulties that have resulted. The two most important reasons for the increase in contamination that often accompanies use of antibiotics are (1) a tendency for the individual to allow his technique to become sloppy because of reliance on antibiotics and (2) the masking of contamination by the presence of antibiotics that suppress organisms without eliminating them. A corollary of the latter is the suppression of most contaminants and the emergence of those, such as mycoplasmas, that are resistant to the antibiotics being employed.

In addition to the points made by Coriell, it is well to mention the ease with which contaminants can be spread through cell culture lines when antibiotics are used routinely. In the absence of antibiotics most contaminants multiply rapidly and their presence is manifest by the development of turbidity, pH changes, and/or a marked cytopathogenic effect. Such cultures are discarded quickly and the trouble is eliminated. When antibiotics are employed, such contamination is suppressed. In a significant percentage of cases, the suppression is incomplete and a resistant form of the organism persists in low numbers or the organism is phagocytized and develops as an intracellular parasite where it is not reached by antibiotics. In either case, organisms may persist and may be spread uniformly throughout the line upon subculture.

VIII. Effects of Contaminants on Cell Cultures

The effects of contamination are thoroughly covered by Armstrong, Kenny, Barile, Fogh, Holmgren, and Sanders (Chapters 3, 5, 6, 7, 8, and 11) for the respective agents. As noted, the effects may vary from complete cell destruction to subtle biochemical effects that are almost inapparent and very difficult to demonstrate. It is obvious that the more careful, observant, and critical a person is, the more likely he is to be aware of these changes and to recognize the possibility of dealing with a contaminating agent.

It is not surprising that viruses present the greatest problem because some of the agents listed as indigenous or cryptic by Sanders may represent the ultimate in parasitism. However, as pointed out by Sanders (Chapter 11), the very fact that indigenous and cryptic "viruses" appear to be almost ubiquitous in continuous cell lines from some species ought to cause us to be hesitant in drawing conclusions as to their origin.

Fogh (Chapter 7) presents a very interesting and very complete analysis of the effects of one particular strain (HT) of *Mycoplasma fermentans* on one strain (FL) of human amnion cells. This study not only provides very important information concerning some of the more subtle effects of a contaminant of a cell culture but also serves as a pattern for the analysis of other systems. This type of information is much too scarce. It is not clear to what extent it is safe to extrapolate specific information from the particular system to any other system.

IX. Concluding Remarks

During the past 25 years, tissue culture has developed from a "mystical" art to a scientific method that has permeated every area of cell biology. To a very large extent, this has been possible because of the developments in our understanding of contaminants and the development of protective measures as documented in this volume.

Some of the major developments have been: (1) the recognition that viruses are common tissue contaminants and development of methods for their detection and quantitative study; (2) the recognition of mycoplasmas as agents of contamination, with the subsequent elucidation of their biology and epidemiology; (3) the introduction of antibiotics and the eventual realization of their attendant problems; (4) the improvements in filtration methods for sterilization; (5) the development of laminar flow systems; (6) the general improvement in methods for detecting, isolating, and identifying contaminants; (7) the development of characterized reference strains (American Type Culture Collection), and (8) the development of methods for freezing and storing "clean" seed stocks.

In spite of these advances, the use of contaminated cultures is much too common and too much time, effort, and material are expended needlessly. There are several reasons for this: (1) the failure to develop and practice good technique, often accompanied by dependence on antibiotics; (2) the lack of awareness of the problems and how to avoid them; (3) the failure to start with characterized materials; and (4) the necessity of using serum and other biologicals that frequently contain mycoplasmas, viruses, or other agents.

The first three of these difficulties can easily be corrected. Were the procedures and recommendations set forth in this volume followed by the majority of those doing cell and organ culture, more than 90% of the contamination currently encountered could be eliminated.

Two areas of further research can help to eliminate most of the problem remaining. Continuing effort ought to be made to develop more sensitive and more rapid means of detecting, isolating, and identifying contaminants. Any improvements in these areas permit more prompt recognition of problems and elimination of contaminated cultures at the earliest possible moment. They also make possible more effective sterility test procedures.

The use of natural products, such as serum, continues to pose serious problems. This can be eliminated either by finding adequate means of sterilization that preserve essential properties or by developing knowledge that makes it possible to eliminate their use.

Finally, it seems appropriate to raise the questions: At what point in control and elimination of contamination do we make our systems so artificial that they no longer have reference to the original tissue as it existed *in vivo*? Should we strive for a completely axenic situation?

Reference

Kourany, M., and Kendrick, P. L. (1966) *J. Infect. Dis.* **116**, 495–513.

Author Index

Numbers in italics refer to the pages on which the complete references are listed.

A

Aaronson, S. A., 117, *128*, 140, *162*, 207, 221, *229*, *231*, 247, 249, 250, 252, *254*
Abinanti, F. R., 237, *242*
Achong, B. G., 4, *24*, 221, *229*
Ackermann, W. W., 154, *156*
Adams, R. A., 4, 15, 22, *23*, *24*
Adler, H. E., 150, *156*
Afshar, A., 153, 154, *156*
Ahl, R., 214, *230*
Albino, A., 247, *254*
Allan, R. D., 68, *76*
Allerdice, P. W., 22, *26*
Aluotto, B. B., 109, *126*, 135, *156*
Andersen, W., 32, *48*
Anderson, D. R., 109, 116, 118, *126*, 133, 134, 151, *156*, 164, 165, 217, *229*
Anderson, H. C., 184, *193*
Anderson, T. F., 228, *229*
Andrewes, C. H., 154, *156*, 246, 247, *254*
Aoki, T., 211, *229*
Archard, L. C., 154, *156*
Archibald, R., 253, *254*
Armstrong, D., 32, *48*, 53, 61, 62, 63, *64*, 133, 151, 154, 155, *156*, *159*, *162*
Armstrong, J. A., 154, *156*, 196, 197, *203*
Ashworth, J. N., 239, *241*
Atler, M. J., 236, *242*
Attridge, J. T., 221, *229*
Auersperg, N., 18, *23*
Augusti-Tocco, G., 211, *229*
Aula, P., 123, *126*, 152, *156*, 182, 183, *193*
Axelrad, A. A., 2, 5, 7, *26*

B

Bailey, G. D., 134, *162*
Bailey, J. S., 144, *156*
Bak, A. L., 135, *156*
Baker, J. A., 236, *241*
Balduzzi, P., 149, *156*
Baltimore, D., 251, *254*
Bang, F. B., 46, *48*
Baranska, W., 21, *25*
Barbanti-Brodano, G., 21, *25*
Barile, M. F., 37, 44, *48*, 52, *64*, 72, *76*, 110, 111, 113, 115, 116, 117, 123, *126*, *128*, 132, 133, 134, 135, 138, 139, 140, 141, 142, 144, 145, 146, 147, 149, 150, 151, 152, 153, 154, 155, *156*, *157*, *158*, *160*, *161*, *162*, 164, 165, 166, 168, 169, 243, *254*
Barker, L. F., 125, *126*, 132, 138, 150, 153, *157*, *160*, 169
Barr, Y. M., 4, *24*
Barrnett, R. J., 134, *160*
Barron, A. L., 14, *25*
Barski, G., 21, *23*
Basilico, C., 21, 22, *24*, *25*,
Basnight, M., 244, *255*
Bassett, C. A. L., 7, *23*
Bauer, A. W., 53, *64*

Baumann, G. W., 22, *25*
Beacham, L. H., 122, *127*
Beale, A. J., 8, *23*
Bean, H. S., 40, *48*
Bearn, A., 17, *24*
Becker, Y., 154, *157*, *158*, *161*
Beinstein, A., 21, *23*
Benirschake, K., 3, 16, *25*
Bentvelzen, P., 227, *231*
Benyesh-Melnick, M., 154, *160*, 183, *193*, 249, *256*
Berezesky, I. K., 251, *255*
Berg, R. B., 132, 135, 150, *157*, 174, *193*
Berman, L., 2, 9, *26*
Bernhard, W., 211, 221, 225, *229*, *230*, 250, *254*
Berry, D. M., 154, *157*
Bessis, M., 221, *229*
Beswisk, T. S. L., 132, *157*
Betts, A. O., 154, *159*
Beveridge, W. I. B., 139, *157*
Biberfeld, G., 151, *157*, 217, *229*
Biberfeld, P., 151, *157*, 217, *229*
Biedler, J. L., 246, *254*
Biesele, J. J., 5, *25*
Billingham, R. E., 5, 6, 7, *24*
Bird, H. H., 7, *23*
Black, F. T., 135, *156*
Black, P. H., 250, *254*
Blacklow, N. R., 251, *255*
Blakemore, W. S., 38, 43, *48*
Blevins, A., 53, *64*
Blinzinger, K., 225, *229*
Bloom, D., 253, *254*
Boatman, E. S., 109, 118, *126*
Bode, H. R., 135, *160*
Bodey, G. P., 43, *48*, 135, *156*
Bogden, A. E., 211, *229*
Bohl, E. H., 237, *242*
Bonar, R. A., 221, *230*
Bonifas, V. H., 123, *128*, 132, 154, *161*
Booe, I. M., 211, *231*
Boone, C. W., 37, *48*, 144, *160*
Boothe, A. D., 221, *230*
Bopp, W. J., 221, *229*
Borman, I., 138, *159*
Borrel, A., 132, *157*
Boshivitz, C., 138, *159*
Boulger, L., 225, *229*
Bouteille, M., 225, *229*, *230*
Boyse, E. A., 15, *26*
Brackett, R. G., 22, *25*
Bradt, C., 22, *25*
Bradvarova, L., 132, 152, *160*
Brancato, P., 123, *128*, *161*
Brand, I., 20, *23*, *26*
Brand, K. G., 2, 5, 7, 8, 20, *23*, *26*
Breg, W. R., 22, *26*
Breton-Gorius, J., 221, *229*
Brightwell, A. S., 221, *229*
Brindle, S. A., 57, 63, *64*
Brinkley, B. R., 226, *229*
Brooks, R. E. 225, *229*
Brown, A., *127*, 149, *159*, 217, *229*
Brown, D. T., 228, *229*
Brown, R. G., 235, *241*
Brown, R. W., 122, *127*
Brown, T. McP., 144, *156*
Brownstein, B., 154, *157*
Brunner, H., 125, *126*
Bucciarelli, E., 225 *229*
Bullock, W., 43, *48*
Buoen, L. C., 20, *26*
Burke, C. N., 244, *254*
Burris, C., 109, *129*, 133, *162*
Buskirk, H. H., 124, *127*
Butel, J. S., 251, *255*
Butler, M., 121, *127*, 132, 134, 135, 142, 144, 151, 153, 154, 155, *157*, *159*

C

Cabasso, V. J., 235, *241*
Campbell, D. H., 7, *23*
Carbone, P. P., 153, *159*
Carpenter, R. R., 125, *128*
Carski, T. R., 73, *76*, 124, *127*, 133, 135, 138, 144, 145, 148, 149, 151, 152, *156*, *157*, *158*, *162*
Caruso, T. D., Jr., 37, *48*
Carver, D. H., 252, *255*
Cary, S. G., 121, *129*
Casemore, D. P., 195, 198, 200, *203*
Caspersson, T., 22, *23*
Castellani, A., 197, *203*
Castor, L. N., 5, *26*
Castrejon-Diez, J., 153, *157*
Cerva, L., 198, 202, *203*
Cesarini, J. P., 214, *229*
Chagnon, A., 153, *159*
Chandra, S., 226, *229*
Chang, R. S., 46, *48*, 195, *203*
Chang, S. L., 197, *203*

Chanock, R. M., 109, 111, 115, 119, 121, 125, *126*, *128*, 132, 134, 135, 139, 140, 144, 150, 153, 154, *157*, *159*, *160*, *161*, *162*
Chapman, A. L., 221, *229*
Chapman, J. A., 132, *151*
Charbonneau, J. R., 149, *156*
Charney, J., 6, 7, 14, *24*
Chatterji, J., 132, *157*
Cheatham, W. J., 235, *241*
Cherry, J. D., 135, 150, 151, 152, 154, *157*
Chessin, L. N., 247, *254*
Chi, L., 195, 198, *203*
Chiu, S. Y., 20, *23*, 26
Cholon, J. J., 22, *26*
Chopra, H. C., 211, *229*
Christiansen, C., 135, *156*
Christofinis, E., 8, *23*
Chu, E. H. Y., 5, *23*
Chu, H. P., 134, 149, *160*
Chumakov, M. P., 132, 153, *159*
Chung, R. A., 122, *127*
Cieciura, S. J., 4, *26*, 248, 249, *255*
Clapper, W. E., 43, *49*
Clark, G., 68, *76*
Clark, H. W., 144, *156*
Clarke, J. K.,, 221, *229*
Clarkson, B., 4, *23*
Claude, P., 211, *229*
Clausen, J. J., 8, *23*
Clifford, P., 221, *229*
Cline, W. L., 119, *128*
Clyde, W. A., Jr., 109, 110, 115, *127*, 134, 135, 150, 151, *157*, *159*
Cochrum, K. C., 153, *157*, *161*
Coffman, W. D., 2, *24*
Coggins, L., 236, 237, *241*
Cohen, A., 122, *127*
Cohen, G., 150, *157*
Cohen, H., 221, *229*
Cohn, E. J., 239, *241*
Cohn, M., 211, *231*
Cole, B. C., 118, 124, *128*, 149, 150, *157*, *161*
Cole, R. M., 135, *162*
Collier, A. M., 135, 151, *157*
Collier, L. H., 144, *157*, 243, *254*
Colter, J. S., 2, 7, *23*, *24*
Compeer-Dekker, G., 124, 125, *129*
Conner, G. H., 225, *230*
Consigli, R. A., 154, *162*
Cook, M. K., 144, *161*
Coombs, R. R. A., 2, 8, 9, 15, 21, *23*, *24*
Cooper, P. D., 245, *254*
Copperman, R., 153, *157*
Corbitt, G., 197, 198, 202, *203*
Coriell, L. L., 2, 3, 4, 7, 12, 14, 17, 18, 19, 23, *24*, *26*, 38, 39, 43, 44, 46, *48*, 52, 58, 62, *64*, 126, *128*, 144, 148, *157*, *159*
Cornefert, F., 21, *23*
Cottew, G. S., 142, *157*
Coughlin, J. I., 248, *255*
Cox, H. R., 235, *241*
Crandall, R. A., 235, *241*
Crosbie, R. B., 125, *128*, 149, *161*
Cross, G. F., 124, *127*, 132, *157*
Culbertson, C. G., 195, 196, 197, 198, 199, 200, 201, 202, *203*
Cummings, G., 153, *159*
Cupp, E. W., 14, *25*

D

Dales, S., 206, *229*
Dalton, A. J., 225, *229*
Danes, S., 17, *24*
Daniel, M. R., 2, 8, 15, *23*, *24*
Darby, N. V., 236, 237, 238, *242*
David, G. B., 68, *76*
Davidson, M., 118, *129*, 151, *162*
Dawson, A., 20, *24*
Defendi, V., 2, 5, 6, 7, 21, *23*, *24*, 120, *127*
DeForest, A., 40, *48*
de Harven, E., 4, *23*, 206, 207, 211, 217, 221, 227, *229*, *230*, *231*, 246, *254*
Deinhardt, F., 4, *24*
Delarue, J., 225, *229*
DelGiudice, R. A., 133, 134, 135, 138, 140, 141, 142, 144, 145, 146, 150, 151, 152, *156*, *157*, *158*, *162*, 166
Denny, F. W., 134, 135, 151, *157*, *159*
De Petris, E., 211, *229*
de Thé, G., 225, *229*
Diamond, L., 249, *254*
Diehl, V., 4, *25*, 247, *254*
Dienes, L., 217, *230*
Dittmar, S. L., 246, *255*
Dmochowski, L., 140, *158*, 207, 210, 217,*231*
Donner, L., 248, *255*
Douglas, M., 197, *203*
Dowling, A. M., 189, 190, *193*
Dreyer, D. A., 140, *158*
Dubes, G. R., 154, *158*
Dujardin-Beaumetz, E., 132, *157*, *158*

Dulbecco, R., 46, *48*, 245, 247, *254*, *256*
Dumont, A., 225, *230*
Dunkel, V. C., 246, *255*
Dunlap, R. C., 244, *256*
Dunnebacke, T. H., 197, *203*
Dupuy-Coin, A. M., 225, *230*
Dy kman, L., 153, *157*

E

Eagle, H., 2, 3, 15, *24*, 120, 122, *127*, 132, 137, 152, *158*, *159*, 238, *242*
Earle, W. R., 2, 7, *23*, *24*, *26*, 34, 46, *48*, 249, *254*
Early, R. L., 45, *48*
Easterbrook, K. B., 228, *230*
Eaton, M. D., 132, 150, 151, *158*, *162*
Edward, D. G. ff., 109, 113, *127*, 133, 134, 135, 139, 144, 146, 150, *158*
Edwards, G. A., 73, *76*, 118, *127*, 151, *158* 174, *193*, 214, *230*
Eisinger, M., 246, *254*
Ellaway, W. J., 135, 151, *157*
Ellison, S. A., 183, *194*, 250, *254*
Elsner, V., *127*, 149, *159*
Emmons, J., 152, *160*
Enders, J. F., 21, *26*, 250, *256*
Ensminger, P. W., 198, 199, *203*
Ephrussi, B., 21, *24*, *27*
Epstein, M. A., 4, *24*, 221, *229*
Erlandson, R. A., 225, *230*
Ershov, F. I., 154, *159*
Eskeland, J., 21, *25*
Estela, L. A., 235, *242*
Evans, V. J., 2, 7, *23*, *24*, 32, 46, *48*
Eveland, W. C., 72, *76*, 138, 139, *156*

F

Faber, J. E., 109, 112, *126*, *128*, 135, 148, *156*, *160*
Fabrizio, D. P., 144, 148, *157*
Fahey, J. L., 20, 21, *24*, *25*
Falk, H. L., 253, *255*
Farber, S., 4, 15, *24*
Farnes, P., 17, *24*
Farnham, A. E., 151, *158*
Fedorko, M. E., 227, *230*
Feldman, H. A., 195, 196, 197, 198, 199, *203*
Felts, R. W. 144, *156*
Fenner, F., 246, *254*
Fincher, E. L., 40, *48*
Findlay, G. M., 135, *158*
Finegold, I., 20, *24*
Fischman, H. R., 244, *255*
Fisher, E., Jr., 153, *157*
Fisher, T. N., 153, *157*
Fitzgerald, E. A., 123, *128*, 154, *161*
Flanagan, J. F., 20, *25*
Fleer, G. P., 132, 153, *159*
Fogel, M., 3, 15, 20, *24*
Fogh, A., 32 *48*
Fogh, H., 66, 67, 70, 72, 73, 74, 75, *76*, 116, 121, 122, *127*, 132, 140, 151, 152, *158*, 174, 175, 176, 178, 179, 180, 181, 182, 188, 189, 190, 191, *193*, 214, *230*
Fogh, J., 22, *24*, 32, *48*, 52, *64*, 66, 67, 70, 72, 73, 74, 75, *76*, 116, 118, 121, 122, 123, *127*, 132, 140, 149, 151, 152, *158*, 174, 175, 176, 178, 179, 180, 181, 182, 183, 184, 185, 186, 187, 188, 189, 190, 191, *193*, 214, *230*, 251, *254*
Foley, G. E., 3, 4, 15, 22, *23*, *24*
Fong, C. K. Y., 244, *255*
Ford, C. E., 21, *25*
Ford, D. K., 5, *24*
Ford, M., 155, *161*, 168
Fowler, R. C., 144, *156*
Fox, H., 121, *128*, 134, 153, *160*
Fox, S. M., 246, *254*
Foy, H. M., 113, *127*
Franklin, H. A., 4, *26*
Franks, D., 3, 8, 9, 14, 15, 20, *24*
Frei, E., 43, *48*
Freireich, E. J., 43, *48*
Freundt, E. A., 109, 113, *127*, 133, 134, 135, 144, *156*, *158*
Friedlander, M., 211, *230*
Friedman, M., 183, *193*
Friedman, M. H., 53, *64*
Friend, C., 124, *127*, 149, 150, *158*, 207, *230*, *231*
Friis, R. R., 247, 252, *256*
Frommes, S. P., 221, *231*
Frothingham, T. E., 132, 135, 150, *157*, 174, *193*
Fudenberg, H. H., 153, *157*, *161*
Fullmer, H. M., 195, 196, 199, *203*
Furness, G., 134, *158*

G

Gaeta, L., 233, 235, *242*
Gaffney, E. V., 132, 152, *160*, 179, 188, 190, *193*

Gafford, L. G., 122, *128*, 154, 155, *158*
Gajdusek, D. C., 244, *255*
Gale, J. L., 125, *127*
Gangal, S. G., 14, *24*, *25*
Garcia, F. G., 244, *255*
Gardner, M. B., 22, *25*, 251, *255*
Gartler, S. M., 3, 15, 17, 18, 20, *23*, *24*
Gaskill, H., 14, *24*
Gay, F. W., 221, *229*
Geantet, P., 132, *157*
Gehle, W. D., 227, *231*
Gentry, G. A., 122, 123, *128*
Gerber, P., 22, *24*, 246, 249, *254*
German, J., 253, *254*
Gerner, R. E., 4, *26*
Gesner, B., 132, 151, *158*
Gey, G. O., 2, *24*, *26*, 45, 46, *48*
Gey, M. K., 45, *48*
Gibbs, C. J., Jr., 244, *255*
Gilden, R. V., 207, *230*
Giles, N. H., 5, *23*
Gill, P., 120, *127*, 154, *158*
Gillespie, J. H., 236, *241*, *242*
Girardi, A. J., 144, 153, *158*
Giuffre, N. A., 57, *64*
Glade, P. R., 247, *254*
Goldblum, N., *158*
Goldenberg, G. J., 15, *27*
Goldstein, M. N., 7, *24*
Gonzales, F., 132, 152, *160*
Goodburn, G. M., 151, *158*, *160*
Goodman, M. R., 124, *127*, 132, *157*
Goodwin, D. F. W., 132, *158*
Gori, G. B., 124, *127*
Grabowski, M. W., 134, 135, 145, *156*, *157*, *158*
Grace, D. M., 21, *26*
Grace, J. T., Jr., 140, *159*, 246, *255*
Grace, T. D. C., 5, *24*
Graham, A. F., 154, *157*
Granger, H., 20, *24*
Gratzek, J. B., 236, *242*
Gravelle, C. R., 221, *229*
Grayston, J. T., 113, *127*
Green, H., 21, 22, *24*, *25*, *27*, 248, 249, 253, *256*
Green, M., 247, 248, *254*
Green, R. H., 244, *255*
Greene, A. E., 6, 7, 14, 22, *24*, *25*, *26*, 39, 43, *48*
Greenspan, F. P., 41, *48*
Gregg, N. C., 7, *25*
Grey, C. E., 140, *158*
Groyon, R., 236, *242*
Groyon, R. M., 236, 238, *242*
Grumbles, L. C., 153, *159*
Guoen, L. C., 20, *23*
Gurner, B. W., 2, 8, 9, 15, 20, *23*, *24*
Gutekunst, D. E., 236, *242*
Györkey, F., 226, *230*
Györkey, P., 226, *230*

H

Habel, K., 16, *24*, 245, *256*
Habel, L., 7, *25*
Hacker, C., 123, *127*, 149, *158*
Haguenau, F., 206, 225, *230*
Hahn, E., 72, 73, *76*, 121, *127*, 174, 175, 176 189, *193*, 214, *230*
Hakala, M. T., 122, *127*, 132, *159*
Hales, R., 140, *158*
Hall, B., 183, *194*, 250, *255*
Hall, C. F., 153, *159*
Hall, W. T., 32, *48*, 211, *230*
Hallauer, C., 243, *254*
Halliburton, B. L., 154, *158*
Hammerling, U., 211, *229*
Hampar, B., 183, *194*, 250, *254*
Hamparian, V. V., 144, 153, *158*
Hanafusa, H., 251, *254*
Hanafusa, T., 251, *254*
Handler, A. H., 3, 15, *24*
Handy, A. H., 237, *242*
Hargreaves, F. D., 123, *127*, *159*
Harley, E. H., 122, *127*
Harris, H., 17, 21, *25*
Harris, M., 3, 16, *25*
Hartley, J. W., 207, *229*, *230*, 244, 247, 249, 250, 252, *254*, *255*
Harvey, J. J., 211, *229*
Haschemeyer, R. H., 228, *230*
Hay, D., 211, *230*
Hay, R., 248, *254*
Hayes, S. F., 125, *128*, 149, *161*
Hayflick, L., 4, *25*, 112, 113, 114, 115, 116, 118, 120, 121, 122, 123, 124, 125, 126, *127*, *128*, 132, 134, 135, 139, 140, 144, 149, 152, 153, *157*, *158*, *159*, *161*, 183, 192, *194*, 214, *231*, 238, *242*, 248, *254*
Hearn, H. J., 127, 149, *159*
Heberling, R., 243, *255*
Hedley-Whyte, E. T., 225, *230*

Hehlmann, R., 253, *254*
Heine, U., 221, *230*
Hellerstein, E. E., 22, *23*
Hellstrom, I., 15, *25*
Hellstrom, K. E., 15, *25*
Helmboldt, C. F., 244, *254*
Henle, G., 4, *24*, *25*, 246, 247, *254*
Henle, W., 4, *24*, *25*, 246, 247, *254*
Henry, K., 225, *230*
Henry, M., 211, *231*
Herderschee, D., 118, 124, 125, *127*, 141, 144, 149, 154, *159*
Herrick, P. R., 22, *25*
Hersch, E. M., 153, *159*
Hetrick, F. M., 149, *161*
Hinz, R. W., 225, *230*
Hirsch, J. G., 118, *127*, 151, *159*, 214, 227, *230*
Hirschman, S. Z., 63, *64*
Hirshaut, Y., 247, *254*
Hlinka, J., 195, *203*
Hlozanek, I., 248, *255*
Hodges, R. T., 154, *159*
Hoerline, A. B., 237, *242*
Hoggan, M. D., 251, *255*
Hok, K., 183, *193*
Holland, J. F., 122, *127*, 132, *159*
Holland, J. G., 207, *230*
Hollande, E., 211, *231*
Hollmann, K. H., 206, *230*
Holmes, D. H., 199, 200, *203*
Holmgren, N. B., 14, 22, *24*, *25*, 52, *64*, 134, 152, *158*, *159*, 251, *254*
Hopps, H. E., 145, *157*
Horneff, J., 43, 46, *48*
Hornibrook, J. W., 7, *25*
Horoszewicz, J. S., 122, *127*, 132, 140, *159*, 246, *255*
House, W., 116, 117, *127*
Howatson, A. F., 206, *229*
Hsiung, G. D., 243, 244, *255*
Hsu, T. C., 3, 5, 16, *25*, 44, *48*, 183, *193*, 249, *255*
Huck, R. A., 236, *242*
Huebner, R. J., 207, *230*, 237, *242*, 244, 252, 253, *255*
Hughes, W. L., 239, *241*
Hull, R., 41, *48*, 195, *203*
Hummeler, K., 32, *48*, 62, *64*, 151, *159*
Humphrey, R. M., 249, *255*
Hunt, R. D., 244, *255*
Huygelen, C., 244, *255*

I

Ibrahim, A. N., 14, *25*
Igel, H. J., 253, *255*
Imagawa, D. T., 144, *160*, 240, *242*
Inglis, V. B. M., 154, *159*
Inoue, R., 21, *25*
Inui, F., 45, *48*
Ishi, Y., 225, *230*
Ito, E., 4, *26*

J

Jacobs, J. P., 122, *128*, 132, 152, *160*, 182, *194*
Jahnes, W. G., 195, 196, 199, *203*
James, W. D., 119, *128*
Jarret, O., 211, *230*
Jensen, E. M., 246, *255*
Jensen, F. C., 21, *25*, 183, *194*, 247, 250, *255*
Jensen, K. E., 134, *159*
Jézéquel, A. M., 197, *203*, 214, *230*
Johanson, B., 21, *25*
Johansson, C., 22, *23*
Johnsen, M. A., 41, *48*
Johnson, E. Y., 251, *255*
Johnson, R. W., 124, *127*
Joncas, J., 153, *159*
Jones, F. S., 146, *161*
Jones, T. C., 118, *127*, 151, *159*, 214, *230*
Jouan, C., 132, *157*
Juang, I., 122, *127*

K

Kadota, K., 221, *230*
Kagan, G. Y., 132, 153, 154, *159*, *161*
Kagnoff, M. F., 53, *64*
Kajima, M., 207, *230*
Kaklamanis, E., 138, *159*
Kakuk, T. J., 225, *230*
Kalica, A. R., 125, *126*
Kalifat, S. R., 225, *229*, *230*
Kalter, S., 243, *255*
Kamrin, A. A., 2, 7, *24*
Kaneseki, T., 221, *230*
Kano, K., 20, 21, *25*
Kao, F., 3, *26*
Kaplan, A. S., 244, *255*
Kaplan, H. S., 249, *255*
Kaplow, L. S., 244, *255*
Kasza, L., 154, *159*

Katz, E., 247, 252, *256*
Katzen, S., 154, *159*
Kazam, E., 37, *48*
Keeler, R. F., 125, *128*, 149, *161*
Kelloff, G. J. 207, *230*
Kelton, W. H., 134, *159*
Kelus, A., 2, 8, 15, *23*
Kendrick, J. W., 235, *242*
Kendrick, P. L., 261, *269*
Kenny, G. E., 109, 110, 111, 112, 113, 115, 118, 119, 120, 121, 122, 123, 124, 125, *126*, *127*, *128*, 132, 141, 149, 152, *159*, *160*
Kern, J., 37, *48*, 138, 142, 147, *156*
Kilbourne, E. D., 245, 252, *255*, *256*
Kilham, S., 228, *230*
Kim, K. S., 134, *159*
Kindig, D. A., 206, *230*
King, D, W., 44, *48*
King, J. R., 135, *162*
King, N. W., 244, *255*
Kingston, D., 38, *48*, 196, 197, 198, *203*
Kirby, W. M. M., 53, *64*
Kirk, R. G., 135, *160*
Kirschstein, R. L., 123, *128*, 132, 154, 155, *161*, 168
Kirsten, W. H., 206, *230*
Kitamura, H., 226, *230*
Klein, E., 21, *25*
Klein, G., 15, *25*
Klien, M., 40, *48*
Kleinberger, E., 135, *158*
Klement, V., 252, *255*
Kline, L. B., 41, *48*
Kniazeff, A. J., 2, 3, 4, 7, 12, 17, 18, 19, 23, *26*, 32, *48*, 144, *160*, 233, 235, 236, 237, 238, *240*, *242*
Knowles, B. B., 21, *25*
Koch, R., 43, *48*
Koestner, A., 225, *230*
Kohn, D. F., 151, *159*
Kondratick, J., 221, *230*
Koprowski, H., 21, *25*, 183, *194*, 247, 250, *255*
Korol, W., 211, *231*, 246, *255*
Koshimizu, K., 141, *160*
Kotin, P., 253, *255*
Kourany, M., 261, *269*
Kraemer, P. M., 120, *127*
Krishan, A. 225, *230*
Kronauer, G., 243, *254*
Krooth, R. S., 3, 16, 17, *25*
Krueger, R. G., 211, *231*
Kubicek, M. G., 2, *24*
Kudo, R. R., 197, *203*
Kufe, D., 253, *254*
Kuff, E. L., 211, *230*, *231*
Kusana, T., 21, *24*
Kutsakis, A., 7, *25*

L

Laird, H. M., 211, *230*
Lambertenghi, G., 227, *230*
Landing, B. H., 22, *25*
Lane, W. T., 207, *230*
Langford, P. L., 140, *158*
Langham, R. F., 225, *230*
Latta, H., 7, *25*
Law, L. W., 15, *25*
Lawn, A. M., 225, *230*
Lawson, L. A., 122, *128*
Lazar, P., 225, *230*
Lazarus, H., 4, 15, 22, *23*, *24*
Leach, R. H., 121, 123, *127*, 132, 134, 142, 144, 153, 154, 155, *157*, *159*, *161*
Lee, D. Y., 124, *127*
Lee, K. M., 236, *242*
Le Gacé, S., 40, *49*
Leibovitz, A., 238, *242*
Leighton, J., 66, *76*
Lemcke, R. M., 111, *127*, 133, 135, 144, *158*, *159*
Leonard, M. J., 249, *256*
Levan, A., 3, 5, *25*, *26*, 183, *194*, 250, *255*
Levashov, V. S., 141, *159*
Leventhal, B. G., 132, 153, *156*, *159*
Levine, E. M., 122, *127*, 132, 152, *159*
Levinthal, J. D., 151, *158*
Levitt, J., 154, *157*
Li, C. P., 195, 196, 199, *203*
Lidwell, O. M., 38, *48*
Like, A. A., 244, *255*
Likely, G. D., 2, *26*, 46, *48*
Lillie, R. D., 199, *203*
Lindberg, R. B., 121, *129*
Lindell, K. F., 43, 46, *49*
Litsky, W., 40, *48*
Litsky, Y. B., 40, *48*
Littlefield, J. W., 21, *25*
Liu, C., 151, *159*
Long, J. F., 225, *230*
Loveless, J. D., 188, 190, *193*
Lowy, D. R., 207, *230*, 247, 249, 252, *255*

Ludovici, P. P., 22, *24*, 52, *64*, 152, *158*, 251, *254*
Lueders, K. K., 211, *230*, *231*
Luftig, R. B., 228, *230*
Luginbuhl, R. E., 244, *254*, *255*
Lund, R. O., 66, *76*, 174, *193*
Lundgren, D. A., 43, *49*
Lussier, G., 153, *159*
Lutsky, I. I., 118, *128*, 151, *160*
Lynch, P. M., 3, 15, *24*
Lyon, G., 225, *230*

M

McAllister, R. M., 22, *25*, 251, *255*
MacCallum, R. O., 135, *158*
McClain, M., 236, *242*
McCombs, R. M., 249, *256*
McCulloch, E. A., 2, 4, 5, 7, *25*, *26*
McGarrity, G. J., 38, 39, 43, 44, 46, *48*, 52, *64*, 126, *128*, 148, *157*, *159*
McGee, Z. A., 132, 133, 135, 150, 153, *159*, *160*, 169
McGinniss, M. H., 132, *161*
McGregor, D., 122, *127*, 132, 152, *159*
McHale, J. S., 6, *26*
MacKenzie, R. D., 135, *158*
Macpherson, I. A., 4, *27*, 111, *128*, 132, 152, *159*, 249, *255*
Macy, M. L., 15, *26*
Madin, S. H., 44, *48*, 236, 237, 238, *242*
Magrath, D., 225, *229*
Malizia, W. F., 110, 111, 113, 115, 116, *126*, 140, 141, 142, 144, 151, *156*
Malloy, E., 45, *48*
Malmgren, R. A., 211, *230*
Malmquist, W. A., 221, *230*, 236, *242*
Mamura, T., 249, *255*
Manaker, R. A., 20, *24*, 151, *156*, 217, *229*, 247, *254*
Manchee, R. J., 119, *128*, 132, 135, 150, 151, *159*, *160*
Mandle, R. J., 39, 43, *48*
Maniloff, J., 109, *128*, 134, *160*
Mann, D. L., 21, *25*
Mansell, P. W. A., 221, *229*
Mansfield, M. E., 237, *242*
Manson, L. A., 120, *127*
Mantel, N., 37, *48*
Marcus, P. I., 252, *255*
Maré, C. J., 132, *160*
Markov, G. G., 132, 152, *160*
Marmion, B. P., 151, *158*, *160*
Marsden, H. B., 197, 198, 202, *203*
Martin, C. H., 150, *157*
Martin, G., 253, *256*
Mascoli, C. C., 195, *203*
Matsuda, K., 154, *159*
Matsuoka, Y., 20, *25*
Matsuya, Y., 22, *25*
Maurer, B. A., 249, *255*
Mazzali, R., 117, *128*
Medrek, T. J., 246, *255*
Meier, H., 207, *230*
Melendez, L. V., 244, *255*
Melin, M., 239, *241*
Melnick, J. L., 251, *255*
Melnick, P. J., 17, *25*
Merchant, D. J., 14, 16, 22, *24*, *25*
Merker, P. C., 184, *193*
Meryman, H. I., 44, *48*
Metzgar, R. F., 20, *25*
Meyer, H. M., Jr., 132, 153, *160*, 169
Meyer, R. D., 63, *64*
Meylan, C., 248, *255*
Middlekamp, J. N., 225, *231*
Migeon, B. R., 22, *25*, *26*
Miki, G., 244, *255*
Milgrom, F., 14, 20, 21, *25*
Miller, D. A., 22, *26*
Miller, G., 152, *160*
Miller, O. J., 22, *26*
Millers, R., 14, *25*
Minamishima, Y., 154, *160*
Minner, J. R., 195, 198, *203*
Minowada, J., 249, *255*
Mintcheva, A., 132, 152, *160*
Mitchell, G. B., 236, *241*
Mitchell, M., 40, *49*
Mizutani, S., 251, *256*
Modest, E. J., 22, *23*
Molander, C. W., 144, *160*, 237, 240, *242*
Monneron, A., 221, *230*
Monroe, J. H., 249, *254*
Montes De Oca, F., 15, *26*
Moore, A. E., 195, *203*
Moore, D. E., 134, *158*
Moore, D. H., 207, 211, *230*
Moore, G. E., 4, 20, *25*, *26*, 226, *230*
Moorhead, P. S., 4, 5, 6, 7, *24*, *25*, 183, *194*, 238, *242*, 250, *255*
Moraillon, A., 225, *230*
Morgan, H. R., 44, 45, *48*

Morgan, J. F., 152, *160*
Mori, M., 225, *230*
Morowitz, H. J., 109, *128*, 133, 134, 135, *160*
Morris, J. A., 135, 144, 152, *156*
Morton, D. L., 211, *230*
Morton, H. E., 110, 111, *128*, 141, 152, 153, 157, *160*, *161*
Morton, V., 53, *64*
Moses, H., 247, *254*
Mountain, I. M., 7, *26*
Moyer, K. B., 154, *158*
Mulford, D. J., 239, *241*
Munroe, J. S., 225, *230*
Murphy, P., 226, *229*
Mussgay, M., 214, *230*
Myers, B., 207, 210, 217, *231*
Myers, D. D., 207, *230*
Myers, R. J., 228, *230*

N

Najarian, J. S., 153, *157*
Nakamura, M., 154, *160*
Nardone, R. M., 132, 152, *160*
Nathenson, S. G., 21, *25*
Neff, J. M., 21, *26*
Negroni, G., 132, *160*
Neimark, H. C., 109, *128*, 133, 135, *160*
Nelson, J. B., 124, *127*, 132, 149, 150, 153, *158*, *160*
Nelson-Rees, W. A., 236, 237, 238, *242*, 251, *255*
Nesburn, A. B., 244, *255*
Newberne, J. W., 236, *242*
Newman, J. T., 244, *256*
Newnham, A. G., 134, 149, *160*
Nichols, W. W., 5, 6, 22, *25*, *26*, 123, *126*, 152, *156*, 182, 183, *193*, *194*, 250, *255*
Nielsen, A. H., 221, *229*
Noble, W. C., 38, *48*
Nocard, E., 132, *160*
Nomarski, G., 68, *76*
Novak, K., 198, 202, *203*
Nowinski, R. C., 207, *230*

O

O'Connell, R. C., 112, *128*, 148, *160*
O'Conor, G. T., 251, *255*
Ørskov, J., 134, *160*
Östergren, G., 183, *194*
Offers, S., 227, *231*
Officer, J. E., *127*, 149, *159*, 217, *229*
Ogata, M., 141, *160*
Old, L. J., 15, *26*, 207, 211, *229*, *230*
Olshevsky, V., 154, *157*
O'Malley, J. P., 132, 150, 153, *160*, 169
O'Neill, E., 45, *48*
Onen, M., 122, *128*, 132, 152, *161*, 183, 192, *194*, 214, *231*
Oppenheim, J. J., 153, *160*
Organick, A. B., 118, *128*, 151, *160*
Orgel, L. E., 248, *255*
Orlando, M. D., 124, *127*
Oroszlan, S., 207, *230*
Osterberg, H., 68, *76*
Ostergren, G., 250, *255*
Overton, W. M., 198, 199, 200, *203*

P

Page, Z., 8, *23*
Paley, A., 144, *160*
Palmer, A. E., 244, *256*
Pan, J., 120, *127*, 154, *158*
Parker, R. C., 2, 4, 5, 6, 7, *25*, *26*
Parks, W. P., 207, 221, *230*, *231*
Pasieka, A. E., 152, *160*
Pasteur, L., 43, *48*
Paton, G. R., 122, *128*, 132, 152, *160*, 182, *194*
Patrizi, G., 225, *231*
Patt, J. K., 125, *126*, 138, *157*
Patuleia, M. C., 124, *127*, 149, 150, *158*
Paucker, K., 154, 155, *156*
Paul, F. J., 251, *255*
Pavia, R., 135, 150, *158*
Pavilanis, V., 153, *159*
Peetermans, J., 244, *255*
Pereira, H. G., 154, *156*
Pereira, M. S., 196, 197, 198, 202, *203*
Périer, O., 225, *231*
Perkins, F. T., 121, 122, 123, *128*, 132, 152, *160*, *161*, 182, 183, 192, *194*, 214, *231*
Perlman, D., 57, 63, *64*, 124, *128*, 149, *160*
Peters, R. I., 207, *230*
Peterson, W. D., Jr., 2, 3, 9, 17, 18, 22, *26*
Petrov, P., 132, 152, *160*
Petts, V., 225, *230*
Petursson, G., 248, *255*
Pfefferkorn, E. R., 228, *229*
Pipes, F., 140, *158*
Pirie, N. W., 133, *160*
Pollack, R., 21, *24*
Pollar, M., 207, *230*
Pollock, M. E., 110, 111, 113, 118, 119, 120,

121, 122, 123, 124, 125, 126, *127*, *128*, 132, 141, 149, 152, *159*, *160*
Pomerat, C. M., 45, *48*
Pomeroy, A. P., 132, *158*
Ponten, J. A., 183, *194*, 250, *255*
Popoff, N., 225, *231*
Porter, K. R., 221, *231*
Pothier, L., 22, *23*
Potmesil, M., 217, *231*
Powelson, D. M., 132, *160*
Prescott, B., 119, *128*
Pressman, D., 20, *25*
Priori, E. S., 207, 210, 217, *231*
Pritchard, W. R., 236, *242*
Puck, T. T., 3, 4, *26*, 248, 249, *255*
Pugh, W. E., 252, *255*
Pulvertaft, R. J. V., 249, *255*
Purcell, R. H., 109, 115, 119, 121, *128*, 134, 135, 144, 150, 153, 154, *159*, *160*, *161*, *162*

Q

Quersin-Thiery, L., 7, *26*

R

Rabotti, G. F., 225, *229*, *230*
Rabson, A. S., 20, *24*, 251, *255*
Rahman, S. B., 124, *128*, 149, *160*
Rakovskaya, I. V., 132, 141, 153, 154, *159*, *160*
Ralph, P., 211, *231*
Ramos, L., 175, 176, 178, 189, 190, 191, *193*
Randall, C. C., 122, *128*, 154, 155, *158*
Rands, E., 117, *128*, 140, *162*
Rapp, F., 16, *26*, 183, *193*, 251, *255*
Rask-Nielson, R., 138, *162*
Ravdin, R. G., 183, *194*, 250, *255*
Ravid, Z., *158*
Razin, S., 109, 124, 125, *126*, *128*, 133, 134, 135, *160*, *161*, *162*
Recher, L., 140, *158*
Reczko, E., 214, *230*
Reed, S. E., 154, *161*
Rees, K. R., 122, *127*
Reich, P. R., 135, *161*
Reid, B. L., 154, *159*
Reimer, V., 233, 235, *242*
Reisinger, R. C., 237, *242*
Reitman, M., 40, *48*
Reynolds, L. I., 41, *49*
Reynolds, R. K., 149, *161*
Richardson, L. C., 226, *229*
Riggs, D. B., 110, 111, 113, 115, 116, 117, *126*, 134, 135, 140, 141, 142, 144, 145, 151, 152, *156*, *157*, *158*
Riggs, S., 125, *128*
Riviere, M., 225, *229*
Robert, A., 225, *230*
Roberts, D. H., 134, 154, *161*
Robillard, N. F., 135, *158*
Robinson, A. R., 3, 4, 17, 18, *26*, 248, 249, *255*
Robinson, L. B., 108, 110, *128*, 132, 149, *161*, 174, *194*
Robinson, V. B., 236, *242*
Roderick, T. H., 15, *26*
Rodwell, A., 110, *128*
Rogentine, G. N., Jr., 21, *25*
Rogul, M., 133, 135, *159*
Roizman, B., 108, 110, *128*, 132, 149, *161*, 174, *194*
Romano, N., 123, *128*, *161*
Rongey, R. W., 251, *255*
Rose, H. M., 44, 45, *48*
Rose, J. A., 135, *161*
Rosebury, T., 58, *64*
Rosenberg, L. T., 125, *128*
Roth, T. F., 221, *231*
Rothblat, G. H., 110, 111, *128*, 141, *161*
Rothfels, K. H., 2, 5, 6, 7, *26*
Rouhandeh, H., 235, *242*
Rous, P., 146, *161*
Rouse, H. C., 123, *128*, 132, 154, *161*
Roux, E., 132, *160*
Rowe, W. P., 207, *230*, 244, 247, 249, 250, 251, 252, *254*, *255*
Rozee, K. R., 228, *230*
Rubin, H., 251, *254*
Ruchman, I., 135, *162*
Ruddle, F. H., 15, 16, 19, 21, *26*
Russell, W. C., 152, 154, 159, *161*
Ruys, A. C., 118, 124, 125, *127*, 141, 144, 149, 154, *159*

S

Sabin, A. B., 132, 135, 153, *161*, 246, *255*
Sachs, L. 15, *24*
Sakamoto, H., 154, *160*
Saksela, E., 183, *194*, 250, *255*
Salas, J., 21, *24*
Sandberg, A. A., 4, *26*
Sanders, F. K., 246, 247, 253, *254*, *255*

Sanford, K. K., 2, *24*, *26*, 32, 44, 46, *48*
Sarkar, N. H., 207, *230*
Sarkar, S., 211, *231*
Sarma, P. S., 207, *230*
Sasaki, S., 134, *161*
Sato, G. H., 3, *27*
Sato, T., 207, 227, *230*, *231*
Sattar, S. A., 237, *242*
Saturk, M., 237, *242*
Savan, M., 236, *242*
Scala, A. R., 151, *158*
Scaletta, L. J., 21, *24*
Schaedler, R. W., 39, 43, *48*
Scher, W., 207, *230*
Scherer, W. F., 2, *26*
Schidlovsky, G. 211, *230*
Schimke, R. T., 52, *64*, 117, *126*, 138, 139, 140, 141, 152, *156*, *161*
Schlesinger, R. W., 123, *128*, 132, 154, *161*
Schmidt, P. J., 132, *161*
Schoefl, G. L., 21, *25*
Schulte-Holthausen, H., 249, *256*
Schulze, I. T., 154, *161*
Schwartz, A. J. F., 235, *242*
Schwöbel, W., 144, *161*
Scolnick, E. M., 207, 221, *229*, *231*, 247, 249, 252, *254*
Scriba, M., 149, *161*
Sell, E. K., 3, 16, 17, *25*
Seman, G., 140, *158*
Semar, J. B., 124, *128*, 149, *160*
Senterfit, L. B., 134, *159*
Shannon, J. E., 2, 3, 4, 7, 12, 15, 17, 18, 19, 23, *24*, *26*, 44, 46, *48*
Sharon, N., 207, *230*
Sharp, J. T., 125, *128*
Shaw, E. J., 124, *127*
Shearer, M. C., 22, *25*
Shedden, W. I. H., 118, 124, *128*, 149, *161*
Shein, H. M., 250, *256*
Shelokov, A., 195, 198, *203*
Shepard, M. C., 73, *76*, 124, *127*, 137, 148, 149, 151, *157*, 161
Sherer, W. F., 44, *48*
Sherris, J. C., 53, *64*
Shipkey, F., 225, *230*
Shipman, C., Jr., 22, *25*, 211, *231*
Shishkov, N., 132, 152, *160*
Shively, J. A., 140, *158*
Shows, T. B., 15, *26*
Shreffler, D. C., 14, *24*
Shreeve, M. M., 197, *203*, 214, *230*
Shullenberger, C. C., 140, *158*
Siegel, B. V., 225, *229*
Siegesmund, K. A., 118, *128*, 151, *160*
Siegl, G., 243, *254*
Silberman-Ziv, G., 20, *23*, 26
Silverberg, R. J., 7, *25*
Silvers, W. K., 5, 6, 7, *24*
Simberkoff, M. S., 153, *161*
Siminovitch, L., 2, 5, 7, *26*
Simon, J., 225, *229*
Simpson, W. F., 2, 3, 9, 13, 17, 18, 22, *26*
Sinclair, F., 154, 155, *158*
Singer, S. H., 123, *128*, 132, 154, 155, *161* 168
Singh, B. N., 197, 202, *203*
Sinkovics, J. G., 140, *158*, 226, *230*
Smadel, J. E., 132, 153, *160*, 169
Smetana, K., 217, *231*
Smirnova, T. D., 132, 153, 154, *159*, *161*
Smith, C. B., 134, 153, *159*
Smith, G. H., 211, *231*
Smith, J. W., 195, 198, *203*
Smith, K. O., 227, *231*, 244, 246, *256*
Smith, N. H., 134, *159*
Smith, P. F., 120, *128*, 134, 135, *161*
Smucker, E. A., 211, *231*
Snyder, J., 135, *156*
Sobeslavsky, O., 119, *128*
Somers, C. E., 249, *255*
Somerson, N. L., 111, 121, *128*, 134, 135, 144, 150, 153, 154, *157*, *158*, *160*, *161*, *162*
Sorieul, S., 21, *23*, *24*
Spear, G. S., 15, *26*
Speigelman, S., 253, *254*
Spendlove, R. S., 125, *128*, 149, *161*
Spitler, L., 153, *161*
Stanbridge, E., 33, 46, *48*, 112, 115, 121, 122, 123, *127*, *128*, 132, 144, 152, *159*, *161*, 214, *231*
Stanbridge, R., 183, 192, *194*
Stavropoulos, K., 138, *159*
Steger, L., 53, *64*
Steiner, J. W., 197, *203*, 214, *230*
Stenback, W. A., 214, *231*, 244, *256*
Stenchever, M. A., 21, *24*
Steplewski, Z., 21, *25*, 247, *255*
Stevenson, R. E., 8, 9, *24*, 37, 44, *48*
Stewart, S., 225, *231*
Stewart, S. M., 144, *162*
Stinebring, W. R., 118, *127*

Stitch, H. F., 183, 192, *193*, *194*
Stitt, D., 152, *160*
Stock, D. A., 122, 123, *128*
Stocker, M., 249, *255*
Straub, D. C., 235, 242
Strife, A., 4, *23*
Strom, R., 21, *25*
Strong, L. E., 239, *241*
Stuart, A. B., 22, *23*
Stuck, B., 211, *229*
Stulberg, C. S., 2, 3, 4, 7, 9, 12, 13, 15, 17, 18, 19, 22, 23, *24*, *26*, 44, *48*
Sverak, L., 221, *230*
Svoboda, J., 248, *255*, *256*
Swanborg, N. K., 3, 17, 18, *26*
Sweet, B. H., 6, 14, *25*, *26*
Swetly, P., 21, *25*
Swim, H. E., 5, *26*
Switzer, W. P., 132, *160*
Sykes, J., 66, *76*
Syverton, J. T., 2, 5, 7, 8, *23*, *26*, 110, 111, 123, 126, *128*, 132, 141, 149, *160*
Szybalska, E. H., 3, *26*
Szybalski, W., 3, *26*

T

Tachibana, D. K., 125, *128*
Takemoto, K. K., 7, *25*, 245, *256*
Tall, M. G., 14, *24*
Tankersley, R. W., 154, *162*
Tashjian, A. H., Jr., 3, *27*
Tauraso, N. M., 116, *128*
Taylor, H. G., 140, *158*
Taylor, H. L., 239, *241*
Taylor-Robinson, D., 109, 111, 115, 117, 119, *128*, 132, 134, 135, 144, 150, 151, 152, 154, *157*, *159*, *160*, *161*, *162*
Teich, N., 207, *230*, 247, 249, 252, *255*
Temin, H., 251, *256*
Tennant, J. R., 15, *26*, 227, *230*
Tessmer, C. F., 140, *158*
Theodore, T. S., 135, *162*
Thiel, J. F., 244, *256*
Thiéry, J-P., 221, *229*
Thomas, A., 211, *231*
Thomas, L., 118, 122, *127*, *129*, 132, 138, 151, 152, 153, *158*, *159*, *161*, *162*
Thompson, R. G., 236, *242*
Thornbeck, J. G., 153, *161*
Titmuss, D. H. J., 4, *27*
Tjio, J. H., 3, *26*
Tobin, J. O' H., 197, 198, 202, *203*
Todaro, G. J., 117, *128*, 140, *162*, 207, 221, *229*, *230*, *231*, 247, 248, 249, 250, 252, 253, *254*, *255*, *256*
Todd, J., 132, 152, *160*
Toolan, H. W., 184, *194*
Toshima, S., 226, *230*
Tournier, P., 211, *229*, 250, *254*
Trapp, A. L., 237, *242*
Treadwell, P. E., 120, 121, *128*, 132, 152, *160*
Trentin, J. J., 183, *194*, 214, *231*, 244, *256*
Trexler, P. C., 41, *48*, *49*, 154, *159*
Tsanev, R. G., 132, 152, *160*
Tsareva, A. T., 154, *159*
Tsilinskii, Y. Y., 141, *159*
Tully, J. G., 53, *64*, 133, 134, 135, 138, 144, 148, *157*, *158*, *161*, *162*, 174, *194*
Tumilowicz, J. J., 22, *26*
Turck, M., 53, *64*
Turner, H. C., 111, *128*, 134, 135, 153, *160*, *162*, 207, *230*, 244, 253, *255*
Tyndall, J., 43, *49*

U

Ulrich, K., 4, *26*
Usenik, E., 221, *231*
Uzman, B. G., 4, 15, *24*, 225, *230*

V

Valdesuso, J. R., 119, *128*
Van deBogart, R. K., 154, *158*
Van Demark, P. J., 152, *162*
Van Der Maaten, M. J., 221, *230*
Vander Velde, G. C., 211, *231*
Van Herick, W., 132, 150, 151, *162*
Van Hoosier, G. L., Jr., 183, 192, *194*, 214, *231*, 244, *256*
van Rhijin, G. R., 118, 124, 125, *127*, 141, 144, 149, 154, *159*
Vedder, H., 45, *48*
Verna, J. E., 138, 145, *156*, *158*
Vigier, P., 248, *256*
Vilain, C., 211, *231*
Vogel, J. E., 195, 198, *203*
Vogelzang, A. A., 124, 125, *129*
Vogt, M., 46, *48*, 245, *256*
Vogt, P. K., 247, 252, *256*
Volkert, M., 238, *242*
Volkman, L. E., 211, *231*
Vonka, V., 249, *256*

W

Waddell, A., 116, 117, *127*
Wahl, D., 154, *156*
Waite, M. R., 228, *229*
Wallace, R. E., 7, *23*
Waltz, H. K., 2, *24*, 46, *48*
Wang, R., 21, *24*
Wang, S. S., 195, 196, 197, 198, *203*
Ward, J. R., 150, *157*
Warhurst, D. C., 196, 197, 198, *203*
Warren, J., 150, *162*
Warren, S., 184, *194*
Waterman, S., 244, *255*
Watkins, J. F., 21, *25*, *26*, 247, *256*
Watson, J., 211, *231*
Watson, M. L., 217, *231*
Weber, A., 221, *231*
Wedum, A. G., 40, *48*, *49*
Weinberg, A. N., 16, *25*
Weiss, M. C., 21, 22, *27*
Weiss, R. A., 247, 252, *256*
Weissman, S. M., 135, *161*
Werder, A. A., 221, *229*
Westwood, J. C. N., 4, *27*, 40, *49*
Whang-Peng, J., 246, *254*
Wheeler, C. E., 245, *256*
Whipp, S., 221, *231*
Whitby, K. T., 43, *49*
Whitcomb, J. G., 43, *49*
Whitefield, W. J., 43, 46, *49*
Whittlestone, P., 132, *158*
Wichelhausen, R. H., 108, 110, *128*, 132, 149, *161*, 174, *194*
Wilbur, J. R., 207, 210, 217, *231*
Williams, C. O., 109, *126*, *129*, 133, 135, *156*, *162*
Williams, R. C., 197, *203*
Williamson, J. D., 154, *156*
Wilson, A. B., 21, *23*
Wilson, S. R., 144, 148, *157*
Windle, M. S., 225, *230*
Winters, A. L., 154, *162*
Witebsky, E., 14, *25*
Wittler, R. G., 109, 112, 121, *126*, *128*, *129*, 133, 135, 148, *156*, *159*, *160*, *162*
Wivel, N. A., 211, *230*, *231*
Woman, S. R., 3, 15, *24*
Wong, D. C., 121, *128*, 134, 153, *160*, *162*
Wong, S. C., 245, *256*
Woodside, N. J., 211, *229*
Wright, B. S., 211, *231*
Wright, D. N., 134, *162*

Y

Yagi, Y., 20, *25*
Yaguchi, R., 72, *76*, 138, 139, *156*
Yamashiroya, H. M., 138, 145, *156*, *158*
Yasin, R., 15, *27*
Yasumura, Y., 3, *27*
Yeh, S., 151, *159*
Yerganian, G., 5, *24*, 249, *256*
Yershov, F. I., 155, *162*
York, C. J., 235, 236, *241*, *242*
Yoshida, M. C., 21, *24*
Young, D., 144, *162*
Young, L. S., 53, 63, *64*
Yu, B., 53, *64*
Yumoto, T., 140, *158*

Z

Zajac, B., 4, *24*
Zech, L., 22, *23*
Zhdanov, V. M., 154, 155, *159*, *162*
Zirbell, L. W., 235, *242*
Zmijewski, C. M., 20, *25*
Zucker-Franklin, D., 118, *129*, 151, *162*
Zur Hausen, H., 249, *256*

Subject Index

A

Acanthamoeba, 197
Acholeplasmataceae, 57, 109-110
Agar plates, mycoplasma testing, 75-76, 113-114, 137-138
Amoeba, identification, 196-197
Anaerobic organisms, 59
Antibiotics, 45-47, 57-60, 62, 63, 123-125, 142, 149-150, 267-268
  effects, 63
Antigens
  Forssman, 15
  H-2, 14
  heterophil, 15
Antiserums
  cytoxic effects, 7
  fluorescent antibody, 9-14
Arginine, 119-123
Aseptic techniques, 44-45
Aspergillus, 53
  fumigatus, 62
  niger, 71

B

Bacteria, 29-34, 38-47, 53-55, 70-71
Bovine herpesvirus I, 234-235, 240-241
Bovine syncitial virus, 221
Bovine virus diarrhea, 235-237, 239-241
Broth culture procedure, 138-140
Burkitt's African lymphoma, 246

C

Candida albicans, 58, 61, 62, 71
Centriole satellites, 221
Chromosomal aberrations, 16-20, 152
Chromosomal changes
  mycoplasma, 122-123, 178-183
  mycoplasma/SV40, 188-189
Coated vesicles, 221
Complement fixation, 14
Contamination
  culture testing, 52-56
  cytogenetic determination, 2-3, 5-7
  effects of, 268
  extrinsic cell, 1-27
  fetal bovine serum, 233-242
  Hartmannella, 195-202
  identification, 261-264
  immunologic determination, 2
  interspecies, 2-3, 4-16, 22
  intraspecies, 3, 22
  mycoplasma
    incidence, 110-112, 140-147
    source, 112-113, 142-148
  parasites, 195-203
  photomicrographs, 69-72
  sources, 29-41, 58-62, 258-259
  treatment, 45-47, 57, 59, 62-64, 123-125, 149-150, 176, 200-202, 266-268
  viral, 233-242
    cryptic, 247-254
    detection, 252-254
    identification, 237-241

inherent, 243-256
isolation, 237-241
passenger, 244-247, 250-253
Contamination prevention, 264-266
airborne sources, 38-41
amoeba, 200-202
antibiotics, 45-47
bacterial, 29-49
cell culture exchange, 32-33
culture media, 36-38, 41-43
filtered air, 38, 43
fungal, 29-49
glassware and apparatus, 33-36
media processing, 41-43
mycoplasma, 125-126
Coxsackie virus, 245
Cryptococcus neoformans, 61
Cytogenetic analysis, 16-20
Cytoxic dye test, 7

D

Diphtheroids, 57
Disinfection of culture room, 39-41
DNA, 122, 251-254

E

EDTA, *see* Ethylenediaminetetraacetic acid
Electron microscopy, 205-231
amoeba, 196-197
freeze etching, 228
negative staining, 228
Epstein-Barr virus, 246
ERK-1 cell line, 4-5
Escherichia, 54, 57, 62, 70, 199
ESP-1 cell line, 207, 217
Ethylenediaminetetraacetic acid (EDTA), 36

F

FL cells, 66-75, 69, 72, 73, 189-193
mycoplasma infected, 175-182
reduced tumor production, 183-188
Feline leukemia virus, 253
Fibroblasts, human diploid, 17
Flavobacterium, 54
Fluorescein isothiocyanate, 10
Fluorescent antibody, 9-14
Foamy virus, 221, 244
Forssman antigen, 15
Friend leukemia, 207
Fungi, 55
Fusion, 21-22

G

G6-PD, *see* Glucose-6-phosphate dehydrogenase
Genetic changes, 251-252
irreversible, 3-4
reversible, 3-4
Genetic markers, 3
Glassware
cleaning procedures, 34-36
sterilization, 35-36
Glucose-6-phosphate dehydrogenase (G6-PD), 3, 15, 17, 22

H

H antigens, 20
Hartmanella, 195-202
HeLa cell line, 2, 17
Hemadsorption, 150
Hemagglutination, 7-9, 12-13, 150-151
Hemolysis, 150
Herpes simplex virus, 246
Heterophil antigen, 15
HL-A alloantigens, 21
Hodgkin's disease, 246
Hybridization, 21-22

I

Identification
interspecies, 4-16
intraspecies, 16-22
Immunodiffusion, 14, 20
Immunoglobulins, 20-21
Immunologic determinations
complement fixation, 14
immunodiffusion, 14
Inoculums, 60-63
Intracisternal particles, 211
Intracytoplasmic crystalline aggregates, 225-226
Isoenzyme, 3, 15

K

Karyotypes, 3-7, 16, 250-252

L

Lactose dehydrogenase (LDH), 15
Laminar flow, 43
L cell line, 2, 14
L-form bacteria, 52, 53, 55, 57, 60, 63, 112-113, 217

L-M cell line, 14, 16
Lymphoid cell line, 20

M

Markers
  antigenic, 14-15
  genetic, 16-17
  isoenzyme profiles, 17-20
  metabolic, 16-17
Media, 36-38, 41-43
  for testing, 52, 53
  Soy-peptone-yeast dialyzate, 113-114
MCN cell line, 4
Micrococci, 57
Microscopy
  Nomarski interference-contrast, 66-70
  optics, 68-69
  phase optics, 67-68
  preparation of cell cultures, 66-68
Mixed agglutination, 8-9, 12-14
Molds, 71-72
Mucor sp., 71
Mucosal disease, 235-237, 239-241
Murine leukemia virus, 206, 207, 211, 214
Mutants, conditional lethal, 245, 247
Mycoplasma, 53, 56, 57, 63, 67, 68-70, 72-76, 107-129, 131-162, 173-194, 214-217, 259-262, 268-269
  cell association, 190-191
  cellular interactions, 151-152
  characteristics, 133-135
  chromosome changes, 178-183, 190-191
  classification, 133
  contamination prevention, 148-149
  culture procedure, 135-140
  cytopathic effects, 190-191, 214
  decreased resistance of cells, 189-190
  detection, 113-117, 135-140
  effects on cell cultures, 119-123, 152-153
  electron microscopy, 214, 217
  elimination, 123-125, 149-150
  FL cells, 175-182
    reduced tumor production, 183-188
  history, 132-133
  host-parasite relationships, 117-119
  identification, 115-116
  incidence of contamination, 140-142
  isolation, 114-115
  origin, 141-142
  physiology, 110
  preferred attachment, 190-191
  serum, 37-38
  sources of contamination, 142-148
  taxonomy, 109
  testing, 53, 72-76
  ultrastructure, 109-110
  virus-cell interactions, 154-156
Mycoplasma orale, 60, 61
Mycoplasma salivarium, 61

N

Naegleria, 198
Nomarski, 71, 72
Nuclear bodies, 221, 225

O

Oncornaviruses, 251
Orcein stain, 67
  mycoplasma testing, 72-76

P

Parainfluenza 3 virus, 237
Paramyxovirus, 245, 252
Penicillium sp., 71
Perichromatin granules, 217, 221
PGM, *see* Phosphoglucomutase
Phosphoglucomutase (PGM), 15, 17
PK-13 cell line, 16
Poliovirus, 244-245, 247
Proteus mirabilis, 57
Pseudomonas, 54, 70
  aeruginosa, 58, 61, 62, 63

R

R particles, 211, 214
Reovirus, 214
RNA, 122, 251-254

S

Saccharomyces sp., 71
Serratia sp., 70
Serum, 36-38, 142-144
Shipping fever, 237
Simian virus 40 (SV40), 188-191, 193, 247-248
Staphylococcus, 54
  albus, 70
Sterility tests, 32-33, 259-261
Streptococcus viridans, 57

T

T-antigen, 16
Transformation, 3, 16
Trichosporon sp., 71, 72
Trypan blue exclusion test, 227

U

Uridvl transferase, 17

V

Virus
  cell culture growth, 248-250
  contamination procedure, 238, 239
  helper, 253-254
Virus-like particles
  centriole satellites, 221
  coated vesicles, 221
  perichromatin granules, 217, 221
Virus particles, type C, 247

Y

Yeasts, 71